The Roots of American Industrialization

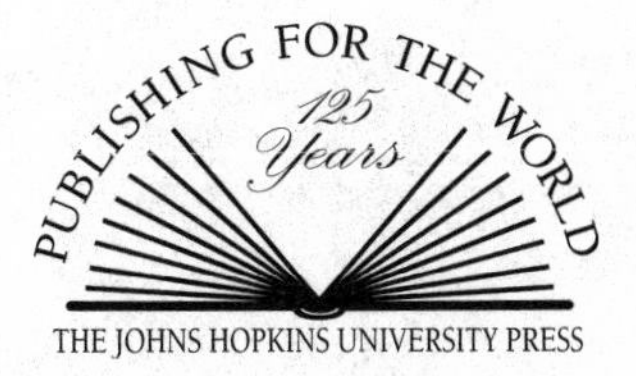

Published in cooperation with the Center for American Places,
Santa Fe, New Mexico, and Harrisonburg, Virginia

The Roots of American Industrialization

David R. Meyer

The Johns Hopkins University Press
Baltimore and London

Printed in the United States of America on acid-free paper
9 8 7 6 5 4 3 2

The Johns Hopkins University Press
2715 North Charles Street
Baltimore, Maryland 21218-4363
www.press.jhu.edu

Library of Congress Cataloging-in-Publication Data

Meyer, David R.
The roots of American industrialization / David R. Meyer.
p. cm.
Includes bibliographical references and index.
ISBN 0-8018-7141-7
1. Industrialization—United States—History—19th century. 2. United States—Economic conditions—To 1865. I. Title.
HC105 .M576 2003
338.0973′09′034—dc21
2002005436

Library of Congress Cataloging-in-Publication Data will be found at the end of this book.

A catalog record for this book is available from the British Library.

Contents

Figures and Maps

Figures

Maps

Tables

Acknowledgments

Writing this interpretation of the roots of American industrialization was a humbling experience, because it would have been impossible without the work of innumerable scholars who provided sweeping syntheses, case studies, econometric analyses, time series, and data compilations covering the nation and the eastern United States for the 1790–1860 period. And I benefited from many scholars' comments on conference papers, lectures at universities, and the exchange of ideas which comes from debating issues in informal settings. Brown University's Department of Sociology, Urban Studies Program, librarians, and research offices provided support at many stages of the writing, and Donna Leveillee, also at Brown, deserves credit for making the maps. Thanks to Edward (Ted) Muller for suggesting the title of the book and to Elizabeth Gratch for applying her editing skills to the manuscript.

CHAPTER ONE

The Puzzle of the Antebellum East

> A people, spread through the whole tract of country, on this side the Mississippi, . . . would probably for some centuries find employment in agriculture, and thereby free us as at home effectually from our fears of American manufactures. . . . It is the multitude of poor without land in a country, and who must work for others at low wages or starve, that enables undertakers to carry on a manufacture, and afford it cheap enough to prevent the importation of the same kind from abroad, and to bear the expense of its own exportation. —BENJAMIN FRANKLIN, 1760

> If there were both an artificer and a farmer, the latter would be left at liberty to pursue exclusively the cultivation of his farm. A greater quantity of provisions and raw materials would, of course, be produced, equal, at least . . . to the whole amount of the provisions, raw materials, and manufactures, which would exist on a contrary supposition. The artificer, at the same time, would be going on in the production of manufactured commodities, to an amount sufficient, not only to repay the farmer, in those commodities, for the provisions and materials which were procured from him, but to furnish the artificer himself with a supply of similar commodities for his own use. —ALEXANDER HAMILTON, 1791

In the late eighteenth century Benjamin Franklin and Alexander Hamilton posed alternative visions of American development. To Franklin ample supplies of farmland supported prosperous agriculture, and manufactures would be imported, but inadequate supplies of land created an impoverished rural population employed as low-wage factory workers producing goods that were cheaper than foreign imports. Hamilton, however, conceptualized a beneficial division of labor between agriculture and manufacturing as workers in each sector achieved greater productivity through specialization and gained from exchanging surpluses. Most discussions of the roots of the antebellum East's industrialization accept Franklin's view that industrialization emerged from impoverished agriculture, not Hamilton's claim that prosperous agriculture and manufacturing benefited each other.[1]

Puzzles in Agriculture and Manufacturing

As late as 1840, agriculture employed most of the nation's workers, and the 500,000 in manufacturing remained so few relative to agriculture that, even if most of them resided in the East, its industrialization seemed meager (table 1.1). Between 1840 and 1860 manufacturing employment tripled, and production soared almost fivefold, ranking this as the fastest period of industrial expansion for the rest of the century, and this growth occurred mostly in the East. By 1860 its manufacturing workers numbered 878,574, three-quarters of the nation's industrial employment.[2] The traditional explanation of the East's industrialization followed comparative advantage logic with specialization of the East in manufactures and commerce, the South in plantation staples (especially cotton), and the Midwest in agricultural food products. Eastern manufactures flowed to the South and Midwest, and the Midwest's food surpluses met deficits in other regions; the South's staples moved primarily to the East and foreign markets.[3] But this explanation collapsed before evidence that the South and Midwest were minor markets for eastern manufactures as of 1840; the South probably took no more than 8 percent of the East's production and the Midwest even less. Over the next two decades the share of the East's total manufactures exported to the Midwest and South together fluctuated between 10 percent and 15 percent, little changed from earlier years. Those shares include products with low value relative to their weight which were not exported to other regions; eliminating those, the share of the East's transportable manufactures exported to the Midwest and South may have reached 25–30 percent by 1840 and possibly somewhat higher by 1860. Nevertheless, shares of eastern production exported to other regions are too low to account for much of eastern industrial growth, except for a minority of manufactures. Foreign markets stimulated little industrial growth, and businesses focused on tariff protection; even during the period of rapid industrialization from 1840 to 1860, exports contributed only 5–6 percent of total value added in manufacturing.[4]

Thus, industrialists looked primarily to eastern markets, yet this is puzzling. As of 1790, virtually everyone in the East lived in rural areas, and, as late as 1820, the rural share dropped imperceptibly; even in 1840, when most observers date the start of rapid industrial growth, 81 percent of the population remained rural, and by 1860, after seven decades of industrialization, about

Table 1.1. The Labor Force in the United States, 1800–1860

Year	Number (thousands)				Percentage Distribution			
	Total	Agriculture	Manufacturing	Other	Total	Agriculture	Manufacturing	Other
1800	1,712	1,274	—	—	100.0	74.4	—	—
1810	2,337	1,690	75	572	100.0	72.3	3.2	24.5
1820	3,150	2,249	—	—	100.0	71.4	—	—
1830	4,272	2,982	—	—	100.0	69.8	—	—
1840	5,778	3,882	500	1,396	100.1	67.2	8.7	24.2
1850	8,192	4,889	1,200	2,103	100.0	59.7	14.6	25.7
1860	11,290	6,299	1,530	3,461	100.1	55.8	13.6	30.7

Sources: Lebergott, "Labor Force and Employment, 1800–1960," 30:118, table 1; Weiss, "U.S. Labor Force Estimates and Economic Growth, 1800–1860," 22, table 1.1.

two-thirds of the East still lived in rural areas (table 1.2). During the period from 1790 to 1860 numbers in New England and the Middle Atlantic climbed continuously, and their combined rural population more than tripled. Even between 1840 to 1860, when industrial production surged almost fivefold, the East added 1.4 million rural residents, a 24 percent increase. As rural dwellers migrated to urban centers, the rural population share declined, comporting with standard interpretations of industrial growth. But growing rural populations remain perplexing, because the conventional view portrays eastern agriculture as continually struggling, first with declining soil fertility and then with competition from midwestern farms.

Sectors of the rural economy—resource extraction (lumbering, mining) and processing (sawmills, smelting)—expanded with industrialization, but they employed small numbers and cannot account for the extraordinary rural growth throughout the antebellum. Standard interpretations highlight New England farms as quintessential losers of market share, first to New York and Pennsylvania farms and then, after 1840, to midwestern farms. By the 1820s most of the East struggled with declining soil fertility, and increasing use of marginal land generated diminishing increments of production. There were exceptions: for example, farms near Boston, New York, and Philadelphia shifted into market gardening, and some areas specialized in dairying (butter and cheese). After 1840 intensive farming near large cities became the norm, and distant areas shifted out of wheat as competition from midwestern grain farms grew fiercer; eastern grain farms shifted farther into dairying and cattle fattening, but commercial farming still had not fully spread throughout the East by 1850.

This interpretation of struggling eastern agriculture provides a weak reed to argue that the farm sector significantly boosted industrialization. Proponents of the "transition to capitalism" explanation of agricultural change argue that farmers resisted market integration and maintained household manufacturing as long as possible, yet farm households directly supported manufacturing as suppliers of rural outworkers or factory operatives. The question remains: who purchased the output of rural workshops and of rural and urban factories that expanded production? We know small nonmechanized factories achieved substantial productivity gains compared to artisan shops, and across diverse industries and in small and large factories labor productivity rose significantly with little machinery investment. But, with nearby anemic rural markets and small urban markets before 1840, why would rural workshops and small factories boost productivity and lower prices of manufactures?[5]

The Philadelphia region's prosperous agriculture in southeastern Pennsylvania and anthracite coal mining northwest of Philadelphia seems to offer a model of transformation of agriculture and manufacturing. From the late 1810s to the late 1830s transportation improvements (especially canals) within Philadelphia's hinterland lowered transport costs and stimulated commercial exchange. Hinterland farmers shifted from wheat and diversified production for Philadelphia's market, and other resource extractors (mining and lumbering) supplied that market; Philadelphia's manufacturers expanded and supplied burgeoning hinterland demand. During the 1840s the Philadelphia region looked more to other regions of the East for markets and as sources of goods.[6] Yet the Philadelphia regional model does not seem to fit other eastern regions, if the conventional interpretation of agriculture is valid. Boston's New England hinterland presumably contained few prosperous farms, and they were threatened by farmers elsewhere. Similarly, New York City's immediate hinterland did not contain much fertile farmland; distant central and western New York held most of the prime agricultural land. Although the Philadelphia model explains the metropolis' industrial growth, it offers little guidance for explaining expansion of rural workshops and small nonmechanized factories in villages distant from Philadelphia, and, by extension, the model does not explain how those firms grew in other eastern regions.

Therefore, roots of eastern industrialization remain obscure. Manufacturers looked to the East for markets, but interpreters of eastern agriculture provide little foundation for identifying farm households as important markets. Before 1840 urban markets were a tiny share of the East's population, and

Table 1.2. Total Population and Rural and Urban Components in the East, 1790–1860

	1790	1800	1810	1820	1830	1840	1850	1860
	Total Population (thousands)							
Northern New England	324	490	661	778	949	1,079	1,215	1,269
Southern New England	686	743	811	881	1,005	1,157	1,514	1,866
Middle Atlantic	1,337	1,808	2,469	3,180	4,112	5,074	6,574	8,258
Total East	2,347	3,041	3,941	4,839	6,066	7,310	9,303	11,393
	Rural Component (thousands)							
Northern New England	319	481	647	762	923	1,011	1,076	1,087
Southern New England	615	651	676	722	757	791	869	900
Middle Atlantic	1,240	1,638	2,191	2,810	3,510	4,133	4,868	5,365
Total East	2,174	2,770	3,514	4,294	5,190	5,935	6,813	7,352
	Urban Component (thousands)							
Northern New England	5	9	14	16	26	68	139	182
Southern New England	71	92	135	159	248	366	645	966
Middle Atlantic	97	170	278	370	602	941	1,706	2,893
Total East	173	271	427	545	876	1,375	2,490	4,041
	Percentage Urban							
Northern New England	2	2	2	2	3	6	11	14
Southern New England	10	12	17	18	25	32	43	52
Middle Atlantic	7	9	11	12	15	19	26	35
Total East	7	9	11	11	14	19	27	35

Source: U.S. Bureau of the Census, *Historical Statistics of the United States*, ser. A195, A202, A203.
Note:
Northern New England: Maine, New Hampshire, and Vermont
Southern New England: Massachusetts, Rhode Island, and Connecticut
Middle Atlantic: New York, New Jersey, Pennsylvania, Maryland, and Delaware

early industrial villages and towns housing low-wage workers in textile and shoe industries offered meager markets to other manufacturers. Discussions of manufacturing reflect this ambiguity in identifying industrial roots—markets are assumed to exist. Researchers claim that reduced transport costs boosted manufacturing, and they denigrate wagons as a transport mode; canals and steamboats benefited industry, but those waterway modes paled next to railroads as contributors to industrialization. Yet evidence shows transportation improvements mostly failed or exerted little impact on agriculture and industry, and, except for the Erie, Champlain, and anthracite coal canals, canals were costly failures and, in the case of the Erie, exacerbated agricultural decline after 1840 as midwestern farm products poured eastward over the canal. Prior to 1850 railroads mostly carried passengers and did little to spur economic development until that decade. Other explanations of industrial growth stay within the manufacturing sector and cover the sweep of industrial expansion, individual manufactures and firms, or case studies of local industrial growth.[7] These interpretations read history backwards from 1860, when the East was an industrial powerhouse and agriculture had declined relatively, but reading history forward generates an alternative hypothesis: agricultural and industrial transformations were integrated, mutually reinforcing processes. Interrelations among local exchange, nonlocal exchange, and market price signals form an essential framework to explain this transformation.

Economic Exchange and Market Price Signals

Between the late eighteenth and early twentieth centuries leading social theorists grappled with conceptualizing contemporaneous economic transformations—improvements in transportation and communication, swelling flows of commodities within and among nations, and industrialization—framing their thinking partly as an analysis of causes and consequences of the division of labor.[8] To contrast observed economic changes, they created the abstraction of the local, self-sufficient economy. Primitive transportation, combined with small-scale production and low consumption, meant that most goods were exchanged face-to-face between producers and consumers, and, because all goods and services were produced and consumed locally, economies of scale and specialization were limited. Isolation retarded technological change and productivity improvements because the small local population generated all innovations, and only local natural resources were used in production. The local, self-sufficient economy epitomized impoverishment, trapping inhabitants at a low level of development, but social theorists' exotic examples demonstrate that they considered these rare cases; the abstraction was simply a reference point to examine economic growth and development. Exchange among local economies—and thus the termination of self-sufficiency—broke them out of their low-level development trap. Each local economy faced a larger market; consequently, opportunities opened to achieve economies of scale and specialization in production, raising labor productivity. Exchange spread information, thus spurring technological change, and prices for goods fell, encouraging individuals to increase consumption. Nonlocal exchange did not occur spontaneously; social theorists attributed breakdowns in local self-sufficiency to intermediaries—merchant wholesalers and financiers linking local economies.

Social theorists provide a rationale for rejecting the claim that any rural community in the eastern United States as of 1790, as well as in colonial America, functioned as a self-sufficient local economy; every rural community had links, no matter how tenuous, to economies across the East and Atlantic world. Country storekeepers were instruments to reach external economies; they collected local surpluses and funneled them to wholesaler-retailers or to general wholesalers in metropolises. General wholesalers controlled lower levels of intermediary exchange among local economies, and they exchanged with more specialized wholesalers handling larger-scale, longer-distance exchange among regions and nations; in reverse, this wholesaler chain reached back to country storekeepers to supply commodities sold to households. Country storekeepers, along with farmers' trips to other local economies to sell surpluses, transmitted prices to local economies.

Price convergence signals greater market integration, but no rural community was isolated from price signals, even feeble ones, from the market economy. In communities with high degrees of local self-sufficiency and low levels of market exchange, farm households exchanged goods, services, and cash with one another through informal creditor and debtor relations, and those exchange relations also held with country stores. Households cooperated to build and maintain local infrastructure, clear land, and construct barns and houses; they shared draft animals and machinery; they worked together during peak labor periods; and they participated in the intergenerational transfer of labor skills. Nevertheless, local communities contained seeds of transformation: retailers and unspecialized wholesalers provided links to external markets, and, as their business increased, they boosted dependence of farm communities on those markets. Market exchange altered opportunity structures as some types of local exchanges and occupations disappeared and new occupations emerged to support market exchange. Deeper participation in external exchanges required greater use of financial capital; thus, differential wealth accumulation introduced more local social and economic stratification. Alterations in communal relations among neighbors and within kinship groups and alterations in gender roles within families sometimes were contested locally and with political and economic actors outside the local economy.[9]

Nonetheless, strict distinctions between nonmarket exchange within the local community and market exchange outside introduce a false dichotomy. As a local community increases integration with the market economy—prices move synchronously and converge—those prices influence amounts and types of external exchanges and impact internal economic relations. Regardless of the degree of local self-sufficiency and extent of external exchange, all exchanges in a market economy are embedded in social structures and social relations. A web of obligations surrounds transactions and transcends prices for all exchanges within a local community whether based on barter, reciprocity, or cash, and social networks involving friendship groups, families and relatives, churches, social clubs, and professional organizations reinforce obligations. Economic actions within and among cohesive groups assume pseudo-contractual forms based on notions of fairness and reciprocity, thus imputing prices even if not formally stated, and social control mechanisms enforce sanctions for malfeasance substituting for legal enforcement of contracts.

Intermediaries (e.g., wholesalers, financiers) controlling exchange among local communities operate in "communities" of mutual trust, and their network ties provide information about trustworthy exchange partners; social relations among intermediaries are mechanisms to enforce sanctions against malfeasance. Therefore, as eastern agriculture was transformed, local farm communities increased their engagement with the market economy. Wholesalers and retailers, among others, transmitted through price signals information about changing nonlocal demand for agricultural products. Farmers strategically determined supply responses: they reorganized labor tasks, changed crops, altered methods of growing crops and raising animals, and assessed the risks of producing low or high shares of their own foodstuffs versus producing for external markets. Their decisions did not negate social bonds; reciprocal agreements, sharing information, and ties within social organizations remained as powerful as ever. Successful responses to opportunities in the eastern economy meant rising per capita incomes for farm families; in turn, growing farm prosperity supplied capital and provided lucrative markets for local, subregional, regional, and interregional manufacturers.[10]

Roots of the Antebellum East

The antebellum East's agricultural and industrial transformation was rooted in economic, social, and political ferment from late colonial times through the Revolution to the early decades of the new Republic.[11] Robustly growing populations provided prima facie evidence that people enjoyed prosperous lives, relative to the standards of the time; New England and the Middle Atlantic colonies collectively grew at a compound annual rate of 2.8 percent from 1680 to 1780. They ranked among the world's healthiest, and real incomes increased at annual rates of as much as 0.6 percent; the pace accelerated from the 1740s to the Revolution. Aggregate economic output of the continental colonies soared about sevenfold in real terms between 1720 and 1774, widening opportunities for trade, finance, farming, and manufacturing. Greater population, production, and trade enhanced the division of labor and encouraged new efficiencies throughout the economy, and colonists diversified production, supplying ever-larger shares of their consumption. They shifted their import mix to higher-quality goods and increasingly purchased lower-quality goods from domestic sources, rather than importing them; this enlarged markets for local craft shops and small workshops, and their numbers proliferated during the late colonial period. Growing prosperity was exhibited in larger, more elaborate houses; greater per household consumption of clothing, dishes, silverware, and furniture; elegant public architecture; flourishing literature; greater numbers of clubs and associations; and founding colleges.

Although the New England and Middle Atlantic economies consumed most of their own production, they benefited from rising trade during the last several decades before the Revolution. New England and, to a lesser extent, Middle Atlantic merchants garnered substantial profits from wholesaling, shipping, and insurance through trade with West Indies' slave plantations, whereas trade with Europe often meant sharing or even taking a secondary position vis-à-vis European merchants. Colonial merchants invested profits in more trade, shipbuilding, and resource processing (sugar refining, rum distilling, lumbering), generating multiplier effects. Fishing and whaling contributed large profits to New England merchants, and trade with Europe gave them experience running sophisticated trade services. New England farmers and rural manufacturers supplied diverse production—barrel staves, shingles, potash, livestock, beef, and pork—for the West Indies trade. Boston and New York merchants profited from growing imports of manufactures from Great Britain and greater trade with the West Indies, and they engaged in coastal trade. Drawing on hinterland grain (mostly wheat), Philadelphia merchants controlled swelling grain trade with southern Europe. On the eve of the Revolution port merchants, interior storekeepers, craft workers, farmers, and resource extractors (fishing, lumbering) experienced rising prosperity, and this economy not only profited from foreign trade but also colonies increasingly integrated and the interior benefited. Although the Revolution disrupted this prosperity, accumulated capital, trade networks, infrastructure (buildings, roads, improved farmland), and skills stood ready to vault the economy to a higher level.

Social and political changes immediately preceding and following the Revolution strengthened the antebellum East's roots. The Republican Revolution

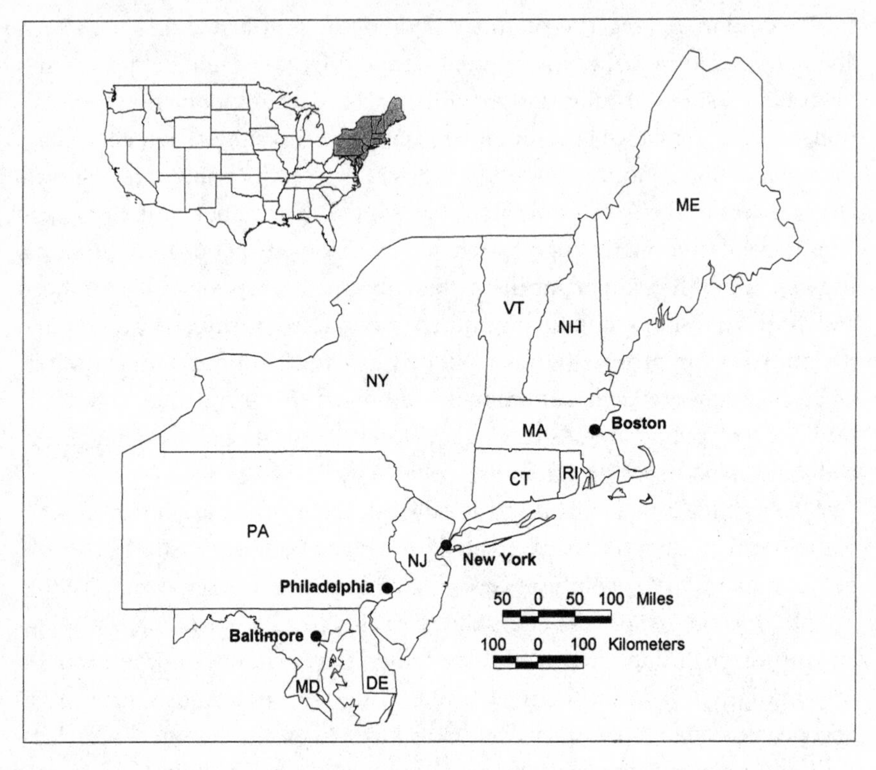

Map 1.1. The East and Its Leading Metropolises of Boston, New York, Philadelphia, and Baltimore

terminated colonial monarchical society; no longer would patronage and kinship of the "gentlemen" elite govern social, political, and economic relations. Workers were extolled for the dignity of their labor, and private interests became recognized in legal affairs as contracts shifted from declaring preexisting rights and duties to focusing on individual transactions. This opened flexibility in law; now business transactions could be negotiated as contracts, if parties wanted to formalize them that way. Equality meant more than "equality of opportunity"; it meant every person was as good as everyone else. That ideal was combined with praise for ordinary people's consumption; their desire to consume motivated their industriousness and productivity. These changes produced the political, legal, and social milieu liberating entrepreneurial and commercial dynamism in the antebellum East from the coastal ports to the frontier.

A Hypothesis

The puzzle of agriculture and manufacturing in the antebellum East can be resolved, but it requires revision of previous interpretations and reconsideration of evidence. Instead of decline, eastern agriculture built on colonial success and thrived from 1790 to 1860. Farm productivity rose, permitting surplus labor to enter factories and providing swelling volumes of farm products for growing urban populations. Because agriculture's competitiveness increased, many farms on poor soil distant from markets declined, but other farms on good soil and with market access—including those near urban centers—thrived as farmers specialized in high-value farm products. This farm population demanded manufactures, and combined rural and urban demand for manufactures in the East supported diverse industrial development. Manufacturers reduced production costs, and the lower prices of goods stimulated demand further. Transportation improvements exerted developmental impacts, but most benefits came from lowly wagons and road improvements, because wagons sufficed for transporting high-value agricultural goods and high-value manufactures.

The East sustained this developmental process as capital accumulation in prosperous rural areas and burgeoning cities supplied growing investment funds. This progress unfolded from 1790 to 1840, when most markets were in the East. As midwestern markets, primarily, and southern markets, secondarily, expanded after 1840, they complemented eastern markets and spurred larger-scale industrial growth in the East and a greater shift of farmers into production for urban markets. This hypothesis argues that each metropolitan regional hinterland of the East—around Boston, New York, Philadelphia, and to a lesser extent Baltimore—experienced similar transformations of agriculture and manufacturing (map 1.1). Their specialties diverged based on resource endowments, concentrations of innovative firms, and the decisions of investors, but overall developmental processes within each region were similar. These metropolitan regions did not grow in isolation before 1840; instead, businesspeople forged growing networks of interregional linkages within the East starting in the 1790s. Thus, this region experienced integrated, mutually reinforcing transformations of agriculture and industry.

Part 1 / The Early Republic, 1790–1820

CHAPTER TWO

Prosperous Farmers Energize the Economy

> The verdure, which here overspreads a great part of the whole region, . . . produces the most cheerful sense of fruitfulness, plenty, and prosperity. . . . Orchards, also, everywhere meet the eye. Herds of cattle are seen grazing the rich pastures. . . . Neat farm-houses, standing on the hills; a succession of pretty villages, with their churches ornamented with steeples. . . . The farmers, throughout this tract, are more generally wealthy than those of any other part of Connecticut. Their farms are chiefly devoted to grazing; and their dairies, it is believed, are superior to any others.
>
> —TIMOTHY DWIGHT ON QUINEBAUG RIVER VALLEY, CONNECTICUT, 1800

> Utica is still more improved. In 1799, there were 50 houses in this village; many of them small and temporary buildings. In 1804, the number was 120. The number of stores, also, and mechanics' shops, is very great in proportion to that of the houses. Utica now exhibits the appearance of a handsome town. Its trade has increased . . . and is greater than that of any other town in the state, New-York, Albany, and Troy excepted.
>
> —TIMOTHY DWIGHT ON UTICA, NEW YORK, 1804

Following the successful outcome of the Revolutionary War with England in 1783, the merchant wholesaler elite had no inkling that the economy stood on the precipice of a deep contraction that would last through the rest of the decade. They embarked on an orgy of purchases of British manufactures—textiles, clothing (e.g., gloves and women's hats), china, glass, locks, books, and playing cards—to meet pent-up demand, and British wholesalers eagerly granted ample credit. The flood of credit and imports worked their way to smaller-scale wholesalers, then to retailers, and finally to households purchasing manufactures on credit, but this commercial binge quickly foundered on a new reality: Britain no longer sheltered the United States within the empire. It faced exclusion from some British colonial trade and high duties on goods previously exported freely or at minimal duty to England; plummeting exports of Chesapeake tobacco to England dealt the harshest blow, and other southern staple exports (e.g., rice and naval stores such as pitch and tar) also declined. Likewise, northern states faced market losses: valuable exports of ships to England plunged, the whale fishery declined with lower demand for whale oil, and the fur trade dropped because the British controlled key trading posts. The United States lost access to trade with the British West Indies, which could not import salted fish and meats from the United States, and American merchants lost the carrying trade (sugar) to England. Trade with Mediterranean lands suffered without British naval protection, and the Spanish severely hindered exports of the Mississippi and Ohio Valleys through the New Orleans entrepôt. Finally, domestic prohibitions on the slave trade reduced rich earnings that could be made from importing slaves into the European colonies.[1]

Collapsing export earnings turned the credit binge into a debt crisis. Although England did not take all of the nation's exports, the imbalances in trade between them reveal the problems Americans faced: from 1784 to 1790 cumulative United States imports totaled 17.4 million pounds sterling, whereas exports were 6.6 million. Consumption declined as household income fell, leaving wholesalers, retailers, and producers with large inventories and forcing them to cut prices; this undermined their ability to pay for credit purchases, and falling asset values reduced debtors' ability to repay. Many small wholesalers and retailers went bankrupt, and even leading merchant wholesalers such as Samuel Otis of Boston and Clement Biddle of Baltimore went under. During the years from 1781 to 1785 the price level collapsed 57 percent in New York City and 28 percent in Charleston, South Carolina, and from 1784 to 1789 the debt-led contraction ground prices 18 percent lower in Philadelphia and 20 percent lower in Charleston.

This adverse economic contraction produced a consensus about the direction states should take; their fragmentation, embodied in the Articles of Confederation, hindered economic transformation. A group of visionaries including Benjamin Franklin, George Washington, John Adams, Thomas Jefferson, John Jay, and Alexander Hamilton led the movement to create and ratify an extraordinary document, the Constitution of the United States. In 1789 Washington's first presidential term began. The Constitution established a strong central government with powers to forge a national economy, and the ratification vote indicated this goal had widespread support. Key provisions granted Congress exclusive power to regulate foreign and interstate trade, created monetary union with control of the currency vested in the central government, and established the basis for national laws to protect commercial credit and to permit businesses to issue public securities to finance trade and production. Economic groups would continually try to achieve competitive advantages through appeals for tariff protection, a regular feature of domestic political debates. More significantly, merchants, farmers, and manufacturers gained free access to a national trade area.[2]

The Impact of Foreign Trade

In 1790, however, the national market posed stark challenges: the domestic market, numbering 3.9 million—including 0.7 million slaves who could not freely exercise demands—sprawled over a thousand miles from Maine to Georgia. Most of this low-density population lived east of the Appalachian Mountains, but 110,000 people had crossed into Kentucky and Tennessee, and firms faced immense obstacles selling to them. Rural shares of population averaged 94 percent during the period from 1790 to 1820. From 1790 to the Congress of Vienna in 1815, which brought peace to Europe, the United States took advantage of its "neutral" status in the Franco-British wars to enmesh itself in the world economy and gain considerable shares of trade. Americans did not offer exports of manufactures—except ships—at competitive prices, but they provided commercial trade services and domestic products drawn from natural resources.

Foreign trade expanded rapidly from 1790 to 1807: real value (1790 prices) of exports rose from $20.2 million to $79.5 million, reexports rose from $0.3 million to $43.8 million, and imports retained for home consumption rose from $23.5 million to $68.2 million. European crop failures and large armies prompted rising domestic exports of foodstuffs, which in 1807 constituted over 75 percent of all goods sold overseas; exports of lumber, pot and pearl ashes, and cotton grew, while fish, rice, and tobacco continued apace. Neutral American ships freely roamed the seas for products—sugar, tea, coffee, cocoa, pepper, gold and silver specie, bullion—from the West Indies, South America, and Asia to bring home for reexport to Europe. European and English manufactures of metal goods and textiles led imports, and luxuries and other articles such as sugar, molasses, coffee, wines, tea, spices, specie, and bullion also entered domestic markets. Trade collapsed during the Embargo of 1807 but fitfully recovered for several years before the War of 1812 with Great Britain doomed it again. After the war ended, in 1815, trade surged for several years as an entering flood of European manufactures temporarily crushed domestic producers, and then it stagnated into the 1820s. Booming trade during the neutrality period (1793–1807) seemingly pointed to foreign trade as the economy's locomotive, but it contributed little to income growth; per capita income grew 1.3 percent annually, whereas it would have risen 1.1 percent annually in the absence of foreign trade. Growth of seaport metropolises, merchant capital accumulation, and banking and commercial expansion probably were as consequential for economic development.[3]

Economic Growth and Structural Change

At this time population growth typically signified a healthy economy, because agriculture supported the population, whereas falling food supplies produced malnutrition, disease, rising death rates, and thus smaller populations. Evidence for the United States implies that the economy grew modestly, or, at minimum, food supplies kept pace with demand; population rose 3.0 percent compounded annually from 1790 to 1820, while the East grew more slowly, at 2.4 percent, as people moved west of the Appalachians after the turn of the century. Just three states—Ohio, Kentucky, and Tennessee—added 1.2 million people from 1800 to 1820, equal to one-fourth of the East's population in 1820.[4] Still, the East more than doubled in size from 1790 to 1820, offering farmers, merchants, and manufacturers swelling markets and opening possibilities for specialization and economies of scale. Measures of the East's economic growth are unavailable, but national estimates provide a proxy because the East's share of national population remained over half. Even with the contraction of the 1780s, gross domestic product (GDP) almost doubled from 1774 to 1793, but economic turmoil kept GDP per capita stagnant; during the neutrality years, however, from 1793 to 1807, GDP grew 4.5 percent annually, and GDP per capita grew 1.3 percent. The economy struggled from the start of Jefferson's embargo in 1807 until 1820 as GDP growth fell to 2.8 percent annually, barely matching population growth; thus, incomes stagnated. Nevertheless, since 1793 the economy of 1820, measured in GDP, had almost tripled in size, and per capita incomes increased by one-fifth. Just over half (55 percent) of the rise in GDP per capita during the years 1800 through 1820 came from a labor force shift out of farming into other sectors.[5] After spending for basic needs such as food, shelter, home manufactures, and farm improvements, consumers possessed considerable sums for other goods and services; about 25 percent of GDP per capita remained after necessities had been purchased (table 2.1).

Table 2.1. Gross Domestic Product and GDP per Capita, 1774–1860 (1840 Prices)

Year	GDP ($ Billions)	GDP/Capita Total $	GDP/Capita Above Basic Necessities[a] Subtotal $	GDP/Capita Above Basic Necessities[a] % of GDP/Capita
1774	0.17	70	—	—
1793	0.30	70	—	—
1800	0.41	78	19	24
1807	0.56	84	—	—
1810	0.59	82	21	26
1820	0.81	84	23	27
1830	1.16	90	27	30
1840	1.73	101	36	36
1850	2.58	111	44	40
1860	4.25	135	60	44
		Annual Rate of Growth[b] (in Percentage)		
Pre-1820 Period				
1774–1793	3.0	0	—	
1793–1807	4.5	1.3	—	
1793–1820	3.7	0.7	—	
1807–20	2.8	0	—	
Ten-Year Period				
1800–1810	3.6	0.5	1.0	
1810–20	3.2	0.2	0.9	
1820–30	3.6	0.7	1.6	
1830–40	4.0	1.2	2.9	
1840–50	4.0	0.9	2.0	
1850–60	5.0	2.0	3.1	
Twenty-Year Period				
1800–1820	3.4	0.4	1.0	
1820–40	3.8	0.9	2.2	
1840–60	4.5	1.5	2.6	

Sources: Weiss, "U.S. Labor Force Estimates and Economic Growth, 1800–1860," 27, 31–32, tables 1.2 (variant C), 1.3, 1.4 (broad definition); U.S. Bureau of the Census, *Historical Statistics of the United States*, ser. A7.

Notes:

[a] GDP per capita above basic necessities is the residual after subtracting the per capita perishable output, shelter, home manufacturing, and farm improvements from GDP per capita.

[b] Percent rate of growth computed by compound interest method.

Thus, a budding market for manufactures existed, and, as producers reduced the selling prices of goods previously made in the home, markets expanded even further.

Rising Wage Levels

Across farm and nonfarm occupations the real wages of workers rose from 1780 to 1820, indicating an improved standard of living. Although agricultural wages fluctuated significantly, they rose 1.0 percent compounded annually in Massachusetts and 1.6 percent in Maryland (fig. 2.1). Workers in nonfarm occupations also made modest gains in wages: manufacturing employees in the Brandywine region of Delaware—mostly workers at DuPont, the gunpowder manufacturer—achieved wage gains of about 0.5 percent annually from 1802 to 1820. DuPont workers annually saved about 18 percent of their wages from 1814 to 1820, and their long-term savings rate over the period from 1814 to 1860 averaged 16 percent annually, suggesting that they accumulated moderate amounts of capital. A broader group, the nation's unskilled nonagricultural workers, experienced a solid uptrend in wages of about 1.5 percent annually during the years 1780 through 1820 (fig. 2.2). And, if Philadelphia was representative, urban workers achieved modest annual wage gains of 0.7 percent for laborers, 0.8 percent for joiners who planed wood on ships, and 0.3 percent for house carpenters (fig. 2.3).[6] Thus, wages for farm and nonfarm workers in diverse occupations probably rose at compound annual rates of 0.5 to 1.0 percent from 1780 to 1820. The lower estimate falls within the range of independent growth estimates of GDP per capita, and together these measures confirm the population experienced a rising standard of living.

Capital Investment

Growing capital investment in farms, factories, transportation, and urban infrastructure spurred economic growth, and this investment not only lifted accumulated wealth as each new investment—net of depreciation—added to earlier capital stock but also raised productivity.[7] Domestic capital stock grew 4.2 percent annually during the years 1774 through 1799 and 5.8 percent from 1799 to 1805, substantially exceeding the population growth of 3 percent; thus, each individual accumulated ever-larger amounts of capital stock (table 2.2). Reflecting the difficulties during the Embargo of 1807 and the War of 1812, capital stock increased 2.3 percent annually during the period from 1805 to 1815,

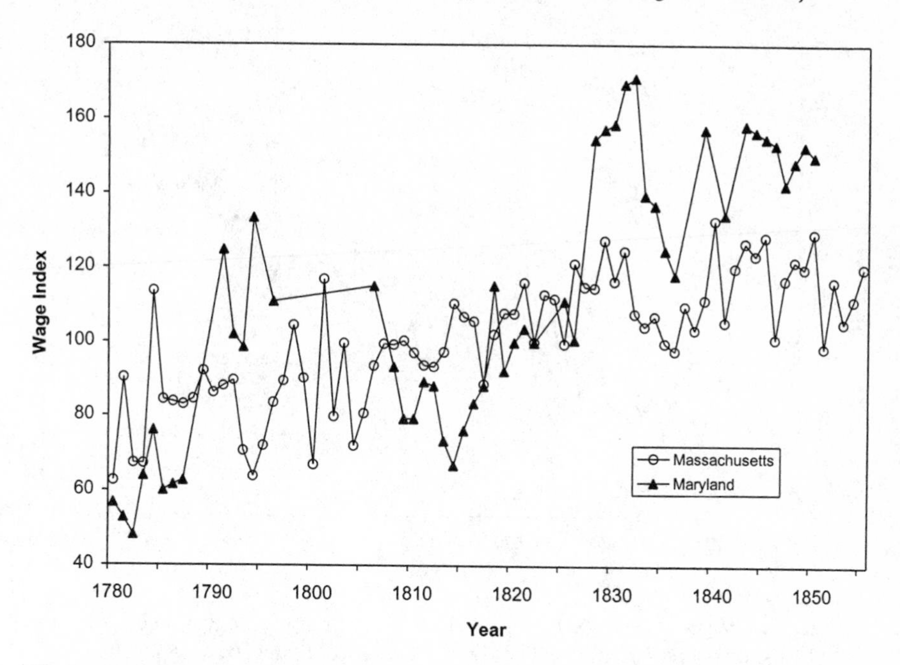

Fig. 2.1. Real Agricultural Wages in Massachusetts and Maryland, 1780–1855 (1822 = 100). *Sources:* Adams, "Prices and Wages in Maryland, 1750–1850," 630–31, table 2; David and Solar, "Bicentenary Contribution to the History of the Cost of Living in America," 16–17, table 1; Rothenberg, *From Market-Places to a Market Economy,* 176–79, app. C.

below population growth. Sectoral changes in capital stock per capita reflect choices about altering productivity.

From 1774 to 1815 the farm sector did not focus on bringing new land into production; the annual rate of change in land clearing and breaking failed to match population growth. Yet the farm sector accumulated capital to accommodate higher levels of food consumption (for meat and dairy products) and to boost productivity through the use of more animal power per capita for farms and transport. Because agriculture dominated the economy, swelling per capita investment in structures indicates that farmers added to their stock (e.g., houses and barns), though this investment also includes urban buildings, mills, workshops, and factories. Growth of per capita investment in equipment during the years 1774 through 1805, along with changes in structures, confirms that the economy acquired greater capacity to grow robustly.

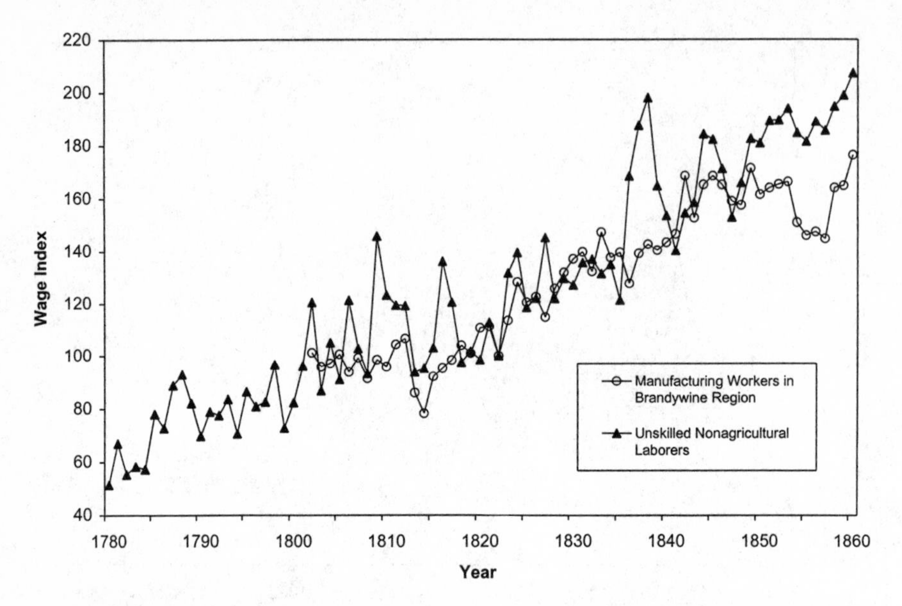

Fig. 2.2. Real Wages of Manufacturing Workers in the Brandywine Region of Delaware and of Unskilled Nonagricultural Laborers in the United States, 1780–1860 (1822 = 100). *Sources:* Adams, "Standard of Living during American Industrialization," 904–5, table 1; David and Solar, "Bicentenary Contribution to the History of the Cost of Living in America," 16–17, 59–60, tables 1, B.1.

Increased per capita stock of inventories likewise comports with economic expansion, because larger inventories were required to support greater production and consumption.

Alterations in the composition of capital stock offer a glimpse into the structural economic changes taking place, and the declining relative significance of the farm sector constituted the greatest shift (table 2.3). As a share of total accumulated capital stock, land clearing and breaking plunged from 56 to 36 percent between 1774 and 1815, and shares of capital stock related to animals fluctuated modestly; the importance of animals as food and power sources and as transportation inputs in the economy neared its antebellum apex around 1815. Substantial gains in structures as a share of capital stock signified broad economic growth, but equipment's share of capital stock remained about 5 percent, implying that the rising output of transportation vehicles, machinery, and tools tracked wider capital investment that was spurring economic expansion. Even as per capita investment in inventories rose to accommodate

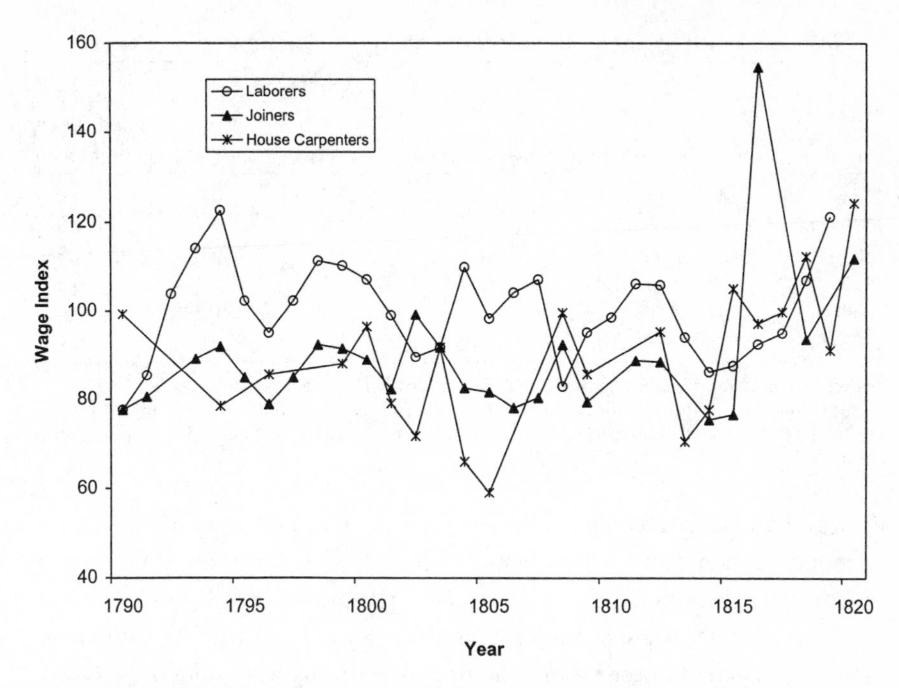

Fig. 2.3. Real Wages of Workers in Philadelphia, 1790–1820 (1822 = 100). *Source:* Adams, "Wage Rates in the Early National Period: Philadelphia, 1785–1830," 425–26, app. table 9.

rising income levels, inventories' share of total capital stock reached its antebellum peak about 1799, confirming an increasingly efficient distribution sector (i.e., wholesaling and transportation) with reduced relative amounts of capital investment in inventories required to support economic growth.

Urban Changes Signify Economic Changes

The availability of agricultural land in the East, even poor-quality land, typically offered most households opportunities to provide for the basic needs of food and shelter; they would not migrate to urban places unless better economic opportunities existed there. Therefore, the East's urban size growth (increased numbers of people in urban places) and urbanization (increased share of the population in urban places) should correlate with economic growth and structural change. Rising agricultural productivity lowers food prices for rural and urban dwellers, and income elasticity of demand for agricultural products

Table 2.2. Annual Percentage Change in Domestic Capital Stock, 1774–1860 (1860 Prices)

		Per Capita				
Period	Total	Structures	Equipment	Inventories	Animals	Land Clearing and Breaking
1774–1799	4.2	0.5	1.4	2.0	0.8	−0.8
1799–1805	5.8	5.5	2.6	1.0	−0.1	−1.5
1805–15	2.3	−0.4	−2.2	−1.7	0.7	−1.5
1815–40	4.0	1.4	1.7	0.7	1.3	0.5
1840–50	4.6	2.2	2.6	2.0	−1.5	−0.4
1850–60	6.6	5.1	6.0	1.5	0.0	0.3

Source: Gallman, "American Economic Growth before the Civil War," 89, 95, tables 2.4, 2.9.

Notes: Total capital stock excludes the value of clearing and breaking land. The capital price index is used as the deflator in the underlying indexes. Annual percentage change computed by compound interest method.

is less than unity (i.e., the percentage increase in the quantity of agricultural products demanded, the numerator, is less than the percentage increase in income, the denominator). Therefore, farm products' total sales grow less than the growth rate of gross domestic product. Surplus farm labor shifts from lower-productivity occupations in rural areas to higher-productivity ones—such as commerce, services, and manufacturing—in urban areas. These activities concentrate in cities because agglomeration economies—economies of scale in supplying concentrated populations and in providing services to businesses and economies of interfirm linkages—exist there, and increasingly specialized activities require exchanges that are better implemented in cities.[8]

The East's urban population more than tripled between 1790 and 1820 (see table 1.2), and the compound annual growth of 3.8 percent significantly exceeded the 2.4 percent growth of population, implying that economic growth, measured in GDP, and rising per capita income (GDP per capita and wages) produced an economic transformation (see table 2.1 and figs. 2.1–2.3). During the rapid economic growth of 1790 to 1810 the urban population increased 4.5 percent annually, whereas during the period from 1810 to 1820, when the economy suffered the most, urban growth fell to 2.4 percent. Urbanization directly indicates structural economic changes, because it measures the transformation from rural to urban occupations. The share of the East's population in urban places rose from 7 to 11 percent during the years 1790 to 1810, a period of rising economic prosperity, but the percentage stagnated during the subsequent decade of economic difficulties (see table 1.2). Eastern regions fared differently during the period from 1790 to 1820: the urban percentage in north-

Table 2.3. Composition of the Domestic Capital Stock, 1774–1860

	Percentage of Domestic Capital Stock (1860 Prices)					
Year	Structures	Equipment	Inventories	Animals	Land Clearing and Breaking	Total
1774	17	4	12	11	56	100
1799	19	5	20	13	44	101
1805	25	5	20	12	39	101
1815	27	5	19	14	36	101
1840	29	6	17	15	32	99
1850	33	7	19	12	28	99
1860	42	9	17	9	22	99

Source: Gallman, "American Economic Growth before the Civil War," 94, table 2.8.

ern New England stayed at 2 percent, whereas southern New England's share rose from 10 to 18 percent, and the Middle Atlantic's rose from 7 to 12 percent.

Finer territorial filtering reveals sharper differences in structural change from rural to urban occupations. Rapid urbanization progressed in Massachusetts, eastern New York, eastern Pennsylvania, and Maryland from 1790 to 1820, and Connecticut's small, steady rise in urban growth and Rhode Island's ascent during the years 1790 to 1810 confirm that their economies had been transformed (table 2.4). New Jersey's urbanization from 1810 to 1820 suggests that changes in nearby eastern New York and Pennsylvania swept along its economy. Western New York's urbanization spurted in the decade from 1810 to 1820, but the quick drop-off afterward implies that economic transformation remained incipient.

Businesses in the regional metropolises of Boston, New York, Philadelphia, and Baltimore led the economic transformation. Merchant wholesalers dominated foreign trade, and from 1790 to 1807 the ports grew on the basis of expanding exports; within the first ten years each witnessed increases in real per capita export values of 60 percent or more. Wholesalers generated powerful local multiplier effects in associated activities such as auctioneering, brokerage, banks, insurance companies, shipping, warehouses, stevedores, and hotels for business travelers. These businesses in turn produced local multiplier effects in construction, and the cumulative growth in employment and households stimulated retailing expansion. This economic growth supported large increases in metropolitan populations; their collective population size soared threefold, offering concentrated, burgeoning markets for hinterland producers of farm

Table 2.4. Urbanization in the East, 1790–1860

	Absolute Change in Percentage Urban						
States and Subareas	1790–1800	1800–1810	1810–20	1820–30	1830–40	1840–50	1850–60
Maine	—	0.7	−0.3	0.3	4.7	5.7	3.1
New Hampshire	−0.4	3.3	−0.2	2.0	5.0	7.1	5.0
Vermont	—	—	—	—	—	—	0.0
Massachusetts	1.9	5.9	1.5	8.3	6.8	12.8	8.9
Connecticut	2.1	1.0	1.5	1.8	3.2	3.4	10.5
Rhode Island	1.7	2.6	−0.4	8.3	12.6	11.8	7.7
New York	1.2	−2.4	−0.9	3.2	4.5	8.8	11.1
New Jersey	—	—	2.5	3.0	4.9	7.0	15.1
Pennsylvania	1.2	−1.5	0.2	2.3	2.6	5.7	7.2
Maryland	3.5	4.5	4.1	4.1	3.9	8.0	17.0
Eastern New York	5.1	2.4	3.5	6.8	5.6	7.9	4.2
Eastern Pennsylvania	3.3	2.4	2.4	3.2	5.2	7.8	18.0
Western New York	—	—	3.2	1.7	1.8	4.4	5.0
Western Pennsylvania	—	—	−2.8	1.1	1.6	4.1	1.9
United States	0.9	1.2	−0.1	1.6	2.1	4.5	4.5

Source: Williamson, "Antebellum Urbanization in the American Northeast," 600, table 1.

Table 2.5. Population of Eastern Metropolises, 1790–1860

Year	Population Size					Total as % of Urban in East
	Boston	New York	Philadelphia	Baltimore	Total	
1790	18,320	49,401	44,096	13,503	125,320	72
1800	24,937	79,216	61,559	26,514	192,226	71
1810	38,746	119,734	87,303	46,555	292,338	68
1820	54,027	152,056	108,809	62,738	377,630	69
1830	85,568	242,278	161,271	80,620	569,737	65
1840	118,857	391,114	220,423	102,313	832,707	61
1850	136,881	696,115	340,045	169,054	1,342,095	54
1860	177,840	1,174,779	565,529	212,418	2,130,566	53

Sources: Taylor, "American Urban Growth Preceding the Railway Age," 311–15, table 1; U.S. Bureau of the Census, *Thirteenth Census, 1910,* 1:82–83, table 57; *Seventeenth Census, 1950,* 1:146–47, table 23; *Historical Statistics of the United States,* ser. A195, 202.

Note: Population of New York based on consolidated boroughs.

Table 2.6. Combined Population Sizes of Eastern Metropolises and Their Hinterlands, 1790–1860 (in Thousands)

Year	New York	Philadelphia	Boston	Baltimore
1790	591	585	851	320
1800	862	772	1,066	342
1810	1,257	1,006	1,297	381
1820	1,695	1,261	1,476	407
1830	2,278	1,586	1,755	447
1840	2,822	1,989	2,029	470
1850	3,589	2,649	2,482	583
1860	4,524	3,354	2,828	687

Source: U.S. Bureau of the Census, *Historical Statistics of the United States,* ser. A195.

Note: Combined population size of each metropolis and its hinterland computed from state population sizes.

New York: ⅔ Connecticut, New York State, ½ New Jersey
Philadelphia: Pennsylvania, ½ New Jersey, Delaware
Boston: New England minus ⅔ Connecticut
Baltimore: Maryland

products, construction materials, fuel, and manufactures (table 2.5). But a decline in their urban share from 72 to 69 percent indicates that hinterland urban places grew faster, buttressing markets for hinterland products. Hinterland urban population more than tripled from 48,000 to 167,000, and some cities—Portland, Maine (7,169 people); Newburyport, Massachusetts (7,634); Providence, Rhode Island (10,071); New Haven, Connecticut (5,772); Albany, New York (10,762); and Lancaster, Pennsylvania (5,405)—had sizable populations as early as 1810. Reconfiguring the eastern population to territories approximating metropolises and their regional hinterlands indicates aggregate market size and growth (table 2.6).[9] The Boston region's population led the East in 1790, but by 1820 New York's region was larger; nevertheless, Boston's region remained a close second. Philadelphia's region maintained third rank during the period 1790–1820, whereas Baltimore's region remained the smallest. The regional markets of Boston, New York, and Philadelphia, especially, and Baltimore, secondarily, headed the transformation from agriculture to manufacturing, and their panoply of farmers, manufacturers, retailers, professionals, and merchant wholesalers recognized that improved transportation spurred regional growth.

Waterway Transportation

Many eastern farmers and resource extractors in lumbering, mining, and construction materials (e.g., stones and bricks) accessed superb natural waterway transport systems of rivers (the Connecticut, Hudson, Delaware, and Susquehanna), bays (the Narragansett, Delaware, and Chesapeake), and coasts. Sailing vessels provided low-cost transport along the Atlantic Coast and in large bays, and they navigated inland on major rivers such as the Connecticut to Hartford, the Hudson to Albany, and the Delaware to Philadelphia; navigation of arks, flatboats, and rafts on inland rivers offered mundane, low-cost transportation. The outlets of most major rivers at tidewater gave hinterland producers of low-value, bulky products access to markets in the metropolises, and inland waterway improvements increased their effectiveness as arteries. Because canals offered low-cost shipment of bulk commodities, they seemingly offered benefits to commercial agriculture and other natural resource activities; nevertheless, the United States failed to construct a successful canal prior to the Erie in the 1820s. Some observers attribute this lag to inadequate supplies of capital, limited knowledge of canal engineering, and financially unsuccessful indigenous examples, but these were not critical factors.[10] Prior to 1820 most major farming areas and other natural resource activities operated either in densely populated areas near large East Coast metropolises, where they relied on road transport, or the areas were near natural waterways; this made canals, with their large fixed capital costs, uncompetitive.

Wagons Provide Effective Transportation Services

Roads and Turnpikes

The ordinariness of wagons disguises their critical integrative position in interior transportation; they bound most origins and destinations directly or as adjuncts to other transport modes. Following 1790 economic prosperity created pressures to improve road transport for passengers and freight: local communities tended their own roads, but meeting increasing demand for subregional, intraregional, and interregional travel required companies and political jurisdictions with financial capacity to build longer roads. State governments jumped into the fray to build the longest roads; however, myriad demands for road improvements overwhelmed governmental capacities to organize, finance, and manage construction. States turned to chartered private or local public companies to build turnpikes and charge tolls to defray costs, which allowed interested parties—such as merchants, farmers, and local governments (principally commercial centers)—to lead road improvement efforts. The East's turnpike boom started slowly in the early 1790s, but completion of the Lancaster Turnpike in 1794 quickly brought attention to turnpikes as solutions to poor roads. Because this turnpike traversed rich, high-density agricultural land between Philadelphia and Lancaster, Pennsylvania, it became an immediate success, but that foundation of heavy demand often would be forgotten in future turnpike ventures.

Construction accelerated after 1795 in New England and after 1800 in the Middle Atlantic; by 1814 the boom started winding down and construction slowed markedly during the 1820s. The federal government contributed to one important project, the "National Road"; the 130-mile section from Cumberland, Maryland, to Wheeling, West Virginia, on the Ohio River was completed in 1818 at a cost of $13,000 per mile. High-quality turnpikes such as the National Road, with beds of crushed stone and gravel to depths of twelve inches or more, typically cost from $6,000 to $15,000 per mile to build. Most turnpike companies, however, built simpler, cheaper roads costing between $800 and $1,500 per mile; they cleared trees and rocks, graded, and sometimes laid a thin gravel layer. By 1820 the East's dense turnpike network linked all leading regional and subregional metropolises and tied hinterland communities to them. Southern New England built the densest turnpike network: radial networks of Connecticut's hubs—Hartford and New Haven—disappeared in the latticelike road system connecting virtually every town by 1814; Providence's radial network, mostly completed by 1820, crossed Rhode Island and reached to eastern Connecticut and neighboring parts of Massachusetts; and Boston's radial network reached every New England state by 1814, crowning it the regional hub.[11]

New York's turnpike construction commenced about 1800, and it built the most mileage: about 900 miles were completed by 1807, and this mileage soared to over 4,000 by 1821. Turnpikes punched westward from New York City and the Hudson River ports of Newburgh, Kingston, and Catskill into the upper valleys of the Delaware and Susquehanna Rivers and on to western New York. Albany, the state's turnpike center, built three trunk lines to western New York, and turnpikes reached north to the Lake Champlain district and east to the Berkshire Mountains of Massachusetts; the Hudson River ports south of Albany forged connections to western Connecticut, and New York City and Albany were connected on both sides of the Hudson. Following the Lancaster Turnpike, other construction in Pennsylvania lagged until after 1800; the state started subsidizing turnpikes through subscriptions to their stock in 1806. Early turnpikes radiated from Philadelphia to major towns east of the Susquehanna River, and Easton, on the Delaware River, formed a secondary hub with turnpikes to the Susquehanna and northeastern Pennsylvania. Construction of trunk lines to western Pennsylvania commenced after 1811, but the 1,807 miles of completed roads in Pennsylvania still lay mostly east of the Susquehanna by 1821. Turnpike companies finished 2,400 miles, including major trunk lines, by 1832: two stone-surfaced roads arced north and south from Harrisburg to Pittsburgh, and from Philadelphia two roads reached north into New York, and one road angled to western New York. Maryland's turnpike construction started around 1805, and the state's turnpikes formed a radial network from Baltimore by 1825: west to Frederick and Cumberland, joining the National Road; and north to York, Hanover, and Gettysburg in the rich farmlands of southern Pennsylvania. New Jersey benefited from its crossroads position between Philadelphia and New York, and there turnpike construction commenced shortly after 1800; most of the mileage, totaling about 550, was completed by 1816. New York merchants helped underwrite turnpikes across northern New Jersey to the Delaware River; they were built to attract the farm and mine products of New Jersey to New York City and to connect with other turnpikes drawing similar goods from northeastern Pennsylvania and adjacent New York State. Several turnpikes serviced the corridor from New Brunswick, at the southern edge of New York harbor, through Trenton to Philadelphia.[12]

Few eastern turnpikes profited owners, and they rarely averaged dividends of more than 2 to 3 percent, far below those of safe government bonds at 5 to 6 percent. The fact that construction continued after unprofitability quickly became apparent suggests that other factors governed investors' decisions. Even the best stone-laid turnpikes were not exorbitantly expensive, and intersections of turnpikes with local roads conferred widespread benefits on communities, which made turnpikes appealing. Merchants, professionals (e.g., lawyers and physicians), and other wealthy individuals supported them and convinced local governments to help underwrite projects, thus spreading costs. These individuals risked limited capital, with little hope of adequate returns on investment, but they gained indirect benefits—such as easier travel and greater business. Evidence from studies of English roads, similar to those in the East, suggests road and turnpike investments generated substantial economic benefits; better surfaces and easier gradients together offered the most benefits, boosting the amount of goods each wagon horse conveyed. From the 1730s to 1838 transport productivity along a major road between Leeds and London rose 1.1 percent annually.[13]

Transport Costs

Wagon transport on turnpikes and other roads in the East proved to be formidable competition for carriage on rivers, and subsequently, on canals and railroads. These modes usually required wagon transport at both origins and destinations, but road transport cost plus the cost of two transshipments did not appear directly in transport rates of other modes, yet transshipment expense—including the use of warehouses, managers, and laborers to transfer and store goods temporarily—added substantially to costs. Commercial wagon rates more directly reflected total transit costs between final origins and destinations because wagons provided door-to-door service, eliminating or sharply reducing transshipment expenses. Over shorter distances wagon transport cost less than waterways, canals, or railroads because low transshipment costs for wagons compensated for higher variable costs per mile, whereas over longer distances other modes cost less than wagons because higher transshipment costs were averaged over greater mileage; thus, lower variable costs per mile dominated total transit costs. Weight, bulk, and value of goods also entered into the transit equation. Over short distances waterways and canals offered cheaper alternatives than wagons for carrying heavy, bulky, low-value commodities such as cordwood (for fuel), bricks and stone (for construction). Yet wagons competitively transported high value per unit weight agricultural commodities—such as eggs, butter, cheese, vegetables, and fruit—over distances of up to one hundred miles, and they competitively carried light, low-bulk, high-value manufactures—items such as pins, clocks, tinware, shoes, textiles—over distances greater than one hundred miles. Wagons offered much greater speed, reliability, and efficiency for moving those goods.[14]

Estimates of commercial wagon rates provide a standard to evaluate the competitiveness of wagons versus waterways, canals, and railroads. Before 1820 ton-mile wagon rates ranged from twenty-five to fifty cents for long-distance carriage from eastern port cities across the Appalachians, and twenty-seven cents per ton-mile was a frequently cited rate for wagon carriage from Buffalo to New York City. Those rates applied to long-distance transport over poor roads between the East Coast and the Midwest when manufactures and other supplies moved westward, but only small volumes of low-value, bulky commodities moved eastward; therefore, high westbound rates embodied the costs of lightly loaded eastbound wagons. East-bound commodities such as wheat and flour could not withstand the cost of a three hundred–mile wagon carriage, because commodity prices (including transport costs) exceeded market prices in eastern cities. Wagon transport on better roads within one hundred miles of East Coast ports, however, cost from one-half to as little as one-third of long-distance carriage across the Appalachians before 1820. Wagon shipment on the sixty-two-mile turnpike between Philadelphia and Lancaster cost about fourteen cents a ton-mile in 1807, and Massachusetts farmers' real costs to transport goods to markets, computed as if they hired commercial haulers, was fifteen to eighteen cents a ton-mile from 1750 to 1820.[15]

Within inner hinterlands of eastern ports farmers' commercial wagon rates remained about fourteen to eighteen cents per ton-mile before and after 1820, yet these rates assume that farmers employed teamsters or implicitly charged themselves commercial costs for trips as if they hired oxen (or horses), wagons, and drivers. Farmers could choose an alternative to commercial rates: if they reduced leisure time and boosted labor contributions of other family members, they could compensate for the opportunity costs of labor time to drive and repair wagons and raise oxen. This strategy cut out-of-pocket transport costs by 50 percent or more to as low as seven to nine cents per ton-mile before

1860, because almost half of inputted costs represented charges for drivers, and the remainder consisted of charges for oxen and wagons, including some labor time; on longer trips farmers carried food and slept next to wagons to reduce costs. Most Massachusetts farmers took goods to market rather than hiring others, and their average distance traveled was almost twenty miles; numerous farmers drove fifty miles, and some traveled up to eighty miles. They rarely used turnpikes, indicating that even country roads served their needs adequately, but, with rising agricultural prosperity, farmers switched to hired drivers. In Worcester County, about forty-five miles west of Boston, most farmers took products to Boston before 1820, and only the most prosperous hired others; by the 1820s, however, many farmers hired teamsters to truck products to markets in Boston and other urban places.[16] Regardless of the transport costs paid by eastern farmers, they employed another strategy to lengthen distances for wagon transport: processing farm commodities to raise value per unit weight before shipment to market, thus reducing the share that transport costs comprised of final selling prices at markets. This explains the ubiquity of gristmills, distilleries, and other processing mills in rural areas; they served local consumption and preparation for market sale. Farmers transformed wheat to flour, corn to whiskey, apples to cider, cattle and pigs to salted meat, and milk to butter and cheese, and, as urban market demand rose, they shifted from grains to higher-value crops such as hay, fruits, and vegetables.

Until about 1850 wagons offered fierce competition to waterways, canals, and railroads for transporting farm products and most manufactures. Concentration of the East's population within one hundred miles of ports, navigable rivers (including the Connecticut, Hudson, Mohawk, Delaware, and Susquehanna), and coastal waterways—modes with low fixed transport costs—exacerbated the competitive positions of canals and railroads. Distances up to fifty miles from markets or navigable waterways, easily within range of wagon transportation, encompassed the East's economic core: southeastern Maine and New Hampshire; all of Massachusetts, Connecticut, and Rhode Island; the Hudson and Mohawk Valleys of New York; eastern Pennsylvania; New Jersey; Delaware; and virtually all of Maryland.[17] Beyond this fifty-mile zone canals and railroads effectively competed with wagons, but the population densities of farmers declined, and revenue remained low, and this recipe for financial failure generated little or no social savings of canals and railroads versus the wagon alternative. To succeed, canals needed to run long distances through rich agricultural farmland distant from markets and from competitive water transport, or they needed to carry huge volumes of low-value, bulky commodities (e.g., construction stone, cordwood, or coal). In the early years of railroads success depended on tapping numerous passengers from high-density inner hinterlands of metropolises and other large cities. Agriculture thrived in this environment of efficient transportation.

Agriculture Thrives

Between 1790 and 1820 numerous households moved across the Appalachians to establish farms in Ohio, Kentucky, and Tennessee, but this outmigration did not produce a widespread rural decline in the East; most areas losing population had poor agricultural land, and rural numbers increased every decade (see table 1.2). The rural population grew 2.3 percent annually, but this figure disguises the different rates in northern New England (2.9 percent), southern New England (0.5 percent), and the Middle Atlantic (2.7 percent); the latter accounted for the majority of the net increase in rural population (74 percent). Rural dwellers mostly worked on farms, but others labored in lumbering and mining and associated processing industries (e.g., sawmills and smelters) and in agricultural processing (e.g., gristmills, flour mills, flax mills, and wool carding). From 1800 to 1820 the farm labor force grew 1.9 percent annually; northern New England states (Maine, New Hampshire, and Vermont) witnessed large increases in farm laborers, whereas southern New England states (Massachusetts, Connecticut, and Rhode Island) maintained stable numbers even as numerous residents emigrated after 1790 to northern New England, central and western New York, northern Pennsylvania, and northeastern Ohio (fig. 2.4).[18] In the Middle Atlantic region New Jersey, Delaware, and Maryland also maintained a stable farm labor force, but Pennsylvania's—the East's largest in 1800—grew 1.9 percent annually between 1800 and 1820; nevertheless, New York's farm laborers surged 4.1 percent annually, and it became the greatest agricultural state in the East. Extrapolating from national figures, farm labor productivity probably increased modestly, and by 1820 productivity was 3 percent higher than in 1800 (table 2.7). National figures, however, include farmers in western New York, northwestern Pennsylvania, Ohio, Kentucky, and Tennessee, who devoted extensive labor to clearing land and constructing fences, houses, and barns, and, initially, they faced little incentive to increase production for distant markets because transportation costs were too high. The rise in productivity in the East's prime agricultural

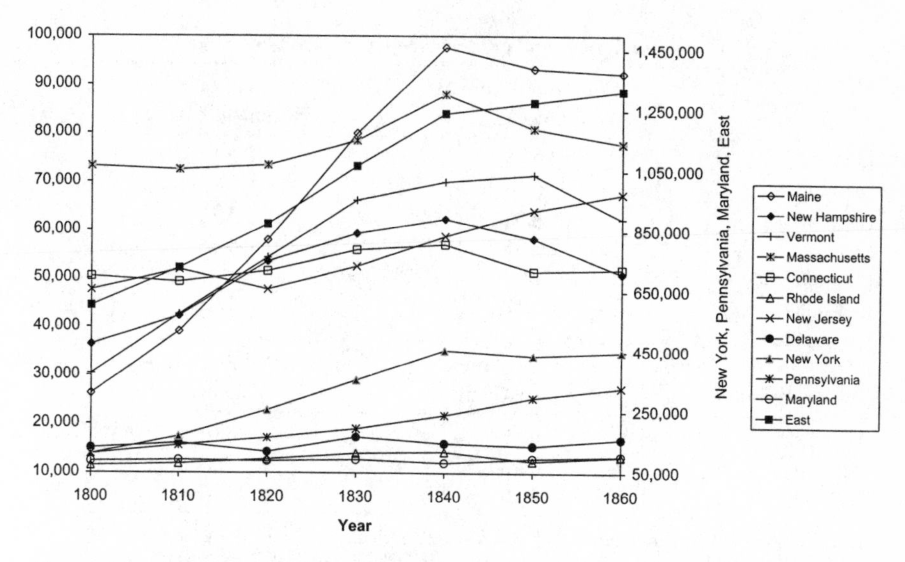

Fig. 2.4. Number of Farm Laborers by State and for the East, 1800–1860. *Source:* Weiss, "U.S. Labor Force Estimates and Economic Growth, 1800–1860," 51, table 1A.9.

areas—within fifty to one hundred miles of Boston, New York, Philadelphia, and Baltimore—certainly was far greater. The urban share of the East's population rose from 7 percent in 1790 to 11 percent during the second decade of the nineteenth century (see table 1.2), indirectly confirming that those areas' farm labor productivity increased faster than indicated by the nation's 0.15 percent annual productivity increase for the period from 1800 to 1820 (see table 2.7).

Within each eastern region farmers responded to local opportunities and to competitors from elsewhere in their region and from other regions; their agricultural responses were textbook examples of Von Thunen principles that posit a large urban center (the market) and agricultural land use based on distance to the city. Nearby farmers faced lower transport costs than those farther away; therefore, those who were closer specialized in bulky and/or perishable, high-value products such as fluid milk, vegetables, fruits, and hay. Distant farmers specialized in low-value grains (wheat) transported cheaply in bulk or processed form (e.g., flour), or they converted grains into higher-value products such as meat (e.g., beef and pork) or dairy (e.g., butter and cheese), whose transport costs constituted small shares of final market prices. Because farms near the city occupied high-value land, they had incentives to raise productivity to maximize returns per acre, and they had a complementary

Table 2.7. Change in Agricultural Productivity per Worker, 1800–1860

10-Year Period	Compound Annual Percentage Change	Period Percentage Change
1800–1810	0.24	2.42
1810–20	0.06	0.59
1820–30	0.35	3.53
1830–40	0.55	5.68
1840–50	0	0
1850–60	1.12	11.83
20-Year Period	**Compound Annual Percentage Change**	**Period Percentage Change**
1800–1820	0.15	3.03
1820–40	0.45	9.41
1840–60	0.56	11.83

Sources: Geib-Gundersen and Zahrt, "New Look at U.S. Agricultural Productivity Growth, 1800–1910," 685, table 6; Weiss, "Long-Term Changes in U.S. Agricultural Output per Worker, 1800–1900," 331, table 2.

relation with nearby city markets because cities provided fertilizer as night soil and as horse manure from draft animals.

Eastern farmers did not simply follow rote Von Thunen principles; between 1783 and 1820 they instituted supply responses to growing market demand. Increasing numbers of farmers developed remedies for soil exhaustion and enhancements to fertile soil, boosting productivity. They applied fertilizer (e.g., manure or gypsum), rotated crops, improved pastures through planting "English grasses" (e.g., timothy or clover), and irrigated meadows and drained marshes to expand pastures. Breeding and other improvements in raising livestock commenced, although breeding remained primarily the domain of elite farmers. To control procreation and to protect livestock lines, breeding required heavy capital investment in fences and barns, and labor costs for new livestock breeds were higher because they needed greater controls on activities. Fences and barns improved livestock production, regardless of breeding, because livestock could be protected and fattened more efficiently.[19]

Massachusetts

Being at the forefront of eastern agricultural transformation from 1770 to 1800, Massachusetts farmers instituted farm improvements, and they shifted from grains to hay two to four decades before facing competition from farmers in New York and the Midwest. Their shift into hay, especially the cultivation of English grasses, signified their commitment to boosting productivity, because hay constituted the core feed input for dairy products (e.g., milk, butter, and cheese), meat, and hides, and increased animal manure fertilized hay and other crops. Farmers near cities held a competitive edge in supplying bulky, heavy hay to urban liveries, which supplied fertilizer to nearby market gardens. Farmers instituted management improvements dealing with livestock (stall feeding), woodlots, fruit orchards, and meadows and other pastures, and they restructured farm space through land clearing, fencing, and connecting barn buildings. Most improvements involved better-utilized labor, and benefits accrued from applying ideas; productivity gains did not rely on changes in farm technology (equipment changed little before 1820) nor on extensive capital. These changes spread statewide between 1785 and 1800, and greater agricultural productivity permitted a shift into other economic activities; thus, agricultural taxes declined as a share of total taxes. The real wage gains of about 1 percent annually during the period from 1780 to 1820 for farmworkers indirectly confirms farm productivity increased (see fig. 2.1).[20]

Agricultural improvements both contributed to, and were a response to, greater market integration in Massachusetts. In the 1790s farm product prices across the state started converging (price variability declined), and the transmission of commercial information produced increased market integration; transport cost reductions exerted minimal impact because costs varied little between 1790 and 1820. Farmers also were integrated into East Coast markets; farm gate prices in rural Massachusetts moved synchronously with wholesale prices in New York City and Philadelphia. Wages of farmworkers began to converge in the 1790s, and limited labor mobility meant that information, similar to farm prices, operated as the mechanism of labor market integration. Between 1791 and 1820 market wagon trips (measured in one-way distances) of Massachusetts farmers and of some farmers in nearby states demonstrate that they participated in large market areas. Excluding the longest trips, farmers' trips averaged about twenty-six miles, with a median of about nineteen; few traveled five miles or less, whereas the most common trips extended fifteen miles, and the second-rank trips reached twenty-five miles. The substantial frequency of sixty-mile trips (fourth rank)—about one-third of the most common distance (fifteen miles)—confirms that farmers looked to distant markets, and Boston accounted for around 25 percent of trips; the important destinations of Salem (a port) and Northampton (in the Connecticut Valley) each accounted for about 6 percent of trips.[21]

Near Boston

Massachusetts farmers' response to market opportunities depended on market access. Brookline, just three miles from Boston's merchant core, exemplifies how metropolitan growth encouraged agricultural intensification in the innermost Von Thunen ring from 1771 to 1821 (map 2.1). The number of farmers remained stable, but cultivated land more than doubled. Grain production—mostly corn for feed and human consumption—more than doubled, and hay output rose by 20 percent, averaging eighteen tons per farm by 1821; as a share of all hay harvested, high-quality English grasses rose from 58 to 76 percent. Although cattle numbers declined, farmers probably shifted to dairy cows to produce fresh milk, thus consuming their production of feed corn and hay. Brookline farmers leveraged their proximity to Boston to sell hay to liveries, and farmers increased the use of manure from those sources. Befitting their location on high-value farmland, farmers engaged in labor-intensive production and boosted the use of hired labor. Fruits and vegetables became greater components of output; the 39 percent rise in cider production confirms the shift to more intensive production.[22]

Agricultural change also swept up Concord farmers, located seventeen miles west of Boston; its land values per acre fell below Brookline's, but Concord farmers incurred higher transport costs (see map 2.1). Production choices reflected access to local markets and the need to transport goods to Boston and ports northward—Salem, Marblehead, and Gloucester; comparisons with Brookline highlight these choices. Concord's farm count remained around two hundred from 1771 to 1826, or four times as many farms as in Brookline. Corresponding to Concord's greater distance from Boston and other commercial cities, woodland and unimproved land constituted almost 40 percent of its land up to 1811, and in that year woodland alone accounted for 26 percent of land use. Tilled land amounted to only 8 to 11 percent of land in use during the years 1781–1821, and grassland accounted for half of land use; most farms kept 80 percent of their improved land (i.e., tilled plus grassland) in grass. Concord farmers raised productivity: during the twenty years from 1781 to 1801 they planted 25 percent of their hay fields with English grasses, and by 1821 they boosted that amount to one-third. Nevertheless, they engaged in less-intensive

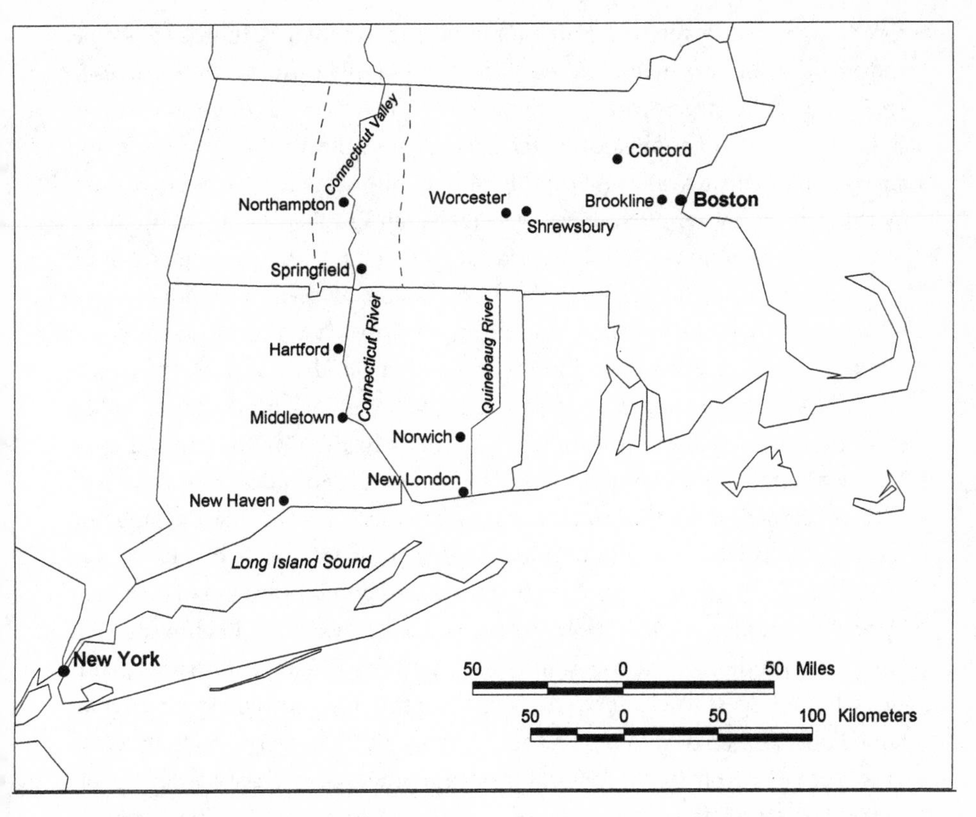

Map 2.1. Massachusetts and Connecticut

farming than those in Brookline; typical farms produced two-thirds as much hay and half as much cider. Milk's characteristics (heavy, bulky, and with a low value per unit weight) and the need to maintain freshness through rapid transit to market meant that Concord farmers could not compete with farms near Boston. Thus, few farms shifted to milk production, and they also produced little butter and cheese, because the profits fell below those from selling hay, corn, and oats as feed and, perhaps, corn for human consumption. Except for cider, little direct evidence exists on these farmers' production of fruits and vegetables, but these probably constituted small shares of sales.[23]

Farmers Farther from Boston

In contrast, the Ward family farm in Shrewsbury, in Worcester County, was over twice as far west of Boston as Concord farmers (thirty-six miles), and before 1820 the nearby town of Worcester housed less than three thousand people, too small to support many farmers. The Ward family owned over 275 acres in 1798, one of the largest farms in the county; thus, their production exemplifies the behavior of sophisticated farmers. By 1794 the Wards drove oxen and sheep to markets in Watertown and Roxbury, immediately adjacent to Boston. Within a decade after 1800, they engaged Nathan Pratt, a friend and neighbor, to transport beef on the hoof, dressed meats (beef, pork), grain, cider, butter, and cheese to Boston on a regular basis. Many of their neighbors shifted to hiring carriers during the decade from 1810 to 1820, and most Worcester County farmers copied them by 1820, following standard Von Thunen principles and transforming output to higher–value added products, including fresh produce, for the Boston market. Farmers in the Connecticut Valley of Massachusetts—about 90 miles west of Boston and 160 miles from New York City—occupied one of the largest fertile agricultural areas in New England, but their products had to withstand much higher transport costs than farmers in Worcester County (see map 2.1). Farmers shipped goods by wagon to Boston or by boat on the Connecticut River and then by coastal waters to Boston and New York City. Productive farms supplied far more food than local markets consumed; the largest towns in 1790—Northampton and Springfield—each contained only sixteen hundred people. By 1790 farms no longer were competitive producers of wheat (for flour) for eastern markets or overseas; wheat farms in southeastern Pennsylvania and Maryland had lower production costs and lower transport cost to major ports (e.g., Philadelphia and Baltimore). Valley farmers continued growing wheat and a variety of other crops—including rye, corn, hay, beef, pork, dairy products, fruits, and vegetables—for their own consumption, but only the largest farms diversified; farmers often exchanged surpluses to meet their needs.

During the period from 1790 to 1820 valley farmers boosted the value of farm goods to withstand long-distance shipment, and beef, pork, butter, and cheese became leading products. Growing markets in Boston and New York City and the robust West Indies markets (up to 1807) pushed food prices higher, spurring changes in farming; by 1800 and continuing to 1820, real land values rose substantially. Farmers raised productivity: they cleared land for pasture, converted pastures to tillage and for planting hay, planted more English grasses to raise hay quality, improved tillage and meadowlands, increased the use of manure for fertilizer, and improved their methods of seed drilling. The valley shifted toward specializing in livestock raising and fattening cattle and hogs, and farmers drove animals on the hoof to Boston or sent them as beef and pork by wagon; they also increased their production of butter and cheese for wagon shipment to Boston. The proliferation of village and country stores after 1800 reflected the greater involvement of farmers in markets outside the valley: these stores collected rural produce and sold manufactures and food products from outside the valley, and stores exchanged with larger retailers in Northampton and Springfield. These retailers and the wholesaler-retailers in Springfield organized the shipment of farm products outside the valley, some of which exited down the Connecticut River.[24]

Thus, from 1790 to 1820 Massachusetts agriculture formed Von Thunen rings of more intensive agriculture close to Boston and less intensive farther away. Boston's and, to a lesser extent, other coastal ports' growth boosted demand for agricultural products, and farmers in each ring increased productivity and supplied larger volumes of goods. Higher rural incomes supported more commercial services in small towns and villages, adding to the demand for farm products. In contrast, Connecticut's agricultural transformation raises a conundrum: observers dismissed the significance of wagon transportation prior to 1820, concluding that most farmers had no access to profitable markets, and the state did not house a regional metropolis of Boston's stature—most farmers were more than fifty miles from either Boston or New York City. Long-standing views of well-known analysts such as Grace Fuller held sway: she described early-nineteenth-century Connecticut as "a loose collection of largely isolated and nearly self-sufficing rural communities."[25] This depiction contrasts, however, with evidence that its citizens ranked among the nation's wealthiest.

Connecticut
Stagnation or Structural Change?

From 1790 to 1820 large numbers of Connecticut's citizens registered views on their economic prospects—they had emigrated and followed paths dating from the 1750s. Population increased only 16 percent in thirty years, whereas nearby Massachusetts and Rhode Island grew by 38 and 20 percent, respectively, and the nation's population exploded, rising 145 percent. The nation's growth reflects a net natural increase (births minus deaths), because during that period few immigrants arrived; applying this growth rate to Connecticut, the net loss of population through out-migration during the years from 1790 to 1820 equaled the state's entire population in 1790 (238,000). Emigrants moved to Vermont and New Hampshire, but most headed to central and western New York, northern Pennsylvania, and the Western Reserve of Ohio (i.e., the northeast quadrant). Most were farmers, and large numbers of them drained from hill towns with poor soil. Yet Connecticut also suffered a "brain drain," as merchants, lawyers, doctors, and clergy left, rising to distinction as business leaders, judges, governors, and college presidents in their adopted states.[26] This emigration should have produced economic stagnation or decline; structural changes, however, supported economic growth.

The roots of this emigration date back to the period from 1690 to 1750, when Connecticut's economy shifted from a focus on local markets to increasing bonds with other East Coast regions and the Atlantic economy. The British West Indies purchased cattle, horses, and staves, and the French West Indies became a market after 1717; Newfoundland bought these provisions, as well as timber, and markets in Nova Scotia expanded in the 1720s; and fishing fleets found markets in Europe, the South Atlantic, and Caribbean. Coastal and West Indies commerce accounted for most of Connecticut's trade, but export earnings constituted a minority of economic activity. Nevertheless, they contributed to agricultural wealth based on the small production of numerous goods, and they generated local multiplier effects: farmers hired workers to clear land; gristmill operators expanded to handle growing agricultural output; consumers demanded goods from craft workers and stores; and country storekeepers, town retailers, and wholesalers demanded transportation improvements. Export earnings paid for imports of consumer and capital goods, and this economic expansion encouraged families to have children; the population grew by 2.2 percent annually during the years 1690 to 1790, close to New England's growth rate (2.3 percent). This growth increased the labor supply, permitting greater production from existing and new farms, and it raised local demand for goods and services. By 1760 new farmers moved into areas of hilly, poor soil. Pivotal merchants in Hartford, New Haven, Middletown, Norwich, and New London (see map 2.1) combined wholesaling and retailing: they collected surplus farm and other natural resource products from their retail operations and from country storekeepers and forwarded them to Boston or New York or directly exported them overseas; in return, they sold imports at retail to consumers and at wholesale to country storekeepers. Eastern Connecticut exported goods to Boston, New York, and Providence, and western Connecticut forged ties by road and coastal vessels to New York.[27] These wholesale-retail merchants were poised to facilitate expansion following the 1780s contraction.

Agricultural Prosperity

Foreign exports of Connecticut's agricultural and other resource products to the West Indies and Central America thrived from 1793 to 1807 then declined erratically but recovered after 1815. Similar to colonial times, exports ranged from cattle, horses, beef, pork, vegetables, and grain to lumber products (e.g., staves, hoops, and boards), but the coasting trade provided steadier outlets because Boston and New York housed large, growing populations and Charleston, South Carolina, was an entree to southern markets (see table 2.5). Foreign and domestic markets encouraged Connecticut farmers to devote effort and capital to improving land, adopting better agricultural techniques, forming agricultural societies, and organizing fairs and thus boosted agricultural productivity, similar to their Massachusetts neighbors.[28] During travels through New England and New York between 1795 and 1815, Timothy Dwight, the peripatetic president of Yale College, chronicled the prosperity of Connecticut's farmers.[29] On his trip in 1796 the Connecticut Valley from Middletown to Suffield elicited warm praise: about Suffield he wrote, "The houses on both sides of the street are built in a handsome style; . . . and in the midst of lots universally covered with a rich verdure and adorned with flourishing orchards, exhibit a scene uncommonly cheerful."[30]

In 1800 he portrayed the Quinebaug Valley of eastern Connecticut as a wealthy farming area specializing in fruit, cattle raising, and dairying (see map 2.1).[31] On a trip in 1798 through northwestern Connecticut, Litchfield and Goshen drew raves: "Litchfield is a handsome town, . . . The houses are well built, and the courthouse is handsomer than any other in the state. . . . [Southwest from the center is a valley] with a collection of good farmers' houses, luxuriant fields, and flourishing orchards. . . . [Goshen] is, perhaps, the best grazing ground in the state; and the inhabitants are probably more wealthy than any other collection of farmers in New-England." On a trip in 1811 that included the shore towns of Fairfield County, Dwight admired the farms: "The soil is better than that of any other in the state, being generally rich, and producing everything which the climate will permit. The pastures and meadows are fine, and the crops of grain are abundant."[32]

Dwight's chronicles cannot be dismissed as uninformed hyperbole. From his perch as president of Yale, this astute observer was a hub in an extensive social network of elite reaching across religious, social, economic, and political spectrums of Connecticut, providing him extraordinary access to information. John Pease and John Niles confirmed Dwight's observations, though less eloquently, and detailed the market achievements of Connecticut's farmers in 1819.[33] Boston's market attracted farmers in northeastern Connecticut, and most of the state—except the northwest corner—produced surpluses that merchants exported to the West Indies or sent in coastal ships directly to southern markets; however, market signals from New York City dominated farmers' decision-making calculus. Specialties corresponded to Von Thunen production and confirm Dwight's portrayals. Producers near New York City and along waterways specialized in heavy, high-value products—such as vegetables, fruits, hay, and construction materials—whereas those farther away specialized in products that withstood long-distance shipment: high-value butter and cheese or low-value grains that could be transported cheaply in bulk, processed (e.g., flour), or converted into higher-value products, such as meat (beef or pork) and liquor (gin or cider brandy). Producers along Long Island Sound from Greenwich to New Haven possessed superior advantages: their nearness to New York City (twenty to eighty miles), the availability of cheap water transportation, and the existence of the trunk line coastal turnpike. Numerous sloops and schooners plied the Sound between New York and the coastal ports of Greenwich, Stamford, Norwalk, Stratford, and New Haven; the latter also had steamboat service three times weekly to New York, New London, and Norwich. Many shore town farmers produced corn, rye, beef, pork, butter, and cheese for New York City as well as for the West Indies and the South. Most towns specialized in potatoes for New York, and Greenwich grew roots and vegetables, Fairfield produced fruit, Norwalk sent lumber, and New Haven forwarded oysters. Producers in coastal towns from New Haven to Stonington accessed vessels active in the coasting trade, and many vessels made regular New York runs; New London hosted round-trip steamboat service of two days. Besides farm surpluses, fish (e.g., oysters, smack, and shad), timber, and ships (from Stonington) headed to New York. Farmers in western Connecticut twenty or more miles north of the shore relied on the dense turnpike network (Danbury and Litchfield were hubs) to reach New York City either directly or via shore ports; their specialties—dairying, cattle raising, and beef and pork packing—withstood costly transportation.

The Connecticut River provided low-cost access to distant markets, and vessels operating from Hartford and Middletown conveyed farmers' goods to New York City and other coastal ports. Apple orchards covered hilly land bordering the valley from Middletown to the Massachusetts line, and fresh fall apples provided large profits; farmers also converted apples to cider and cider brandy for sale. Farms on rich alluvial soils of the valley north of Hartford raised Indian corn for large local gin distilleries, and gin went to coastal markets—in New York, Boston, and Providence—and to southern states. Tobacco growing for export markets commenced in the valley towns of East Hartford and East Windsor, and dairy and cattle farms dotted hilly land bordering the river from Middletown to the shore. Shad fisheries generated profits for residents of valley towns, and quarry stone and wood found ready urban markets, especially in New York City.

In the Quinebaug Valley of eastern Connecticut farmers reacted to similar market signals as farmers in western Connecticut; they specialized in products—such as dairy, beef, and pork—which could withstand long-distance shipment. A turnpike traversed the valley, and Norwich served as the commercial outlet; it housed twelve vessels active in the coasting trade, principally to New York City, as well as a line of steamboats that traveled to that city. The northern valley also accessed Boston's market (about fifty miles away), and numerous turnpikes reached to Providence (about twenty-five miles away); its population grew from 6,380 to 11,767 during the period from 1790 to 1820, offering a sizable market. New Haven (with 8,327 people) and Hartford (with 6,901), both leading subregional metropolises, cast market shadows over nearby farms, replicating on a smaller scale New York City's impact on Connecticut's agriculture. Farmers in New Haven outside the built-up area grew fruits and vegetables, while nearby farmers in Milford and North Haven also raised vegetables, and those in Milford sent small meats to New Haven; farmers in East Haven and Milford raised hay for teamsters and horses owned by wealthy city residents. Farmers in Hartford outside the built-up area raised large quantities of vegetables for local residents, similar to New Haven farmers, and East Hartford, across the river, grew vegetables (e.g., peas and beans) and fruit (e.g., watermelons) for Hartford; farms within ten miles of the city raised hay for its horses. New Haven's and Hartford's market shadow extended beyond ten miles because both demanded extensive dairy products, beef, and pork; farmers accessed these markets as supplements to the huge New York market.[34]

Instead of an agricultural backwater awaiting industrialization, Connecticut built a prosperous farm economy from 1790 to 1820. Farmers raised productivity and leveraged their proximity to New York City, especially, and to subregional metropolises in the state, secondarily, to specialize in high-value products that could withstand wagon and coastal transportation, and the state's prosperous farmers and commercial businesses contributed to the demand for high-value farm products. Like Connecticut's farmers, those in New York State accessed the New York City market, but farmers in central and western New York faced long market trips and limited local markets until about 1810; nevertheless, farmers statewide raised productivity and sold increasing amounts of their output.

New York

Prior to 1790 most of the state's population lived either in New York City, nearby Long Island, immediately north of the city, or in the Hudson and lower Mohawk Valleys (map 2.2). Frontier settlers west of the Hudson initially engaged with distant markets during the first twenty years of the nineteenth century, and, simultaneously, Hudson Valley farmers boosted production to capitalize on the burgeoning New York City market. The Hudson-Mohawk region comprised a twenty-one-county area extending about two hundred miles up the Hudson Valley and about one hundred miles west of the Hudson. Its population soared 57 percent during the 1790s, to 402,313, and much of this growth occurred west of the valley; Oneida County, with its new urban settlement of Utica, grew over tenfold, to 20,839. Then the region's growth slowed to 32 and 17 percent, respectively, during subsequent decades as new settlers looked to western New York.

Farmers moving into the Hudson-Mohawk region, especially the western part, mostly cleared land and improved it for pasturage or tillage during the decade from 1790 to 1800, and many continued those tasks into the next decade. They capitalized on cheap water transport to New York City and expanded their production of wheat and potash, a by-product of burning trees during land clearing; they used more fertilizer to raise productivity; and Hudson Valley farmers added gypsum to the soil to compensate for its lime deficiency. Farmers near New York City accessed large manure supplies for wheat fields and pasturage. Starting in the 1790s, farmers planted more English grasses in pastures and meadows, and by 1810 the practice became widespread. Throughout the region farmers produced butter and cheese for New York City, but two groups stood out. Farmers near the city, especially in Orange County, specialized in butter because their proximity provided a competitive advantage; butter did not retain its freshness and quality for long-distance transport, except during winter months. In contrast, Oneida County farmers compensated for the 250-mile trip to New York City via the Hudson-Mohawk Valleys

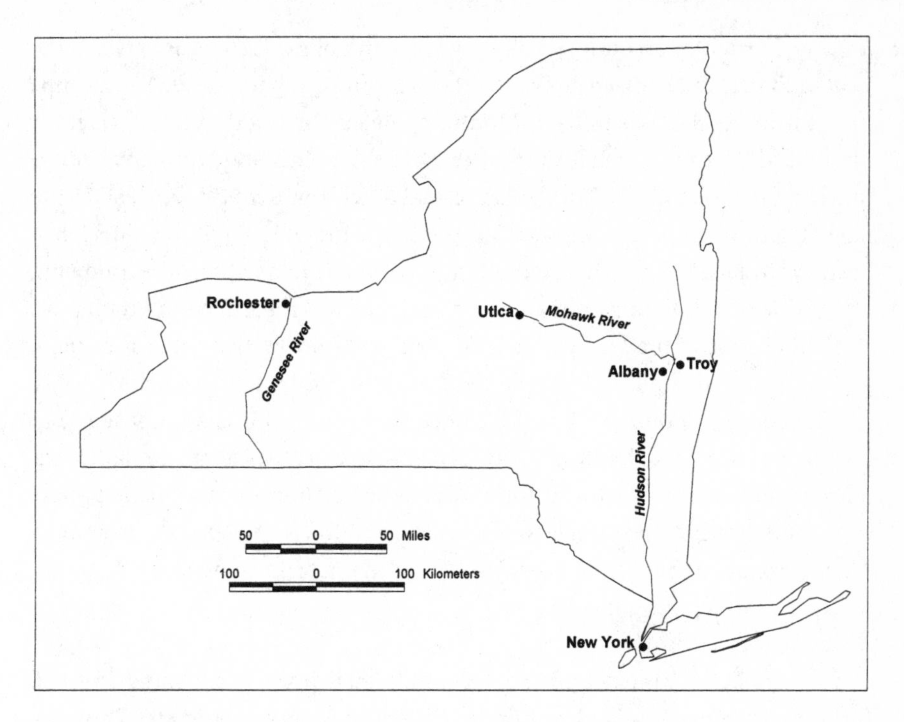

Map 2.2. New York State

and specialized in cheese during the decade from 1810 to 1820; cheese factors contracted for farmers' cheese and sold it in the city by 1815. Hudson River ports—at Albany, Troy, Catskill, Kingston, Poughkeepsie, and Newburgh—grew rapidly during the period from 1790 to 1810 as farmers boosted productivity and a greater share of them sold their output for shipment to New York City. After 1800 river ports aggressively built turnpikes east into Connecticut and Massachusetts and west of the Hudson River, because wagons provided the best transport away from the Hudson artery; rapids and falls along the Mohawk River hindered its transport effectiveness.[35]

During the early 1790s settlers entered the Genesee Valley in western New York, but farmers were four hundred miles from New York City and over two hundred miles by wagon to Albany (see map 2.2). They remained diversified producers for their own consumption, and wheat production for sale as flour outside the Genesee Valley remained modest until the late 1810s. The Susquehanna flowed toward Chesapeake Bay and the Philadelphia and Baltimore markets; however, arks floated for only a few high-water months. Although

Montreal provided an outlet for flour, uncertain political relations between the United States and Canada prevented farmers from relying on it. Flour shipments from Rochester's mills to Montreal totaled about a thousand barrels in 1810 and did not rise much until after 1816; shipments surged to sixty-seven thousand barrels by 1820, but the Erie Canal quickly undercut Montreal. Cattle raising became important in the Genesee in the first decade of the nineteenth century, and early drives headed to Baltimore and Philadelphia; the following decade New York became a major destination. The Holland Land Purchase—stretching west of the Genesee to Lake Erie—was not settled until after 1800, and only fifteen thousand people lived there by 1810. The population grew rapidly after 1810, but, excluding Genesee County, the Purchase housed only fifty thousand people by 1820. Early farmers raised cattle for the Philadelphia market, produced some wheat for sale as flour in Montreal and made potash from land clearing; nevertheless, their three hundred–mile wagon distance to Albany excluded them from most East Coast markets until the 1820s.[36]

The Philadelphia Region

Farmers in the Philadelphia region joined their peers in the inner rings of Boston and New York and responded to the metropolis' markets. Those in southeastern Pennsylvania, however, diverged in one major respect—they occupied one of the largest, most fertile agricultural areas in the East. Because Philadelphia's merchants possessed substantial capital and extensive trading contacts throughout the East and in foreign markets, the region's farmers earned significant financial returns as external markets grew, and they gained from Baltimore's emergence as a competitive mercantile center, because they could choose the market offering the best prices (see table 2.5). Philadelphia region farmers produced wheat cheaper than other eastern farmers before 1790 and maintained their competitiveness over the next several decades, and numerous flour mills converted wheat to high-value flour for wagon transport to the city. During the 1790s wheat remained the leading tillage crop for off-farm sale, and most farms also raised diverse grain crops for feed—corn to feed hogs, oats to feed horses, and hay to feed cattle and dairy cows; farmers also grew rye, vegetables, and fruit for consumption, and larger farms raised livestock, thus meeting their meat requirements. Like peers elsewhere near large metropolises, farmers in southeastern Pennsylvania and nearby areas embraced English grasses to improve pasturage and hay fields, but only those near Philadelphia used manure for fertilizer, because few farms produced enough manure to employ it; rotation cycles for land became more important as ways to boost productivity.[37]

Rings of decreasing intensity of land use with distance from Philadelphia reached about one hundred miles by the 1790s, although differential land quality and ethnic-religious skill levels distorted this figure. Vegetable and fruit production and dairying (mostly fluid milk) thrived near Philadelphia, and cattle driven from other states and elsewhere in Pennsylvania were fattened on rich hay meadows along the Delaware River south of Philadelphia and in Lancaster's environs about sixty miles west of Philadelphia. Chester County farmers—about forty miles from Philadelphia—produced cheese and butter and gained a reputation for their cheese by 1800. Because cheese can be transported long distances, this specialty typically did not occur near metropolises, but Chester producers probably competed elsewhere in the East through Philadelphia wholesalers. Extensive wheat farming within sixty miles of Philadelphia also is unexpected, but wheat was a cash export crop and farmers cultivated large, fertile acreage; in contrast, farmers in Massachusetts, Connecticut, and the lower Hudson Valley occupied less-fertile land. Farther from Philadelphia farmers also grew wheat on fertile land because they were highly competitive vis-à-vis most other eastern farmers. The rapid growth of the upper Susquehanna Valley in northeast Pennsylvania and adjacent New York State in the mid-1790s created a prosperous outer ring of Philadelphia. Farmers diversified production to meet their consumption and to exchange with other farmers and through country stores. They shipped wheat down the Susquehanna to river ports on its lower reaches, and Middletown's flour mills collected large amounts and forwarded flour by wagon to Philadelphia; upper Susquehanna farmers shipped surplus whiskey and pork by wagon and drove cattle to Philadelphia.[38]

Between 1800 and 1820 Philadelphia's population gain equaled the metropolis' size in 1790 (see table 2.5), and foreign commerce contributed to growth up to about 1810; subsequently, domestic business dominated its economy. Philadelphia's merchants served a coastal area along the Delaware River and in nearby Maryland; in turn, the coastal hinterland supplied products for Philadelphia's consumption and for domestic trade and foreign exports. An increasing number of vessels connected the coastal hinterland with Philadelphia, and regularly scheduled trips appeared by 1805, indicating growing output. Because this agricultural area accessed cheap waterway transport, its farmers supplied bulky, low-value products at low delivered prices to Philadelphia. Indirect evidence suggests intensive fruit and vegetable production and dairying expanded near Philadelphia during the first two decades of the nineteenth century, and cattle fattening and dairying increased in the next ring (ten to sixty miles away). Wheat production probably stabilized or even declined within about sixty miles of Philadelphia as farms shifted to higher-value products.[39]

Agricultural Progress

Records of Abraham Hasbrouck Jr., the most important storekeeper in Ulster County (on the west side of Hudson River), vividly document the behavior of agricultural leaders from 1799 to 1820. At Hasbrouck's Kingston store the number of farmers consigning goods for shipment on his sloop to New York City rose eightfold, to 165, and these elite farmers produced sufficient volumes of high-value goods that they contracted with Hasbrouck or other wholesalers in New York City to handle them. Yet the share of farmers consigning goods rose modestly, from 12 to only 27 percent. The absolute increase, from 164 to 436, in the number of farmers not consigning goods exceeded the increase in those consigning goods; Hasbrouck's records reveal that those not consigning achieved a comfortable living standard and participated in the rising agricultural prosperity. Nevertheless, social stratification among farmers rose as those consigning goods for New York City markets responded to market price signals by transforming practices, altering output, increasing productivity, and raising sales. Their incomes rose faster than those of farmers who continued to emphasize self-sufficiency and local exchange and who constituted a significant share of farmers failing to compete under the altered agricultural regime after 1820, thus forming the pool of workers for rural nonagricultural occupations and for manufacturing.[40] Growing agricultural productivity supported the increasing share of the population living in urban places (see table 2.4). Besides migration to large metropolises, rising wealth in farming areas stimulated the development of central places providing retail goods and services for farm populations, and some cities became wholesale-retail centers; these urban places were nodes of demand for farm products and centers of capital accumulation.

Elaboration of the Urban Hierarchy

During the colonial period small dispersed places dominated, and they contained a cluster of several activities—a country store, a craft shop, and perhaps a tavern if the cluster were on a major road; if a place housed the county seat, the government office often stood alone or with one other activity. Dispersed places had little or no centrality because their businesses did not specialize in providing goods and services to people beyond the immediate vicinity; instead, farmers and craftspeople exchanged small surpluses there. The emergence of central place systems indicates farmers' greater engagement with the market economy. Initially, farmers could use their wagons to transport surpluses to markets, but they possessed limited knowledge of distant markets. Local retail merchants with ties to distant wholesalers served as information sources; therefore, farmers profited from selling surpluses to retailers in exchange for goods, and agglomerations of these retailers formed central places. The local emergence of unspecialized wholesalers and freight forwarders in larger central places enhanced marketing efficiencies.[41] Areas of rising agricultural prosperity in the Connecticut Valley of Massachusetts, the state of Connecticut, central New York State, and southeastern Pennsylvania exhibited an elaboration of urban hierarchies during the thirty years from 1790 to 1820.

The Connecticut Valley of Massachusetts

In the late eighteenth century retailers in Springfield, Massachusetts, purchased most of their West Indies goods in Hartford or Middletown, Connecticut, and obtained European goods from Boston or New York. The Dwights, the owners of Springfield's largest store, maintained their strongest connections with Boston. By the 1780s about five Springfield stores operated at both wholesale and retail (although most business was retail), whereas most other stores operated solely as retailers. During the 1790s and early 1800s, coinciding with rising agricultural prosperity, the Dwights established retail branches in valley towns, and by 1800 the Springfield store (which was the main one) advertised regularly at wholesale and retail rates. By 1815 the Dwights opened a warehouse in Boston to supply their retail branches, and within a decade after 1815 more wholesaler-retailers appeared in Springfield. Around 1790 Northampton merchants, located about twenty miles north of Springfield, handled most of the wholesale and retail trade of surrounding Hampshire County. Larger retail merchants imported goods and sold them at wholesale rates to small country stores scattered across the valley. Agricultural expansion, powered first by international markets from the early 1790s until 1807 and then domestic markets after that time, supported the transformation of urban places. Northampton's wholesale and retail business increased, but small villages in the towns (areal political unit in New England) of Amherst, Hatfield, and Williamsburg grew faster, as storekeepers captured growing retail purchases of farmers living nearby. Stores sold more goods that farmers traditionally purchased—such as salt, hardware, notions, books, sugar, rum, and spices—and after 1810 textile sales rose.[42]

Connecticut

Because Connecticut's prosperous farmers served the New York City market, they formed a large market for retail goods and services and supported numerous wholesalers at different levels of specialization; consequently, the state developed an elaborate urban hierarchy. From 1790 to 1820 Hartford and New Haven—leading subregional metropolises under the larger dominance of the regional-national metropolis of New York—experienced population growth rates of 69 and 86 percent, respectively, far outpacing Connecticut's 16 percent growth. As agricultural prosperity boosted trade opportunities, Hartford's merchants extended their reach into Vermont and New Hampshire, and this economic growth supported increasing specialization: some wholesaler-retailers eliminated retailing to operate as general wholesalers in the 1790s, and continued agricultural growth in the next decade encouraged some wholesalers to specialize in selected lines. By 1819 the city's merchants included five wholesale dry goods dealers, twenty-six retailers (broadly defined), and sixty-one grocery, crockery, and provisions stores, and the city housed numerous other businesses, including twenty-one taverns or inns for travelers and a large professional elite (twenty-two lawyers, twelve physicians, and five clergy). New Haven's wholesale merchants focused on the West Indies, coastal, and New York City trade, and a vigorous retail trade of twenty-two dry goods stores and eighty-seven grocery and provision stores served the large local population and surrounding rural dwellers. Wholesalers and major retailers in Middletown (downriver from Hartford) did not effectively compete with Hartford merchants for rich valley trade north of that city; nevertheless, Middletown retailers captured a large business in the lower Connecticut Valley.[43]

The shore towns of Stamford (with fourteen stores), Norwalk (with sixteen), Fairfield (with twenty-five), and Stratford (with twenty-eight stores and fifteen large warehouses) exhibited palpable evidence of successful entry into the New York City market. As key ports and business centers on the coastal turnpike, taverns and inns served travelers, and the towns housed a sizable professional elite. East of New Haven prosperous retail sectors also operated in shore towns, including Guilford (with sixteen stores), Saybrook (with fourteen), Stonington (with twenty); the latter also had a coasting trade to New York City and southern states. Successful interior farmers in western Connecticut accessed several large villages situated on major turnpikes: Danbury's six taverns complemented its eleven mercantile stores, and Litchfield, one of the state's wealthiest towns, held sixteen mercantile stores; both towns housed many professionals. Farmers in eastern Connecticut utilized New London as a commercial outlet, and its merchants traded with New York City, southern states, and the West Indies. New London's twenty houses concerned with navigation included some wholesalers, and it had a large retail sector (with fourteen dry goods stores, forty-two grocery or provision stores, and ten taverns or inns). Norwich, near the mouth of the Quinebaug Valley, possessed a large retail sector (with forty-five stores) and numerous professionals, and it was an outlet for the valley's agricultural produce; its twelve vessels and steamboats provided access to New York City. The towns of Pomfret (with seven stores) and Thompson (with eight) served farmers in the northern part of the valley, and both had small professional classes; Pomfret's four public inns served travelers on the turnpikes. Farmers between the Connecticut and Quinebaug Valleys accessed a sizable central place in Windham, at the intersection of the east-west turnpike connecting Hartford and Providence and the north-south turnpike linking Massachusetts to Norwich; Windham contained fifteen stores and many professionals.[44] By 1819 Connecticut's wholesale centers—Hartford, New Haven, Middletown, New London, and Norwich—signified vigorous agricultural production for markets outside the state, and extensive market purchases by prosperous farmers, craftspeople, professionals, and mill owners supported an elaborate central place hierarchy. Connecticut was fully settled before 1790 and highly accessible to the large New York City market, whereas settlement commenced in central New York State in 1790; nevertheless, rapid inmigration soon produced an urban hierarchy.

Central New York

Inmigration to the New Military Tract of central New York started after 1790, and Utica, at the head of Mohawk Valley, was the gateway and supplier for settlement farther west. When Timothy Dwight, the traveling president of Yale College, visited Utica in 1804, its sizable wholesale and retail sector impressed him. Geneva, on the western side of the Military Tract, was another wholesale and retail supply center.[45] Initially, farmers spent time clearing land and constructing farm infrastructure and thus had little surplus output for sale and limited demand for goods and services from small central places; therefore, few such places existed until several years after 1800. Rapid settlement soon led to the emergence of about a dozen villages during the first decade of the nineteenth century: entrepreneurs from Albany and New York City established some villages to collect surpluses to export and for redistributing imports to nearby farmers, and small-scale local businesspeople founded other villages. By 1820, when the Erie Canal section in central New York opened, the hierarchy consisted of Utica and Geneva as wholesale-retail centers, large central places such as Ithaca, Auburn, and Seneca Falls, new canal villages such as Syracuse, and numerous small villages amid dense agricultural settlement.[46]

Southeastern Pennsylvania

In contrast to central New York's hierarchy built on rapid inmigration of farmers, southeastern Pennsylvania's evolved on a preexisting base of densely settled, prosperous farmers. By 1800 an elaborate central place hierarchy existed; Philadelphia was a regional metropolis (i.e., a wholesaling center) and the highest-level central place, and it housed lower-level retail functions. Within thirty miles of Philadelphia few large central places existed because people in that zone traveled to Philadelphia for high-level goods and services; thus, the zone chiefly contained villages and hamlets. Beyond thirty miles a full complement of large central places, transport centers, towns, villages, and hamlets offered goods and services to surrounding farmers, and upper-level centers also served the urban residents of lower-level centers. Many of the larger urban places—such as Lancaster, York, Reading, and Carlisle—were county seats, and larger towns and cities grew at about the same pace as Philadelphia during the period from 1790 to 1820, indicating rising agricultural prosperity.[47]

The emergence and elaboration of urban hierarchies coincided with rising agricultural prosperity in the East because productive farms released labor for nonagricultural occupations. Farmers demanded ever-larger amounts of goods and services, and, as supply prices fell and incomes climbed, consumption diversified across goods and services, including food, clothing, transport services, craft shop production, and professional services such as doctors and lawyers. Commercial businesses, professionals, and the most prosperous farmers accumulated substantial capital complementing that gained in the mercantile sectors of regional and subregional metropolises. Those capital pools funded manufacturing growth from 1790 to 1820, supporting the development of the East's regional industrial systems.

CHAPTER THREE

Bursting through the Bounds of Local Markets

> A manufacturing spirit was early disclosed in this State; and, with the exception of Rhode Island, there is no State in the Union where it has been cherished with so much attention, or directed to so many objects. . . . Connecticut has superior advantages for manufacturing pursuits; and it is believed that it cannot fail of becoming, at no distant period, an extensively manufacturing community. —PEASE AND NILES, 1819

Within each regional metropolis—Boston, New York, Philadelphia, and Baltimore—pivotal intermediaries such as wholesalers, commodity brokers, and financiers controlled intra- and interregional exchanges of commodity and financial capital. Multiplier effects of their activities made the metropolis the region's largest city, and thus the single greatest agricultural and industrial market. Metropolitan factories expanded to serve intermediary activities and the metropolis' local market, and factories elsewhere—in metropolitan satellites, towns and cities distant from the metropolis, and rural areas—also met demands. In some industries workshops and factories increased production for markets elsewhere in the East, and in several cases—shoes, "Connecticut" manufactures, and cotton textiles—they boosted production for midwestern and southern markets. Thus, regional economic development involved the interrelated processes of agricultural development, industrial expansion, and urban and metropolitan growth, epitomizing the growth of a regional industrial system.[1]

Entrepreneurs in large East Coast metropolises gained from locating factories at the headquarters of intermediaries controlling the intra- and interregional exchange of commodities and capital. As centers for collecting and redistributing resource commodities from wide territories, metropolises were ideal sites for large-scale processing industries (e.g., flour, sugar, and chemicals), and, because commodity shipment often involved transshipment, processing at that point lowered costs. Intermediaries generated demand for manufactures that supported exchange activities: transportation required ships, wagons, and locomotives; shipping called for boxes, barrels, and other packaging; and information processing relied on printing, publishing, and engraving. Intermediary activities made metropolises the first large cities in their regions, and their activities generated ongoing local multiplier effects, stimulating population growth and maintaining their rank as the largest cities (see tables 2.5 and 2.6). Thus, early factories serving local markets often started in the metropolis, and, because they achieved economies of scale before other factories and they possessed superb access to regional markets via the metropolis' wholesaling and transportation services, they dominated regional markets. Factories processing commodities, supporting exchange (e.g., transportation, containers, and information), and serving local and regional consumer markets required intermediate manufactures, especially producer durables (primary metals, fabricated metals, and machinery). Because a metropolis housed all these manufactures, it was the greatest industrial center of its region, but it could not monopolize manufacturing; factories elsewhere competed, and some of their advantages derived from special features of regional market demand.[2]

In the region outside each metropolis household consumers, urban infrastructure, natural resource sector, and intra- and interregional trade generated broad demands for manufactures.[3] Household consumers demanded food (flour, meatpacking, and liquor), clothing (dresses, pants, and shoes), houses and equipment (lumber, nails, furniture, doors, sashes, and stoves), and transportation (carriages and wagons). Expanding urban places required infrastructure: stores, public buildings, offices, and factories needed bricks, lumber, doors, windows, nails, hardware, and furniture; and municipal improvements required lights, gas pipes, water mains, and sewers. Commercial agriculture used barns and sheds (lumber, doors, hinges, and nails), farm implements (plows, hoes, axes, and rakes), and machinery (mowers, reapers, threshers, and harvesters), and some equipment had cast-iron components and machine mechanisms. Processing activities—flour mills, distilleries, lumber mills, and meatpacking plants—were factories; they often used waterpower, but flour and lumber mills also employed steam engines. This arrangement caused machinery technology to spread to processing centers throughout the region, and backward and forward linkages from processing stimulated other regional industries; steam engines required inputs from primary metals, metal fabricating, and machinery firms. Forward linkages from lumber mills contributed to barrel, furniture, and wagon manufactures; flour mills spurred food manufactures; and meatpacking fed by-product manufactures such as soap, candles, and shoes.

Intra- and interregional trade needed printing, publishing, and engraving, and these activities used paper; because paper mills required clean water, they were often located in the hinterland and met large paper demands in regional metropolises. Commodity transport supported the manufacturers of barrels, boxes, and other packaging, and the transport of people and commodities spread demands throughout the region. Yet canal and railroad construction spurred little industry because construction mostly used hand labor or simple lumber products. Steamboats and steamships needed steam engines; however, the construction of vehicles mostly involved hand labor. Locomotive demands spurred heavy equipment manufacturing, including steam engine building and metal fabricating; railroad repair shops operated as large machine shops, contributing to producer durables. Factories produced passenger and freight cars, although craft workers built equipment; vehicles required planed lumber, and metal components such as wheels, axles, and springs boosted metal fabricating industries.

Producer durables appeared in each expanding regional industrial system. Manufactures in the metropolis and elsewhere in the region demanded primary metals (iron bars, rods, plates, and sheets), spurring the growth of the iron industry, and fabricating firms molded, bent, cut, and shaped metal components, which entered into machinery products—textile machines, paper machines, wood-cutting machines, farm machinery, steam engines, and locomotives. The expansion of metal producer durables stimulated a separate industry—machine tools—to produce equipment for cutting, grinding, bending, and polishing metal. Although primary metals manufactures (e.g., blast furnaces and forges) produced key inputs, metal fabricating, machinery, and machine tool industries constituted the pivotal producer durables, and, because they required close contact with other manufactures, most located in larger cities, especially metropolises. Before 1820 pivotal producer durables developed as iron foundries, engine works, and arms firms, but after that year industrial growth spurred a transformation.[4]

Industrial growth ratcheted through each regional industrial system as local and nonlocal multiplier effects. Because regional metropolises housed factories serving exchange and local and regional markets (including producer durables), they had the largest, most diverse industrial sectors, and they could house interregional market manufactures (e.g., textiles, shoes, and apparel), similar to cities elsewhere. Local multiplier effects from these factories powered metropolises to the top industrial position in their regions. Regional growth outside metropolises provided industrial entrepreneurs elsewhere with opportunities to serve subregional markets. Many firms started as craft shops and small nonmechanized factories, and regional growth spurred their expansion in metropolitan satellites, subregional metropolises, cities, towns, villages, and rural areas; factories outside metropolises could serve interregional markets. Local industrial multipliers in each urban place stimulated consumer manufactures, retailing, and government, and industrial multipliers sometimes stimulated sectors providing intermediate inputs or using outputs of expanding industries. Nonlocal multiplier effects ratcheted through regional industrial systems as growing manufacturing firms outside the metropolis required its wholesaling, warehousing, and financing services, thus generating nonlocal multiplier effects in it.

The metropolis benefited from cumulative industrial growth throughout the region, whereas industrial places (such as cities, towns, villages, and rural areas) gained nonlocal multiplier effects from the metropolis only if they produced goods demanded by metropolitan businesses. The regional industrial system did not expand linearly, manufactures grew differentially, technological change varied across industries, and intraregional demand diverged. Firms near the metropolis accessed metropolitan markets, and densely populated inner rings of agriculture and resource extraction constituted large markets; therefore, each regional system possessed industrial satellites (cities, towns, and villages) and rural industry surrounding the metropolis.[5]

Looking beyond Local Markets

Following 1790 the steady rise in agricultural productivity stimulated the growth of local manufactures, and evidence from the Connecticut Valley of Massachusetts suggests this expansion initially came from households producing manufactures for their use and selling small surpluses, itinerant workers adding to household production, and household production for exchange with other households. A farmer's limited capacity to accumulate capital, however, and the farm's few workers and modest division of labor constrained household-based manufacturing. As local markets expanded, craft workers—

such as blacksmiths, shoemakers, and wagon makers—entered part- or full-time production, but they quickly saturated local markets. Local craft shops, grist- and sawmills, and iron furnaces sometimes produced small surpluses for sale to country storekeepers, and these goods might have entered regional or interregional markets; nevertheless, these minor sales had little consequence for industrial growth. Entrepreneurs faced two paths as local economies shifted from an internal focus to greater market integration. Along one path some firms manufacturing goods with low value relative to their weight (e.g., flour, lumber, furniture, and iron products) or producing custom goods (e.g., iron foundries) requiring close contact between buyer and seller expanded into regional markets in the East, and, as transportation improved or economies of scale in production reduced costs, some firms sold to other regions. Flour, saw, and paper mills near Wilmington, Delaware, achieved such fame by 1791 that they were identified simply as the "Brandywine Mills." Manufacturers of goods with high value relative to weight—such as textiles, buttons, arms, tinware, clocks, and hats—followed a second path: they intensified sales to regional markets, and then leading firms assaulted interregional markets in the East. By 1819 John Pease and John Niles noticed numerous Connecticut firms selling tinware, clocks, buttons, and arms outside the state.[6]

Capital for Industry

Demand for Capital

Small craft shops—such as blacksmiths, carriage and wagon works, and tinware shops—selling products to local markets required little fixed capital (land, buildings, and machinery) and working capital (wages and inventory). Although they sold goods on credit and faced payment delays, they did not contend with risks accompanying long-distance sales where they possessed little leverage over buyers. Craft shops grew slowly, and retained earnings funded expansion; small and large workshops and factories required commensurately larger fixed investments, however, and much of their capital went to land and structures. Textile and paper firms allocated larger shares of capital to machinery compared to other firms, yet they invested more fixed capital in land and structures than in machinery. Most antebellum factories invested little capital in machinery; from 1774 to 1815 equipment investment remained at 4 to 5 percent of domestic capital stock, whereas investment in structures rose from 17 to 27 percent (see table 2.3). Factories selling outside local markets required extensive short-term or working capital, and annual sums often exceeded fixed capital. Factories purchased raw materials and paid wages prior to receiving payment for goods, and these payments lagged six months or longer if goods sat in inventory or if buyers purchased on credit. Even textile and paper firms, which had the largest share of capital allocated to fixed investments, required about one-third of capital for working purposes as of the early 1830s, whereas firms making coaches and harnesses, hats, and shoes used about half of their capital for that purpose.[7]

The Supply of Capital

In 1810 Albert Gallatin, the secretary of the Treasury, parroted the supporters of manufacturing who were complaining about a shortage of capital for industry, and he endorsed their solution—to tap government coffers. Nevertheless, capital was widely available, albeit not at interest rates as low as Britain's, and suppliers of fixed and working capital came from diverse sectors. Wholesalers in regional metropolises (Boston, New York, Philadelphia, and Baltimore) supplied fixed capital for factories, and, if they invested cooperatively, sizable sums were available. More commonly, they forged alliances with manufacturers for whom they provided distribution and credit services; wholesalers supplied capital, and manufacturers contributed industrial skills. Of greater significance, wholesalers supplied industrialists with extensive working capital; they guaranteed notes, extended credit for supplies, provided negotiable notes when goods were shipped, and purchased notes from buyers.[8]

Growing farm prosperity generated capital for manufacturing, and the transfer to industry commenced by the 1780s. Farmers shifted toward interest rates for loans, increased their use of debt to finance economic activity, purchased shares of stock in financial institutions and manufacturing firms, and expanded the use of negotiable credit instruments. Wealthy rural residents hiked the provision of credit to others and offered it over wider areas, and they increasingly borrowed in order to lend and shifted assets out of land and into financial instruments, including investments in banks and insurance companies. Although the rural wealthy did not necessarily directly invest capital in industry, the fungibility of capital meant that others made the transfer. The commercialization of rural economies created nodes of capital accumulation at subregional metropolises such as Hartford, Connecticut, and Albany, New

York, where wholesalers, owners of large processing firms, and major retailers were sources of larger sums of capital, whereas successful businesspeople in central places—such as retailers, freight forwarders, feed mill owners, lawyers, and physicians—were leading sources of smaller capital. Central place businesspeople operated in social networks with significant redundancies in business and social ties; they could combine capital to support moderate-sized manufacturing enterprises and could punish malfeasant action by coinvestors.[9]

Financial Institutions

Commercial banks provided modest capital during the period from 1790 to 1820, and their supply pace accelerated after 1820. They accepted few deposits because they mostly made long-term loans or regularly extended short-term loans; deposits that could be withdrawn on short notice were risky sources of capital, and most money for lending was capital stock in the banks. Several individuals formed a commercial bank, and each contributed capital as purchases of shares of bank stock; they usually served as directors, accounting for about one-third of capital stock. Many directors were members of kinship groups, while others were close acquaintances whose bonds of trust emerged through business or friendship ties. Merchant wholesalers led the formation of banks, and manufacturers, large retailers, and owners of mines, lumber mills, and real estate joined them. Insurance companies, savings banks, and other financial institutions, often partly owned by directors, typically purchased large blocks of stock, and other purchasers included retailers, professionals, and prosperous farmers. In New England bank insiders—directors, their relatives, and those with close personal ties to directors—received most loans, and most bank capital entered local and regional economies through individuals rather than firms. In Middle Atlantic banks, however, a wider set of firms and individuals, rather than insiders, received loans, and bankers successfully accessed social networks to confirm the creditworthiness of diverse borrowers. Directors stressed the careful monitoring of loans to preserve capital and maintain the bank's reputation; reputation was critical to attract capital from other community members because they provided the majority of capital.[10]

In New England and the Middle Atlantic the number of banks rose slowly from the early 1780s to about 1800 (fig. 3.1), but this did not indicate limited liquid capital; wholesalers supplied substantial amounts directly as loans or indirectly through credit. After 1800 New England captured the lead in number of banks and continued outpacing the Middle Atlantic until 1812. As the

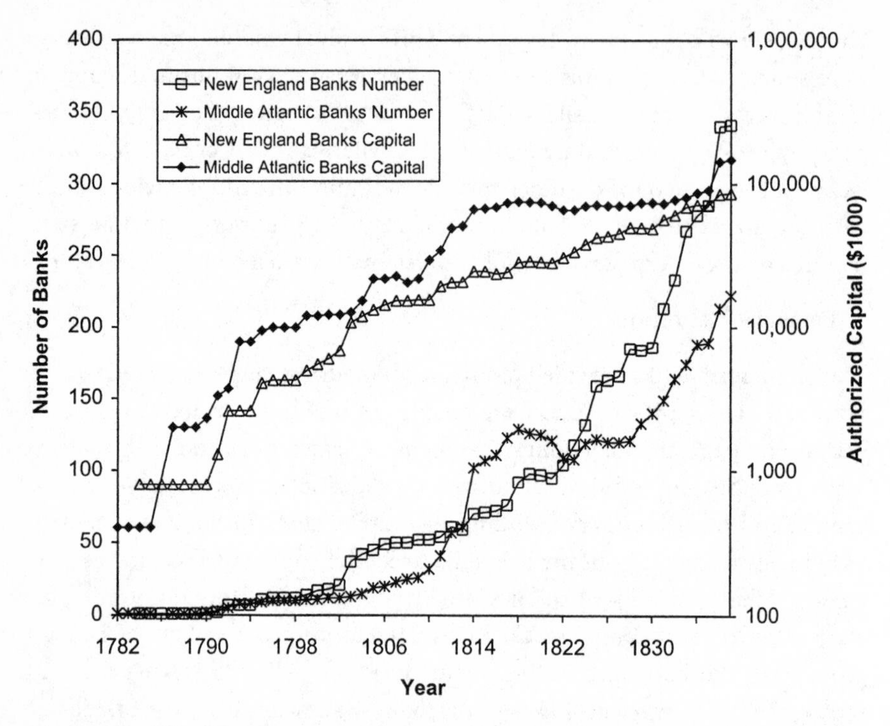

Fig. 3.1. Number of Chartered State Banks and Their Authorized Capital in New England and the Middle Atlantic, 1782–1837. *Source:* Fenstermaker, *Development of American Commercial Banking: 1782–1837,* 13, 77, 80, tables 4, 12, 13.

oldest settled region, New England possessed highly developed rural and urban exchange networks; thus, residents mobilized capital from prosperous farmers outside the regional metropolis (Boston) and in businesses in subregional metropolises (e.g., Providence and Hartford), wholesale-retail centers (e.g., Springfield, Massachusetts, and New London, Connecticut), and central places. Until about 1810 New England had the greatest dispersal of banks in the East, whereas in the Middle Atlantic most agriculture and commerce occurred in close proximity to large banks founded in New York City, Philadelphia, and Baltimore between the early 1780s and early 1790s. And New York, the largest populated state in the East by 1810, remained in the throes of large-scale settlement of new lands, thus retarding bank formation away from the long-settled environs of New York City.[11] After 1812 the Middle Atlantic's number of banks surged, and New England's number kept rising, underscoring the continued capital accumulation and reallocation to investments.

Widespread rising agricultural and commercial prosperity generated similar rates of capital accumulation across the East, as indicated by the close correspondence between the growth rates of aggregate authorized capital of state banks in New England and the Middle Atlantic from the early 1780s to about 1812 (see fig. 3.1). Total capital for the Middle Atlantic stayed above New England after 1786, even though New England's banks outnumbered the Middle Atlantic's after 1791, because the Middle Atlantic housed sizable numbers of highly capitalized banks in New York City, Philadelphia, and Baltimore. The Middle Atlantic's average capital per bank was two to three times that of New England, and this size gap did not diminish until after 1813; as late as 1820, Middle Atlantic banks averaged twice the authorized capital of New England banks. New England's growth rate of aggregate bank capital from 1803 until 1820 remained almost constant, though at a lower pace than before 1803; this matched slower economic growth during the decade from 1810 to 1820. The jump in authorized capital in Middle Atlantic banks after 1803 corresponded to surging bank numbers in the Hudson-Mohawk region and central New York as settlement matured. The stagnation of total capital and the number of banks from 1814 to the early 1830s, however, reflected failed speculative investments such as in transportation infrastructure, which was common during rapid settlement.[12] Consistent growth in the number of banks and aggregate capitalization in New England, along with a wide dispersal of capital nodes outside Boston, supported the funding of manufacturers across the region. Connecticut exemplifies this capital supply.

Connecticut's Supply of Capital

The thriving urban network of subregional metropolises (Hartford and New Haven), wholesaling-retailing centers (Middletown, New London, and Norwich), and substantial mercantile villages (Stamford, Bridgeport borough, Litchfield, and Pomfret) symbolized robust capital accumulation. Merchants acquired capital they could not fully reinvest in their businesses; trade did not expand endlessly because farmers exploited most prime agricultural land before 1790. Similarly, professionals (such as lawyers, physicians, and clergy) required other outlets for profits or inheritances than expanding businesses or investing in houses and libraries. Wealthy farmers whom Timothy Dwight met on his travels invested profits in imposing farmhouses and improved lands, but, after those investments, other outlets for capital beckoned. Large merchant wholesalers, befitting their abundant capital and business acumen, led the formation of financial institutions, and the political and religious elite joined them to found ten banks and nine marine and fire insurance companies during the period from 1790 to 1820, all located in merchant wholesaling centers and ports. Thus, stockholders possessed large pools of capital to borrow for business ventures.[13]

Although craft shops or small nonmechanized factories sometimes generated high rates of return on capital, their small absolute profits did not attract rich investors. Instead, Connecticut's wealthy directed capital to businesses such as wholesaling or to new, large-scale ventures such as shipping firms, transportation companies, western land purchases, and turnpikes, and industrial investments flowed to printing companies and to cotton and woolen textile firms capitalized at over $100,000 or to resource processing firms such as rum and gin distilleries.[14] Few of the elite's industrial firms specialized in goods in which Connecticut firms dominated in the nineteenth century; nevertheless, their investments generated direct and indirect multiplier effects on business activity, improved transportation infrastructure, and created opportunities for other investors—retailers, freight forwarders, feed mill owners, and professionals—to accumulate capital that wended its way to emerging small-scale workshops and factories. Nevertheless, industrialists in prosperous agricultural areas in Connecticut and elsewhere confronted a dilemma: their competitors accessed the same demand and capital. Firms had an option: they could gain a competitive advantage by inventing new products or improving on existing ones and by inventing new ways of making products to boost productivity or improving on those they already had.

Technological Change

An Explanation

Technological change in the form of inventions (i.e., technical advances in products or production processes) and innovations (i.e., commercial implementation of these products or processes) powered growth and change in regional industrial systems. Invention and innovation need not occur conterminously; nevertheless, examining the location of invention offers insights into industrial change. According to standard reasoning, cities were concentrated markets for inventions, and the frictional effects of distance on information flow meant that rural dwellers knew less than urban dwellers about the demands for invention. Because personal interaction generated most information, urban dwellers acquired more and better technical information needed for inventing, and technological disequilibrium and convergence within the growing industries of a city enhanced inventive activity. Therefore, inventions occurred disproportionately in cities rather than in rural areas, and larger cities—regional and to a lesser extent subregional metropolises—possessed greater inventiveness (inventions per capita).

Yet patent evidence for the large-scale, urban industrial growth from 1860 to 1920, when the standard argument should fit, offers mixed support. Inventiveness related positively to the share of population in urban areas, suggesting that urban residents were more inventive than rural dwellers, but inventiveness did not directly relate to city size; inventive rates in many small industrial cities far surpassed those in large metropolises. Their share of inventions declined over time, rather than increasing, as would be expected from their cumulative advantages, and inventive activity did not directly relate to urban growth. The absolute number of inventions related positively to the size of city and of manufacturing employment, which are logical outcomes if inventions occur randomly in a population—that is, larger numbers of individuals produce more inventions. An alternative interpretation views invention as a new combination of preexisting knowledge designed to satisfy demand, and important "breakthroughs" were less significant than small improvements arising from the workaday worlds of inventors or mechanics. Inventors recognized profitable opportunities and/or confronted technological problems needing resolution. Continuous minor improvements following an initial invention accumulated to substantial gains in quality or productivity, and the technical skills that factory workers and managers developed were essential to devising improvements.[15]

Thus, from 1790 to 1820 (and continuing to 1860) technological change proceeded apace. Large regional and subregional metropolises had high inventive rates, because industrialists and other individuals sensitive to market opportunities faced large local markets and possessed optimal access to information about demand and technical change elsewhere. Yet, even if those cities had the highest inventiveness, most inventions occurred outside metropolises. Emergent industrial centers—such as Massachusetts' and Rhode Island's textile towns and Connecticut's consumer goods villages—housed workers and managers facing technical problems; their survival required sensitivity to competitive pressures from local or distant firms. Increasing agricultural prosperity away from large cities boosted the demand for manufactures, and some local craft shops were transformed into workshops or small nonmechanized factories; they were able to improve their productivity by changing work organization and adding simple tools. As agricultural prosperity continued, they increasingly competed with one another in a subregion, forcing firms to seek technological changes that enhanced productivity. Thus, inventiveness rose as firms sought market advantages. Because most technology did not use sophisticated mechanical principles, inventiveness was widespread.

The Nation and the East

Inventive activity, measured by the number of patents, closely followed the business cycle from 1790 to 1820 (fig. 3.2). Robust growth between 1790 and 1810 provided fertile conditions for entrepreneurs to develop new and improved products; inventions soared at one of the fastest rates of the antebellum, and, even during the 1810–20 slowdown, annual inventions remained near the peak set around 1810. This pattern also appeared in the breakdowns of inventions across combined sectors (agriculture, construction, transportation, and manufacturing) and in manufacturing, which was the largest patent sector, separately (table 3.1). Regional rankings of inventiveness stabilized by the period from 1805 to 1811: southern New England (first) and New York (second) stood far above other regions; Pennsylvania ranked third; and rural northern New England and the southern mid-Atlantic shared the bottom. Compared to small-city counties and rural counties, residents of large-city counties exhibited higher inventiveness for all economic sectors and for manufacturing separately, and the absolute gap between large cities and the rest widened considerably between 1791 and 1822. Rural counties compared favorably with small-city counties by the years 1805 to 1822, possibly because many rural counties contained towns with populations of three to nine thousand, which made them important urban places for that period. Although large cities had higher inventiveness, people living elsewhere accounted for most inventions. Between 1805 and 1811, when the distribution of inventiveness among states and regions had stabilized, high rates occurred in leading areas of commercial agriculture: eastern Massachusetts, focused on Boston; the Berkshires, western Connecticut, and the Hudson Valley, focused on New York City; and southeastern Pennsylvania, focused on Philadelphia. This distribution seems to reflect waterway access to large market areas. This interpretation, however, overstates those arteries' significance and downplays the impact of agricultural prosperity

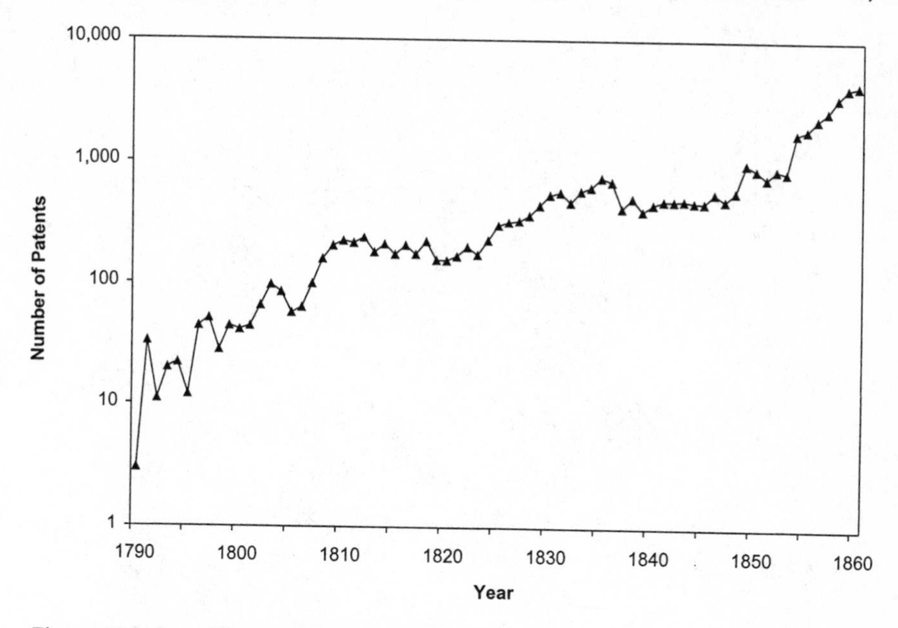

Fig. 3.2. Number of Patents Issued Annually, 1790–1860. *Source:* U.S. Bureau of the Census, *Historical Statistics of the United States,* ser. W99.

on inventiveness. Most firms did not require waterway transport because wagons provided low-cost shipment for manufactures, just as wagons were suitable for most farmers. Even if workshops and small nonmechanized factories sold products in the metropolises, wagons provided efficient transportation.[16]

Inventor characteristics comport with this interpretation of the spread of inventiveness across areas of agricultural prosperity. Between 1790 and 1822 those patenting one invention over their careers accounted for 50 percent of patents, and those with two career patents contributed 20 percent—testimony to widespread invention; those patenting extensively (six or more career patents) accounted for only 12 percent of patents. The elite in commerce and the professional classes in urban areas dominated patenting (holding 50 percent of patents) between 1790 and 1804, but their share dropped to 39 percent between 1805 and 1822. The share of patents held by artisans (e.g., carpenters; shoemakers; and makers of watches, jewelry, and instruments) fell slightly, from 35 to 31 percent, between 1790 and 1822, because their occupations were removed from the competition facing workshops and nonmechanized factories. Over that period the share of patents held by machinists and toolmakers rose from

Table 3.1. Annual Patent Rates per One Million Residents by State and Region for Manufacturing and for All Sectors, 1791–1846

	1791–98	1799–1804	1805–11	1812–22	1823–29	1830–36	1836–42	1843–46
Manufacturing								
Northern New England	1.1	4.5	7.3	6.8	10.5	24.0	14.2	9.6
Southern New England	2.4	11.2	27.8	31.2	31.8	59.9	42.2	49.6
New York	6.0	5.6	26.6	17.5	37.7	32.2	15.8	32.2
Pennsylvania	7.0	10.0	16.2	15.5	13.8	21.3	9.7	20.0
Southern Middle Atlantic	2.1	8.8	12.3	13.2	10.4	14.1	7.8	19.5
Other States	0	2.2	1.0	2.0	3.0	4.0	2.2	2.7
National	2.0	5.7	10.7	9.7	12.5	16.1	8.7	12.9
All Sectors								
Northern New England	1.9	7.5	15.2	15.1	33.0	65.5	32.9	20.0
Southern New England	7.2	26.7	65.2	55.4	60.4	106.4	79.5	74.5
New York	10.9	16.4	62.0	49.9	81.3	95.6	49.6	65.8
Pennsylvania	17.2	14.5	29.7	33.6	32.2	53.3	32.9	42.5
Southern Middle Atlantic	4.1	17.0	23.7	34.9	31.9	41.4	40.8	40.0
Other States	1.2	3.4	3.4	6.1	10.4	13.2	7.7	9.9
National	5.2	11.3	23.9	22.9	30.0	41.8	24.5	27.3

Source: Sokoloff, "Inventive Activity in Early Industrial America," 824–25, table 1.

Note: Regional definitions:
Northern New England: Maine, New Hampshire, and Vermont
Southern New England: Massachusetts, Connecticut, and Rhode Island
Southern Middle Atlantic: New Jersey, Delaware, and Maryland
"All sectors" includes agriculture, construction, transportation, and manufacturing.

4 to 11 percent, and the share held by other producers and dealers of metal products (e.g., stove manufacturers and blacksmiths) rose from 8 to 11 percent. Therefore, the decline from 85 to 70 percent in the share of patents held by the elite and artisans between 1790 and 1822 foreshadowed the shift to individuals attuned to industrial demands among urban and rural populations in areas of agricultural prosperity.[17]

Many manufactures remained oriented to markets within subregions or regions, but firms in some industrial agglomerations expanded into interregional markets after 1790. The mix of infrastructural components differed among agglomerations: some products required unskilled labor, whereas others used highly skilled labor; and some workshops retained the flexible production of diverse goods using simple hand-powered machines, whereas others shifted to the large-scale production of uniform commodities using water-powered machines. Shoes, Connecticut manufactures, and cotton textiles exemplify this rise to national market dominance; each developed in early areas of agricultural prosperity.

The Boston Region Shoe Complex

Except for the small urban population, most colonial households made their shoes, but they could not gain proficiency in shoe manufacturing because each household made few of them, and cutting and sewing leather and designing and fitting shoes required skill; therefore, households produced crude shoes. By the 1750s local shoemakers overthrew most household manufacture, and widespread demand and simple shoe technology created employment for numerous shoemakers who served local demands. Because manufacturing high-quality, expensive shoes required substantial time and skill, a division of labor among workers or shops contributed limited production efficiencies. Local, small-scale retail producers dominated this custom market, and sometimes shoes were made-to-order.[18] To break into local markets, regional and interregional shoe manufacturers needed to enter at the broad, low end, and they achieved cost advantages over local shoemakers: efficiencies from division of labor within or among craft shops, the purchase of low-cost raw material inputs from firms supplying many manufacturers, and efficient wholesale distribution.

The Boston region's specialization for interregional shoe markets rested initially on the broad, low-end market, and large, captive populations guaranteed markets that supported the development of a production and distribution system. The shoe complex's roots date from about 1750 to the Revolutionary War, when Lynn shoemakers supplied cheap shoes to Salem merchants for export to West Indies slave plantations, and after the war shoe wholesalers expanded markets for cheap shoes and boosted supplies from Lynn shoemakers and others in the region. Ebenezer Breed, a Lynn native and member of a Quaker family of shoe manufacturers, typified shoe merchants: in 1786 he moved to Philadelphia to operate as a commission merchant and developed close ties with other Quaker merchants. Merchants in Philadelphia and nearby Baltimore possessed marketing channels to acquire leather and to export goods to southern cities, especially those supplying slave plantations. Amos Rhodes, Breed's agent in Lynn, furnished him contacts with shoe manufacturers, and, by making trips to Richmond, Petersburg, Charleston, Augusta, and Savannah, Breed cemented ties with southern merchants supplying captive slave markets. In retrospect Breed wisely chose wholesale distribution outlets serving Virginia, North and South Carolina, and Georgia. A large slave population cultivated tobacco in southeastern Virginia, and from there to northern Georgia the Piedmont stood on the threshold of an explosive expansion of slave labor cotton agriculture to supply Britain's booming textile industry. From 1790 to 1820 the production of cotton bales soared from 3,000 to 335,000, and the black population (mostly slaves) in the Piedmont states doubled from 551,000 to 1,101,000. Ebenezer Breed and merchants like him profited handsomely from bulk shoe orders that slave owners placed, and he recognized the need to reduce competition from British shoe producers. Breed leveraged his ties to Philadelphia Quaker merchants and became a leading lobbyist for the tariff on imported shoes passed in 1789.[19]

Massachusetts dominated interregional market shoe manufacturing, and the South was the largest market. In 1810 Massachusetts shoemakers produced 6.3 shoes per capita, over three times the national average (2.0), and accounted for 51 percent of shoes entering interregional markets. Boston merchants funded, supplied, and marketed the shoe production of Lynn and other towns in the region, Philadelphia and Baltimore supplied hides, and their merchants distributed shoes from Massachusetts before 1820. Craft shops making complete shoes dominated production until about 1810; termed *ten-footers,* after their dimensions, they housed a master, several journeymen and apprentices, and the master's wife and several children. They employed a simple division of labor: women bound uppers; journeymen and masters cut leather pieces, tacked uppers and soles to lasts, and sewed uppers and soles together; and children and apprentices performed minor tasks. By 1789 Massachusetts exhibited specializations by town: Lynn produced women's and children's shoes, and by 1810 its output of women's shoes reached about 100,000 pairs annually; and towns in southern and western Massachusetts produced men's shoes, brogans (heavy work shoes), and high boots. Increased shoe demand opened opportunities to reorganize production, and from 1810 to 1820 shoe producers experimented with central shops. In its early form this small operation hired workers to cut leather for distribution to ten-footers, and they returned finished shoes to central shops. They gradually experimented with an increased division of labor, possibly enhancing productivity, and the larger capital required to fund materials shifted control over production and distribution to those entrepreneurs with the greatest financial resources.[20]

By 1820 Boston region shoe manufacturers made substantial inroads into markets for standardized cheap shoes, forcing local shoemakers to focus on shoe repair and custom orders. Rising per capita income encouraged entrepreneurs to expand the production of cheap shoes, but income gains were too modest to account fully for growing production. The reduction in shoe prices brought about by new forms of production—especially the division of labor in ten-footers, which eliminated the costly processes of individual shoemakers—probably contributed more to shifting manufacturing to markets for low-cost shoes. Individual shoe firms remained small-scale producers, and increased output came from numerous tiny firms employing a division of labor within firms and a specialization of function among firms; workers used hand tools. Although Massachusetts controlled about half of interstate trade in shoes, it accounted for only 21 percent of total value of national shoe production in 1810—thus, national market dominance was not ensured.[21] In contrast to shoes, Connecticut manufactures do not fall into a single group based on a division of labor or cost-cutting strategy. Because Connecticut contained some of the East's most prosperous agricultural areas, it also exemplifies agriculture's positive impact on industrialization; firms in a large number of different manufactures moved into interregional markets after 1790.

Connecticut Industrialization: Yankee Ingenuity or Archetype?

Connecticut's industrial inventiveness approached mythical proportions, and the facts seem to confirm it: measured by patents per capita, Connecticut led national inventiveness from 1790 to 1900 and exceeded the second-ranked state by about one-third from 1870 to 1900. Commentators coined the term *Yankee ingenuity* to depict the industrial inventiveness of New Englanders, and Connecticut Yankees epitomized ingenious inventors and manufacturers. Joseph Roe, a leading student of mechanical invention and machinery in the nineteenth century, argued that Connecticut mechanics' roots in the English middle class explained their inventiveness, and others claimed that the absence of opportunities in agriculture and commerce encouraged inventiveness.[22] Undoubtedly, Connecticut possessed talented mechanics, yet they could claim no more ingenuity than talented mechanics elsewhere. Prosperous agriculture undergirded Connecticut's industrialization, and this was unexceptional; Connecticut was an archetype of eastern industrialization.

Domestic and Local Market Manufactures

Supervisors and collectors of the Treasury Department contacted officials serving under them as well as prominent local residents for their comments on manufactures, and their correspondences submitted for Hamilton's "Report on Manufactures" convey Connecticut's industrial status in 1791. These astute observers agreed that domestic cloth manufacture held sway; nevertheless, an army of craft workers met the ongoing needs of the farm economy for boots and shoes, saddles, harnesses, farm implements, and wagons. Some crafts sold surplus production outside the state, but those were minor exports, and references to agricultural and other resource-processing manufactures (e.g., flour mills, sawmills, and distilleries) remained subdued. Farmers and other resource extractors expanded and deepened ties to New York City's market, and subregional metropolises and central place villages flourished, and by 1819 a different manufacturing economy existed in Connecticut. Domestic manufactures attained minor importance only in eastern counties (Tolland, Windham, and New London) and were insignificant in central and western counties (Hartford, New Haven, Middlesex, Fairfield, and Litchfield), which were oriented toward New York City. Processing manufactures blanketed Connecticut: most towns housed grain mills, sawmills, distilleries, tanneries, fulling mills, and carding machines serving household consumption and their market production.[23] These firms added dispersed points of capital accumulation to those of village retailers, professionals, and prosperous farmers, thus supporting new manufacturing.

Extremes of individual wealth—tiny concentrations of leading merchants in

Hartford and New Haven or on a larger scale in Boston, New York, and Philadelphia—provided markets for craft workers making unique, exquisite metal and wooden goods. Nevertheless, these craft workers gained no experience with production for larger-scale markets, in which they faced harsh competition. Connecticut's broad-based prosperity, which was sweeping up farmers, small mill owners, and business and professional leaders in cities, towns, and villages, posed different challenges to craft workers who did not serve the wealthy. Competition with other craft workers for this large market required changes in labor organization and in techniques and technology to improve quality and raise productivity. Entrepreneurs met this challenge and developed many thriving firms forming complexes of workshops, and they made substantial progress acquiring dominant national positions in various manufactures such as hats, tinware, clocks, plated ware, buttons, brassware, and hardware. If "Yankee ingenuity" is dismissed as an explanation, a puzzle remains: how did those Connecticut firms achieve success across many different manufactures? In contrast to the large, simple markets of shoes and textiles reached by wholesalers and retailers, Connecticut's firms faced uncertain markets and confronted local craft shops that were more formidable competitors than household shoemakers, spinners, and weavers. And the initial ideas for most products and simple manufacturing technology provided few barriers to entry during the early years of production. By the 1790s, however, Connecticut's entrepreneurs possessed substantial advantages: they occupied a rich information environment, which allowed them to reach complex markets outside Connecticut and to develop organizations, techniques, and technologies of production to undercut distant craft shops.

Information Hubs

Educational Networks

By the late eighteenth century Connecticut's superbly educated, literate citizenry possessed an exceptional capacity to acquire and process information. No less a luminary than Yale's president Timothy Dwight testified to the high-quality education: "schools are everywhere in existence, and are everywhere managed with a good degree of propriety." Schools received widespread citizen support, and from 1730 to 1760 the colony designated large sums from western land sales to bolster them. In 1795 the state established a permanent fund from land sales in the Western Reserve of Ohio, and the legislature designated interest proceeds to support town schools, supplementing state and local tax aid. Citizens supported up to twenty academies offering more rigorous education than public schools, and at the educational pinnacle stood Yale College, the training ground of the social, religious, political, and business elite. Educated together and united as alumni, they formed a seamless cloth of shared interests and information exchange which permitted shifting coalitions to found mercantile enterprises, banks, insurance companies, and industrial firms with tacit, if not direct, government support. The wealthiest had little inclination to found or underwrite new, small workshops in industries such as hats, tinware, or clocks, but the elite shared information about business changes in state, regional, national, and international arenas.

Although the elite were concentrated in the subregional metropolises of Hartford and New Haven, many lived in rural areas as wealthy farmers and in villages working as major retailers, clergy, lawyers, or physicians, because prosperous rural economies offered opportunities. Yale-educated clergy were information hubs in the rural parishes of western Connecticut by 1750, and sizable numbers of them lived in Hartford (forty-five), New Haven (thirty-seven), Litchfield (twenty-three), and Fairfield (twenty-nine) Counties by 1770. Few of them served their native parishes; thus, most had ties elsewhere. Their alumni contacts, travels to New Haven, pulpit exchanges with other parishes, county ministerial association meetings, providing lodging for travelers, and local visits to wealthy farmers, professionals, and businesspeople returning from trips gave the clergy access to diverse information to share with their parishioners. Lawyers complemented clergy as information sources after 1780 and even surpassed them in terms of mobility; they frequently traveled to meet other lawyers, judges, and clients, thus operating as information pivots about local and regional economies. Litchfield village housed a law school founded in 1784 which claimed students from across the nation; it graduated 474 students between 1784 and 1812.[24] Because these social networks had multiple, nonredundant channels among participants, information moved rapidly around the network, and no one monopolized information.

Connecticut's citizens actively engaged in literary pursuits, which were logical extensions of enjoying widespread literacy and a prosperous economy: farmers communicated by letters with friends and relatives, citizens supported reading pursuits so extensively that social libraries appeared in most villages by 1819, and most citizens lived within twenty miles of bookstores. Books were purchased on extended trips to merchant wholesaling centers and ports such as Hartford (with seven bookstores), New Haven (four), Middletown (two),

New London (two), and Norwich (two) as well as on shorter trips to large villages such as Norwalk (one) and Litchfield (two); most of these places had book binderies. Since 1773 every wholesaling center, except Middletown, possessed newspapers, and by 1819 all of them published weekly papers; Norwalk, Bridgeport, and Windham also had newspapers. The postal service established offices in every substantial village, thus ensuring that written communications reached everyone regularly. The dense network of turnpikes and other roads eased the flow of written information through the postal service, and good roads permitted efficient face-to-face contact. Besides many radial turnpikes surrounding Hartford and New Haven, Connecticut's latticelike grid connected most towns by 1814. The travels of Yale's president Dwight provide a window on the period from 1795 to 1815; he rarely complained about road conditions and on any given day comfortably traveled twenty to twenty-five miles on horseback and sometimes covered even greater distances.[25] Thus, Connecticut mechanics and local financial supporters lived within dense information networks effectively communicating ideas about new industrial endeavors, methods of organizing production, and techniques to make goods. Yet mechanics and their financiers also needed information about markets and distribution channels outside of the state to break the bounds of the local market.

Market Information

Connecticut entrepreneurs sat astride the information corridor between Boston and Baltimore and were near New York City, the national pivot of market information. Wholesalers and retailers in mercantile centers and in small ports along Long Island Sound regularly accessed commercial information through selling agricultural and other resource products at the New York market and through purchases made there. By 1794 all of Connecticut, except the northeast corner, received information from New York within five days, placing the state near the center of national information networks. And Connecticut's mercantile firms maintained direct trade ties with large and small ports from Maine to Maryland and with southern ports, led by Charleston, South Carolina; most ties dated from at least 1750 and strengthened up until 1820. Because Connecticut exported diverse commodities, merchants acquired a broad understanding of external market demand, and superb information networks within the state transmitted this market knowledge to manufacturers. State emigration following 1750 and continuing past 1820 created ongoing ties through friends and families to emerging markets in northern New England and, more important, to large population growth areas in central and western New York, western Pennsylvania, and northern Ohio. In contrast, potential manufacturers in rich agricultural areas elsewhere in the East had less access to information about interregional markets, because their information came filtered through wholesalers in Boston, New York, Philadelphia, and Baltimore, which was biased toward goods with proven, large-scale demand (e.g., shoes and cotton textiles). These wholesalers exported regional staples such as wheat, flour, and fish, which were widely demanded in other regions or internationally; they had little incentive to explore the complexities of demand in the hinterlands of other regional metropolises.[26]

In the century from 1750 to 1850 peddlers provided alternative information sources about distant markets and means to reach them. Few peddlers served frontier settlements and areas of low population density and poor roads, because farm household incomes were too low. Peddlers acted as roving country stores, but this local exchange between farm household and peddler operated under severe constraints. Households possessed limited cash or items to barter, and peddlers could not transport heavy, bulky agricultural output; country storekeepers were better equipped to accumulate farm surpluses for long-distance transport to markets. Therefore, peddlers carried little farm produce to market, and they assigned low values to household crafts because these items had limited resale value. As farm households increased their market integration and their incomes rose, they shifted their purchases to village stores offering wider ranges of goods and better prices for farm surpluses; these stores became more accessible with improved roads. Peddlers served a niche, nevertheless, as purveyors of goods with a "high value per unit weight," including new goods that storekeepers did not stock and goods such as clocks and tinware which were demanded in such small quantities that storekeepers stocked few of them; Connecticut workshops developed a specialization in them. During the economic upturn of the 1790s workshops sold mostly to Connecticut's prosperous farmers and village households. Workshop owners sometimes peddled goods, but hired peddlers provided wider distribution channels; after 1800 workshops used peddlers to penetrate markets outside the state. When peddlers carried diverse products, it increased the sales opportunities with each customer and reduced the distribution costs for each good.[27]

Small Connecticut firms became prominent in various consumer goods industries targeting two different markets. Fur hat manufacturers targeted wealthy farmers and urban consumers reached through metropolitan wholesalers, principally in New York City and to a lesser extent Charleston, South Carolina, whereas manufacturers of tinware, spoons, plated ware, buttons, and clocks targeted farmers, village craft workers, mill owners, and professionals living in prosperous agricultural areas. Producers targeting the latter market used Connecticut to test consumer receptivity to their goods, but sales outside the state posed problems. Firms needed production technologies and labor organization undercutting rural and village craft workers, and they avoided competing against craft shops in metropolises because these shops dominated local markets. As rising agricultural prosperity boosted purchasing power, manufacturers required marketing strategies to convince consumers to buy better goods—for example, tinware, metal spoons, plated ware, and buttons—as substitutes for cruder alternatives or to buy goods such as clocks which they could not previously afford.

Hats in the Embrace of New York City

By 1791 Connecticut housed several hat workshops that had recently commenced production for markets outside the state, and Danbury, about fifty miles northeast of New York City, already was famous as the leading producer; its firms broke the bounds of local markets through local finance and distribution through New York wholesalers. Fur hat production predominated, implying that urban consumers, and perhaps wealthy farmers, were the main markets; Danbury's firm of O. Burr and Company, reputedly the largest in the state (with seven journeymen and ten apprentices), sold hats at wholesale in New York City and also planned to open a retail shop there. Danbury's retail merchants occupied the hub of expanding turnpike and road networks in western Connecticut, providing access to nearby beef cattle and dairy farms, to Hartford and the Connecticut Valley, and to New York City. They accumulated capital from selling farm goods in New York and retail goods to prosperous farmers, craft workers, and professionals in Danbury's vicinity. O. Burr and Company continued retailing while expanding hat manufacturing; some retailers followed Burr's example, whereas others remained solely retailers. In either case capital flowed to hat manufacturing, primarily as credit: Danbury's mercantile stores purchased raw materials (e.g., furs) for hat manufacturers and debited accounts, and merchants permitted manufacturers to pay workers with "orders" cashed at their stores for goods. The circuit was completed when manufacturers paid retailers with hats that the latter sold to New York wholesalers.[28]

This symbiosis between mercantile credit and hat manufacturing supported the substantial growth of Danbury's hat industry. Workshops produced as many as twenty thousand hats by 1800, and within less than a decade fifty-six hat shops were operating; most remained tiny and employed perhaps three to five workers, but large firms soon emerged. In 1812 Tweedy and Benedict employed thirty workers, and by 1814 the White Brothers had fifty workers, capital of $50,000, and produced up to twenty-four thousand hats annually. Danbury's merchants and hat manufacturers, like other Connecticut businesses, looked to southern markets for direct sales, and newly wealthy cotton planters probably were targets; firms chose Charleston as distribution center and opened four wholesale outlets between 1802 and 1820. Those southern forays into direct wholesaling remained sideshows, however, to New York merchants' grip on the hat trade; they supplied furs, and before 1812 hats were being finished in New York. Sometimes a Danbury firm such as White Brothers established a warehouse in New York, but that strategy merely moved firms closer to wholesalers. By 1819 Danbury's hat industry had grown considerably; most, if not all, of its twenty-eight hat workshops lined about one mile of the road in town, along with many mercantile stores and other shops, and production spilled over to nearby New Milford and Oxford. Scattered hatters' shops in towns such as New London and Litchfield indicated that producers still carved out local markets, and two workshops in Hartford and one in New Haven probably focused on large local and nearby markets. A Hartford workshop employed thirty-six workers, and another one in East Hartford used waterpowered machinery to make hats, suggesting sales to out-of-state markets, and Bridgeport borough housed workshops that exported to other states. Firms still engaged in direct wholesaling to the South from Charleston, building on Connecticut wholesalers' long-standing ties; nevertheless, New York merchants controlled most hat wholesaling.[29]

Marketing Metal Consumer Goods

Producers of metal consumer durables—items such as tinware, spoons, plated ware, and buttons—confronted a dilemma: consumers demanded them more as incomes rose but purchased them infrequently, and at the modest income levels prevailing in prosperous agricultural areas, country and village stores stocked limited numbers and varieties of these goods. Consequently, wholesalers could not feasibly distribute them, except to large-city markets, and infrequent consumer purchases meant that producers readily saturated markets around workshops. Tinware producers solved the dilemma, and manufacturers of spoons, plated ware, and buttons rode their coattails; they transformed lowly local peddlers into interregional distributors of light, high-value goods to the mass market on farms and in villages. Metal consumer durables manufacturing shared several characteristics. Retailers supplied fixed and working capital, often purchased workshop output, and organized distribution through peddlers; some retailers also entered production. Technology changed somewhat, but most production increases came from hiring additional employees, reorganizing work tasks, and employing a greater number of workshops. Internal production economies of scale remained limited: most shops employed several workers, and few hired more than ten before 1820. External economies of production and distribution developed as clusters of firms emerged in a small area of Connecticut: firms shared the peddler distribution system, copied techniques and styles, spun off new workshops from existing ones, and tapped a growing pool of workers moving among firms.

Tinware

Tinware constituted ideal products for Connecticut's rural and village markets experiencing rising agricultural prosperity. At lower income levels households purchased pots, pans, buckets, cups, and ladles for everyday use, and, as income rose, they bought roasters, trays, candlesticks, trunks, and storage containers (for pills or jewelry); consumers could purchase plain or elaborate, painted versions of these items. British producers monopolized tinplate manufacturing until the late nineteenth century; therefore, international and national wholesale networks distributed tinplate to small tin shops. Boston wholesalers initially controlled tinplate imports, but, with the increasing hegemony of New York wholesalers over markets in central and western Connecticut, they captured control of the tinplate trade by the decade from 1810 to 1820. Tinplate moved by sloops traveling to New Haven, Middletown, and Hartford, and their small-scale wholesalers supplied large retailers near burgeoning tin shops; retailers provided credit to tin shops for purchasing tinplate and producing tinware. Under the leadership of the Pattison family shop, dating from the 1750s, Berlin became the hearth of the tinware industry, and their shop was a training center and a source of information and supplies for the tinware trade. In nearby Meriden the Yale family, who were leaders in other metal trades, operated a tin shop by the 1790s, but major expansion there came in the first two decades of the nineteenth century. Nearby Bristol commenced tinware manufacturing by the 1790s: local retailers such as Elijah Manross and the Mitchell family funded local tin shops with credit and goods and, in return, received tinware they sold through peddlers; these retailers then started tin shops. The Mitchell family owned Bristol's largest retail store and with their peddlers became leaders in the tinware business.[30]

The tinware complex developed a division of labor, which fanned growth. Individual tin workers pounded, bent, and molded tin into utensils, but after 1800 japanning (varnishing) became popular for making better articles. This technique, which permitted painted designs to be fixed to products, required skill to make the varnish and bake the utensil at the proper temperature; the painting was the specialty of a corps of young women traveling among shops. The manufacture of tinners' tools commenced; Berlin's North family ran a mercantile store and added a shop to make tinners' tools. Although no Connecticut workers invented a major tin machine before 1820, shops quickly adopted a machine, invented by Calvin Whiting and Eli Parsons in Dedham, Massachusetts, in 1804, for finishing the rough edges of tin. Seth Peck from Southington, Connecticut, purchased patent rights for manufacturing the machine by 1816, setting the foundation for the state's tinware machinery business. Production scale economies using machinery remained limited before 1820 because few machines existed to make components. Workers increased production through efficiencies acquired when making large volumes of the same product; therefore, small tin shops predominated. Wealthy farmers, mercantile store owners, and tavern owners invested capital in these shops, and high worker mobility among the complex's shops provided labor.[31]

Peddler distribution powered the sales of a rising cornucopia of tinware. Tin shop owners such as Edward Pattison of Berlin sold tinware by horseback or wagon within Connecticut through the economic contraction of the 1780s, when demand was restrained. Local peddlers may have purchased some tinware from shops to sell, but they were not a reliable distribution system when greater numbers of tinsmiths in Berlin, Meriden, and nearby towns started production during the 1790s upturn. Firms experimented with contract peddlers specializing in selling tinware out of state and with directly employing them. Around 1800 tinware peddlers widened travels throughout New England and New York; between 1795 and 1815 Yale president Timothy Dwight encountered them on trips ranging from Cape Cod, Massachusetts, to western New York. From 1800 to 1820 tinware dealers intensified their peddler distribution in New England and New York and expanded to New Jersey and Pennsylvania. Bloomfield's Filley family, leading tinware dealers after 1806, supplied Vermont peddlers until 1809, and beginning about 1810 they opened peddler distribution branches, first in Elizabethtown, New Jersey, then near Troy, New York, and by 1816 in Philadelphia. Southern markets became important during the first decade of the nineteenth century, as the Piedmont cotton boom created a small-planter middle class. Peddlers took wagon loads south at the start of winter, selling goods along the way, and when they reached Richmond, New Bern, Charleston, or Savannah they met tinsmiths sent southward who produced tinware that peddlers sold in the interior; in early spring peddlers sold their horses and wagons and returned by ship to Connecticut. By 1819 peddler distribution blanketed the nation east of the Mississippi River, giving Connecticut workshops a firm grip on the market. Workshops in Berlin, the industry hearth, each employed as many as twenty peddlers to distribute tinware, and one firm reputedly employed forty. Bristol and Meriden each matched Berlin's five workshops, nearby Middletown had six workshops, and a few appeared in surrounding towns, some as far east as New London.[32]

Once tin shop owners and retailers recognized the value in selling tinware through peddlers during the 1790s, new tin shops had incentives to locate near one another to tap into others' sales organization, contributing to demand for peddlers and, in turn, raising the probability that additional tin shops would locate nearby. As tinware production soared and retailers took over the organization of the large peddler corps, tin shops had further incentives to locate in Connecticut's tinware complex. Elsewhere in the East tin shops hoping to sell outside their local areas faced growing barriers: independent peddlers favored supplies from Connecticut, because its shops offered greater volume and variety at lower prices, and its tin shops and retailers possessed an efficient peddler system with greater access to information about markets in the East, South, and Midwest. This army of tinware peddlers offered superb distribution for other manufacturers of metal goods such as spoons, plated ware, and buttons; therefore, firms in these industries agglomerated near tin shops. These manufactures remained less important than tinware before 1820, but later they surpassed it. Tinware faced competition from products made out of different materials, whereas spoons and plated ware generated the silverware and cutlery industries and buttons spawned the brass industry; these manufacturers shared overlapping metalworking skills, including casting, cutting, bending, and polishing metal.

Spoons, Plated Ware, and Buttons

Prior to 1800 the manufacture of spoons, plated ware, and buttons was sporadic. British imports offered formidable competition in large cities, and retailers in rural areas and villages stocked few of these infrequently purchased items; thus, wholesaling distribution was inefficient. Craft shops saturated local markets, but they could not sell in distant ones. Meriden's Yale family of tinware producers was among the earliest button manufacturers; they added pewter buttons in 1794 and probably peddled buttons, along with their tinware, in Connecticut. Following a slow start during the 1790s, the formation of firms making spoons, plated ware, and buttons picked up, aided by the expanding peddler distribution of tinware in New England and New York, and after 1810 growth accelerated, coinciding with intensified peddler distribution in the Northeast and its expansion in the South and Midwest. More firms appeared in Meriden; some made only buttons, whereas others combined buttons with tinware. Charles Yale's success at selling tinware in the South probably motivated the Yale brothers' decision to add Britannia ware to tinware manufacturing in 1815; they imported skilled English workers for the task. Connecticut button inventors had a banner year in 1815, receiving nine patents, and they worked in the industry's core—Meriden, Bristol, Southington, and New Haven.[33]

Button manufacture in Waterbury (near Bristol and Southington) diverged from other towns in 1802 when Silas Grilley joined with Abel and Levi Porter and Daniel Clark to found Abel Porter and Company to produce brass buttons. Using scrap copper from old kettles, stills, and sugar boilers, the firm added zinc to cast brass ingots, transported ingots to an iron rolling mill in nearby Litchfield, and returned sheets to Waterbury for final rolling. By 1811 Abel Porter and Company metamorphosed into Leavenworth, Hayden, and Scovill (known as the Scovill firm), uniting Leavenworth's and Scovill's mercantile capital with the button expertise of David Hayden. Hayden arrived at Waterbury in 1808 after gaining experience in Attleboro, Massachusetts, a leading button manufacturing center about twelve miles northeast of Providence. In 1817 Daniel Hayden, David's brother, joined the firm to manufacture lamps and other brass articles, and he received training in metal and machinery skills: as a machinist in Williamsburg, Massachusetts; as a gunsmith at the

Springfield Armory; and as a textile machinery builder, first with Samuel Slater, then with Wilkinson's machinery firm in Providence, and finally on his own at Haydenville, near Williamsburg.

Thus, the early brass industry possessed strong network bridges to pivotal metalworking centers in New England. Benedict and Burnham, another major Waterbury firm, also started inauspiciously, and, consistent with brass industry growth in Waterbury and vicinity, the firm had close ties to Leavenworth, Hayden, and Scovill. In 1804 Aaron Benedict and Joseph Burton became partners in a mercantile business, and by 1812 they began manufacturing bone and ivory buttons; they possessed an inside track to button manufacturing because Benedict married Charlotte Porter, the daughter of Abel Porter, in 1808. Producers of metal spoons, plated ware, and buttons clustered in Meriden, Berlin, Wallingford, Middletown, Bristol, Southington, Derby, Salisbury, and Waterbury by 1819, and these goods, along with tinware, were sold to the nation east of the Mississippi through well-oiled peddler distribution systems.[34]

Wooden Clocks

Like small metal consumer durables, clocks faced low demand in rural areas and villages as of 1790. Following British tradition, most clockmakers produced brass movements, and the requisite skills and laborious manufacturing methods raised costs to thirty to fifty dollars, and wood encasements added ten to fifteen dollars; at those prices only wealthy farmers, merchants, and professionals could purchase them. Although wooden movements cost about half as much as brass movements, crude wood construction restricted demand to poor rural areas, yet their prices made them luxury items, and both types of clocks including the case weighed over one hundred pounds. Clockmakers produced two or three clocks at a time and peddled them within a one-day round trip of the shop, or more typically they built clocks to order and traveled by horseback to deliver them; clockmakers seldom finished ten clocks annually, and most made fewer than six. Eli Terry began manufacturing brass and wooden clock movements in Plymouth, just north of Waterbury, in 1793, after apprenticing with specialists in each of the movements. Terry operated like other clockmakers, but a disjunction between the growing success of tin shops in nearby towns and the inability of clockmakers to increase their sales frustrated them.

Shortly after 1800 James Harrison experimented with waterpowered machinery in his Waterbury clock shop; simultaneously, Terry started to upgrade equipment and construct new tools and machinery for making wooden clocks. Their efforts imply recognition that technical changes raised production levels and reduced costs; peddlers would distribute clocks. Harrison and Terry targeted the growing market of middle-income households, replicating the approaches of nearby manufacturers of metal consumer durables, but Harrison lacked the business acumen and quickly faded from the scene. Over the next few years Terry's tools and machinery enabled him to start as many as twenty-five clock movements simultaneously, probably with help from several apprentices. He achieved efficiency gains from batch production of "approximately" interchangeable parts and chose wooden movements because they did not require exacting measurements of all the parts, as did brass movements; no metalworking technology existed to achieve that precision.[35]

Terry's dramatic production increases—about eight times the levels of typical clockmakers—suggest that his clocks were higher quality and cheaper than competitors' clocks. Consumers eagerly purchased clock movements pouring from Terry's workshop, and in 1806 he decided to shift to a new, small factory and raise production to two hundred clock movements simultaneously. Opportunity beckoned to shrewd Waterbury retailers and manufacturers—Edward Porter, a Yale graduate of 1786 and retired clergyman, and Levi Porter, one of the principals in the new (1802) button firm of Abel Porter and Company. The Porters had purchased clock movements, including some made by Terry and distributed through peddlers, thus recognizing that falling prices stimulated demand. In 1806 they negotiated a contract with Terry stipulating he would complete four thousand clock movements at four dollars each by 1809, and the Porters would supply the materials and sell the clocks; Terry scrapped plans for his original factory and found a site for a larger factory. It took a year to build new machinery, and by 1809 he achieved annual production of three thousand clocks, an astounding achievement and a death threat to wooden clock craft shops. The Porters readily sold clocks: those that were sold as movements may have been peddled widely in New England and New York at prices as low as ten dollars, but, if they were sold in a case, they probably never left Connecticut.[36] This transfer of surplus capital from retailing and manufacturing to fund new manufactures would typify the emerging industrial complex, as entrepreneurs utilized interwoven social networks to identify new opportunities.

Before Terry completed the order, news of audacious changes in clock manufacturing and sales reverberated throughout nearby towns. In Bristol the

Roberts' brothers were the only traditional clockmakers in business in 1808; but two clockmakers entered the business in 1809, and four or five clock shops operated through 1812. Production soared as clockmakers probably copied Terry's machines and methods. Industrial secrets could not be kept in this rural and village economy with interwoven social networks of manufacturers, retailers, and professionals: people moved easily among towns, and traveling young women painting tinware also painted clock faces, thus serving as information channels. Yet standard wooden movement clocks had severe defects as consumer items. Hang-on-the-wall versions left pendulums exposed to damage or disruption, and, if they were encased in wood, their final prices more than doubled, to twenty dollars, making them too expensive for many households. Their large cases posed other problems: consumers needed woodworkers to make the cases; clockmakers or distributors had to visit homes to make them; or clocks could be sold with heavy cases, but this restricted how far they could be shipped. Such clocks did not mesh with peddler distribution; Terry recognized these limitations and sold the factory to two of his employees, Silas Hoadley and Seth Thomas, in 1810.[37]

During the next several years Terry used profits from the Porter contract to underwrite work on designing a small clock and completed it by 1812. During the next two years he purchased a factory site and probably initiated contact with skilled clock mechanics in the Bristol-Plymouth-Waterbury area to get assistance in designing machinery for making clock parts; on August 22, 1814, six mechanics received patents for that purpose. Early production runs revealed defects; therefore, Terry discarded traditional clock movement design and completed a revolutionary redesign of clock works. He received the first of six patents on June 12, 1816, which acquired the name the "Pillar and Scroll" shelf clock; it became a legend in mass market clocks. Seth Thomas quickly paid a thousand dollars to license manufacture of the clock. Terry hired Chauncey Jerome, a craft worker skilled at making clock cases, and they mechanized the manufacture of cases using new circular saw machinery; the factory employed waterpower to run its machines. While Terry accelerated production, he worked on design improvements of the shelf clock; production quickly soared to twelve thousand clocks annually in both Terry's and Thomas's factories. Prices of these clocks—initially fifteen to eighteen dollars retail—stimulated demand; as production increased and competitors commenced production, prices dropped, and by 1819 peddlers distributed clocks everywhere east of the Mississippi.

By 1810 clock manufacturing centered on towns with active firms: Bristol had several wooden clock factories, one of which made brass clocks; Plymouth and Waterbury each had two factories; and Winchester, north of Plymouth, had one factory.[38] Terry led the transformation of these clockmaker clusters to a dynamic, competitive complex of clock manufacturers reaching the mass market, but he did not act alone; other clockmakers and skilled craft workers contributed ideas to create machinery for large-scale production. This interchange resulted from the integrated social and economic structures of prosperous agricultural towns; information flowed through wide-ranging, nonredundant social networks. Clocks had bonds with such metal consumer goods as tinware, spoons, plated ware, and buttons; a cluster of towns dominated each industrial product, and their clusters overlapped. Peddler distribution instituted by tin shops united these manufactures, and the growth of each enhanced the peddler system, thereby fueling the expansion of other industries.

Archetypal Transition

Emigrants from Connecticut during the years from 1790 through 1820 left behind thriving agriculture increasingly bound to eastern markets, especially New York City, and Connecticut wholesalers and retailers forged well-oiled trade networks, collecting rural surpluses, redistributing some locally and within the state and exporting the rest. The accumulated capital of wealthy residents supported the formation of financial institutions (e.g., banks and insurance companies) and infrastructure projects, which undergirded future growth. Prosperous farms and other natural resource producers (e.g., those involved in fishing, lumbering, and quarrying) stimulated vigorous local market crafts and processing, and these sectors, along with retail merchants and the professional elite, accumulated capital for manufacturing. Rising agricultural prosperity encouraged craft shops to expand sales to local rural and village markets, which they quickly saturated, forcing them to look elsewhere for open markets. Expansion required access to information about markets and technology, and Connecticut excelled as an information network: it invested in human capital and developed a literate population with interchanges within the state and externally. Firms in manufactures such as hats leveraged their ties to New York wholesalers to reach external markets, whereas firms making metal consumer durables and wooden clocks built a peddler distribution system focusing on the mass market of rural and village dwellers. Connecticut did not have unusually ingenious mechanics; rather, its citizens utilized their agricultural prosperity to build an industrial base, and in this sense Connecticut was an archetype of agricultural and industrial transformation.

Standard interpretations of the rise of cotton textiles, the third manufacture exemplifying the bursting through local market bounds in the East, stress the importance of merchant wholesaler capital, the development of technically skilled textile mechanics, the significance of immigrants to founding textile industry, and the availability of low-wage, surplus labor that was ideal for textile factories. Implicitly or explicitly, these interpretations view the cotton textile industry as emerging in a poor agricultural economy. In contrast, the explanation posed next argues that areas of rising agricultural prosperity in the East formed seedbeds for the industry's growth.

CHAPTER FOUR

The Foundation of the Eastern Textile Cores

> It may be announced, that a society is forming, with a capital which is expected to be extended to at least half a million of dollars, on behalf of which, measures are already in train for prosecuting, on a large scale, the making and printing of cotton goods.
>
> —ALEXANDER HAMILTON, 1791

By 1790 rural and village households controlled the production of cotton textiles, making staggering quantities and varieties—dresses, shirts, pants, undergarments, stockings, handkerchiefs, coats, aprons, ribbons, towels, tablecloths, sheets, and blankets. Simple spinning wheels and hand looms sufficed for plain versions of products, and women and young girls learned requisite skills and dexterity producing large amounts of them.[1] Only the few wealthy households and urban dwellers could afford to purchase cloth, and Britain supplied most of it. Increased per capita income and greater numbers of urban dwellers raised demand—shifting the demand curve rightward—for factory-produced textiles, but income gains came slowly. Factory owners could improve production technology and organization, however, in order to reduce textile prices—shifting the supply curve rightward. The vast output of home-manufactured textiles in rural areas, on farms, and in villages signified huge latent markets and thus galvanized entrepreneurs.

Crompton and Arkwright Spinning Technologies

Alternative British technologies—Crompton versus Arkwright spinning systems—confronted American manufacturers of cotton textiles in 1790, and each offered a unified package of labor organization, machines, and production processes; capital was invested in machinery and buildings. Production followed flow, rather than batch or workshop, principles, and managers monitored labor processes and machine output to prevent bottlenecks. Carding machines combed cotton into parallel form, called "sliver," and in the drawing machine rollers drew sliver out, increasing the parallel form; then roving machinery imparted "twist" to sliver and transferred it to bobbins in preparation for making yarn on the spinning machinery. Because cotton spinning mills used complicated machinery with moving parts, owners and managers hired skilled mechanics to build, adjust, and repair machinery.

Crompton and Arkwright spinning systems employed similar machinery technology from the carding through the roving processes, but they diverged at spinning. In the Crompton system operators, called mule spinners, controlled yarn twisting and coiling on to as many as one hundred or more spindles, processes requiring skill and dexterity. Initially, mule spinners used hand power, but part of the motion was adaptable to waterpower; in either case owners paid high wages to hire and retain skilled mule spinners. Spinning jennies, which were simpler machines requiring much less skill and dexterity, also had numerous spindles, but jenny spinning did not offer the flexibility of mule spinning. In contrast, Arkwright spinning substituted capital for labor to produce coarse yarn through an investment in numerous spinning machines, each with multiple spindles; unskilled operatives attended them, and within several weeks they learned tasks requiring little physical strength or dexterity. Only waterpower could run heavy machine spindles at high speeds; therefore, the Arkwright system required a commitment to waterpower. Cotton mill owners spread the fixed investment in waterpower (for the dam, the raceway to carry water, the water wheel, gearing, and belts) over all machinery from carding to spinning. Under the Crompton and Arkwright systems weaving shifted to hand looms, and weavers worked in cotton spinning mills, craft shops, or homes; effective waterpower looms did not appear until around 1815.[2]

Large-Scale Failures

After the worst effects of the economic contraction of the 1780s had passed, the high profits of British textile firms lured numerous American entrepreneurs into textile manufacturing for domestic markets. They copied British techniques, sneaked machines past British customs, and bribed British mechanics to evade emigration laws. America's wealthiest merchants and most astute political leaders sought to emulate British firms, including the Cabot family of Boston merchants, backers of the Beverly Cotton Factory in 1787; leading merchants in Hartford and vicinity, founders of the Hartford Woolen Factory in 1788; the political and business elite in New Jersey and New York

City, backers of a factory city (Paterson) at the Great Falls of the Passaic in 1792 which Alexander Hamilton, the secretary of the Treasury, also supported; and John Nicholson, a wealthy Philadelphia entrepreneur, land speculator, and founder of a manufacturing village at the falls of the Schuylkill River in 1794. These large-scale textile projects quickly failed, and, as information circulated about them, eastern merchant and political elite probably despaired of capturing domestic markets from the British textile behemoth. By 1790 Britain's cotton spinning industry of 2.4 million spindles dwarfed America's, which had fewer than 2,000. From 1793 to 1807 Britain typically sent at least 30 percent of its cotton goods exports to North America, and their value jumped sixfold, a compound annual growth of 14 percent; this increase in British exports benchmarks the growth of the American textile market.[3] To counterattack Britain's onslaught, firms avoided the causes of past debacles—grandiose plans and extravagant expenditures—and relied on prudent business leadership to eliminate blunders. Cotton mill owners targeted markets in which they already competed successfully against Britain and reduced manufacturing costs to encourage households to make the shift from home production to purchasing textiles.

Market Expansion

Transportation improvements had a minimal impact on textile markets between 1790 and 1810, because shipping costs changed little and cotton yarn or cloth had a high value relative to its weight; even large transport cost reductions barely impacted final consumer prices. Few urban residents spun yarn or wove cloth, and many could afford to purchase cotton cloth; thus, they were prime cloth markets before 1810. With only seven thousand urban dwellers by 1810, the "West" contributed trivially to this demand. Between 1790 and 1810, however, the East's urban population grew 5 percent annually, with Boston, New York, Philadelphia, and Baltimore accounting for about 70 percent of the East's total urban population (see tables 1.2 and 2.5). Eastern urban market expansion accounted for up to one-third of the 14 percent annual growth of British textile exports to the United States before 1807. Migrants to the West provided another cotton cloth market because they faced pressures to clear land in order to commence farming, leaving little time for spinning yarn and weaving cloth; between 1790 and 1810 the West's population increased by 11 percent annually, but this figure overstates its market significance. Most of those living in the West purchased cotton cloth only during early settlement, and then, like eastern residents, they reverted to household spinning and weaving; in 1810 the per capita value of household textile manufactures in western and eastern states differed little. Western households' income was inadequate to purchase commercial cloth before 1810: western markets for their agricultural products remained small, and farmers exported minimal goods down the Ohio and Mississippi Rivers, the only feasible arteries. Together, eastern urban and western markets grew by 8 percent annually, accounting for just over half of the 14 percent annual growth of British textile exports from 1793 to 1807.[4]

Between 1790 and 1810 the eastern rural population rose by 2 percent annually, contributing only 15 percent of the increased demand for British textiles, but rural dwellers' importance is understated, because they formed a huge latent market for commercial textiles; they outnumbered urbanites by thirteen times in 1790 and remained eight times larger in 1810 (see table 1.2). Rising income encouraged eastern farmers first to stop hand spinning cotton yarn and then to cease hand loom weaving. The rate at which households dropped home textile manufacturing implies that they had elastic demand for commercial textiles during their early income gains—that is, their consumption grew faster than income. Per capita income gains for farmers, who made up most of the population, possibly reached 1.3 percent annually between 1793 and 1807 (see table 2.1), but this ranks between the extremes of isolated farmers in western New York and Pennsylvania, northern Maine, New Hampshire, and Vermont and farmers in prosperous agricultural areas. The latter's annual per capita income growth exceeded 1.3 percent, and these areas' village retailers, professionals (e.g., lawyers and physicians), and skilled craft workers had income gains equaling or exceeding farmers, thus adding to commercial textile markets; nevertheless, increased income accounted for a small share of the growth of British exports.

The Supply of Cotton Textiles

Compared to hand spinning, new cotton spinning technology embodied in the Arkwright and Crompton systems dramatically reduced the cost of yarn, and yarn price reductions of 20 to 30 percent were achieved faster than income gains; thus, declining textile prices provided a greater impetus to demand. Successful challenges to the British dominance of cotton yarn and cloth markets required that textile firms hire technically proficient managers and skilled mechanics; both needed training in mills applying Arkwright or Crompton spinning technology. British immigrant managers and mechanics were prime carriers of those technologies to American mills, because private-level secretiveness and government protection in Britain eliminated other avenues for technological diffusion until the 1820s. Rapid technological change and need to adapt technology to American conditions of capital and labor placed premiums on the knowledge of skilled managers and mechanics; the Crompton system also required skilled mule spinners, and they likewise carried skills to the United States. Shortages of managers with technical expertise and of skilled mechanics hindered cotton textile expansion until the mid-1830s, and the spread of indigenously trained managers and mechanics spurred the proliferation of successful mills.[5] The elite's cotton textile debacles in the decade from 1787 to 1797 partly resulted from the failure to hire these managers and mechanics.

Arkwright spinning mills produced coarse yarn for sturdy fabric (for plain sheetings and shirting), and the market included farmers, mill owners, and village retailers, professionals, and skilled craft workers in prosperous agricultural areas in the East and in the West's rural areas, where households dropped home spinning as yarn prices fell; those households could not afford to purchase cloth, whose price did not drop. Arkwright mills selling coarse cloth to markets in cities, among the wealthy in prosperous agricultural areas in the East, and among new western migrants faced difficulties until 1815, when early workable power looms appeared. In the meantime cotton mills confronted three options: employ unskilled hand loom weavers in the factory, "put out" yarn to unskilled home hand loom weavers nearby and then sell cloth, or find skilled hand loom weavers willing to work for low wages. The first two options meant that cloth prices reflected productivity levels that were barely greater than households, and a supply of low-wage, skilled hand loom weavers depended on labor market distortions.

As late as 1815, immediately prior to the introduction of power looms, New England mills wove only 14 percent of their yarn in factories, testimony to the limited benefits gained from hiring hand loom weavers.[6] Crompton mills, or Arkwright mills with several mule spinners, produced yarn suitable for weaving ginghams, cotton flannels, denims, and calicoes. The mill owners faced two options: they could hire skilled weavers to work in their factories, or they could sell yarn to other factories employing these weavers or sell yarn to skilled weavers owning craft shops. Merchants, major retailers, professionals, factory owners, and skilled craft workers in Boston, New York, Philadelphia, and

Baltimore, and in subregional metropolises such as Providence, Hartford, and Albany, constituted the markets for this cloth; other markets included the wealthiest farmers, mill owners, and village retailers, professionals, and skilled craft workers in prosperous agricultural areas.

Factory Expansion

The elite's large-scale debacles initially dampened enthusiasm for cotton manufacturing; as few as ten mills started production between 1791 and 1799, whereas twenty-one were built from 1800 to 1807. During the next two years, which included the Embargo, investors anticipated soaring domestic sales because British mills could not compete; the number of new yarn mills surged to 72. By 1809 perhaps 269 mills existed, yet many were probably shops housing a carding machine and a few spinning jennies. In the five years from 1810 to 1815—spanning the War of 1812 (1812–15) and the closing of American markets to the British—the mill count soared from 76 to 140 within a thirty-mile radius of Providence, the national center of cotton mills, but Britain dumped accumulated production on the American market after the war, causing many mills to shutter permanently.[7]

National mill estimates, however, mix numerous marginal operations with cotton yarn mills that were competitive with British mills. Competitive mills hired technically proficient, experienced managers and skilled mechanics, and mills surviving until 1832 provide a surrogate for them. Only five cotton yarn mills commenced production between 1790 and 1804, because British imports presented formidable competition and small supplies of technically proficient managers and skilled mechanics posed entry barriers (table 4.1).[8] The pace accelerated over the next three years, however, when nine mills began production, even though profits, indexed by factory gross margins, did not reach unusual heights; steady gains in per capita income could not cause the surging mill numbers. Rising numbers of experienced managers and skilled mechanics built a base supporting the surge; they moved around, providing the buildings and machinery to establish cotton mills, and each new mill trained more managers and mechanics.

Nevertheless, even with British competition suppressed during the one full Embargo year of 1808, shrewd investors started only three factories. Their reticence seems puzzling because gross profit margins on cloth rose at that time, and cloth production on hand looms in New England yarn factories, the industry leaders, surged. New competitive mills did not emerge in large num-

Table 4.1. Cotton Textile Industry, 1790–1820

	Wholesale Prices Cents/Yard of Cotton Sheeting				New England Cloth Output	
Year	Raw Cotton	Cotton Sheeting	Factory Gross Profit Margin	Number of Competitive Factories Established	Yards (thousands)	% Change from Prior Year
1790	—	—	—	1	—	—
1791	—	—	—	1	—	—
1800	—	—	—	2	—	—
1804	—	—	—	1	—	—
1805	10	21	11	3	46	—
1806	9	22	13	4	62	35
1807	9	21	12	2	84	35
1808	8	23	15	3	181	115
1809	7	25	18	7	255	41
1810	7	22	15	10	648	154
1811	6	19	13	10	801	24
1812	4	19	15	28	1,055	32
1813	5	22	17	23	1,459	38
1814	6	23	17	45	1,960	34
1815	9	20	11	15	2,358	20
1816	12	19	7	6	840	−64
1817	11	18	7	3	3,883	362
1818	10	17	7	4	7,216	86
1819	10	17	7	3	9,941	38
1820	7	16	9	6	13,874	40

Sources: The wholesale price of raw cotton (per yard of cotton sheeting) is computed based on 1.0 pound of raw cotton needed to produce 2.4 yards of cloth. For that ratio and New England cloth output, see Zevin, "Growth of Cotton Textile Production after 1815," 123–24, table 1, 136, 142. Wholesale prices of raw cotton per pound and cotton sheeting per yard come from U.S. Bureau of the Census, *Historical Statistics of the United States,* ser. E126, 128. Factory gross profit margin is cotton sheeting per yard minus raw cotton per yard of cotton sheeting. The number of competitive factories established in a given year is based on those still operating in 1832 in New England (Maine, New Hampshire, Vermont, Massachusetts, Connecticut, Rhode Island), New York, and New Jersey. This number of factories is computed from McLane, *Documents Relative to the Manufactures in the United States.* For the methodology, see chapter 4 discussion.

bers until 1809, but only seven of them were built, while investors founded seventy-two mills over the course of the two years 1808 and 1809 (see table 4.1).[9] Shrewd investors recognized the importance of prosperous local markets in a mill's vicinity and felt that they could penetrate them even if their British competitors returned. The absence of competition during the Embargo did not accelerate business plans to hire managers with technical expertise and skilled textile mechanics, because the embargo severely crimped lucrative non-local markets for yarn and cloth—port cities and their prosperous agricultural environs, which supplied them with food and products for export. These areas suffered because the terms of the Embargo prohibited both British imports and ships from leaving for foreign ports. Less-astute investors focused on the absence of British competition for local markets of prosperous farmers and villagers; they rushed into production without securing professional mill managers and skilled textile mechanics. Accumulating numbers of them in successful mills supported accelerating starts between 1809 and 1811, when investors established twenty-seven competitive factories, whereas from 1790 to 1808 just seventeen such factories were started (see table 4.1). Cotton cloth production in New England factories soared, but mills barely dented latent cloth markets; the 801,000 yards of cloth woven on hand looms in cotton mills in 1811 represented only 5 percent of the 14.9 million yards produced in homes in 1810.[10]

Investors started ninety-six competitive mills between 1812 and 1814, compared to forty-four founded in the prior two decades, but after the war ended, in 1815, Britain dumped cloth, and mill starts fell. This situation seemed to define a watershed in the cotton textile industry (see table 4.1).[11] Several puzzling facts exist, however, and once reconciled suggest that competitive mill formation continued following a similar logic as earlier. Cotton cloth prices rose during the war, but raw cotton prices also climbed; consequently, gross profit margins closely matched those in the period from 1808 to 1810, when investors started far fewer mills. New England mill owners recognized that cloth production did not offer unusually lucrative returns; they increased output by 32 to 38 percent annually between 1812 and 1814, reflecting growth rates similar to those of previous years. Because New England mills sold most output as yarn, factory gross profit margins, like those of cloth margins, could not expand much. Households could spin and weave, and during the war their capacity to do that prevented mills from significantly raising the prices of yarn and cloth. Increasing numbers of experienced, technically proficient mill managers and of skilled mechanics powered greater numbers of competitive cotton factories from 1812 to 1814, regardless of the fact that British competition had been temporarily eliminated. Under the spell of euphoria of the protected domestic markets, numerous other mills initiated production without experienced employees, but they survived neither British dumping after the war nor the shift to power loom weaving. Cotton mill production recovered after the impact of British dumping dissipated; gross profit margins collapsed, however, when raw cotton prices jumped in response to British and other European demand following the end of the Napoleonic Wars and as prices for cotton sheeting declined. Cloth output rose rapidly, even though few new mills entered production, because older competitive mills started installing power looms for weaving. The cotton textile industry developed from three cores—Providence, Philadelphia, and Boston—which followed distinctive industrial paths rooted in separate constellations of merchant capital, mechanic skills, technology, products, and labor organization.

Cotton Textile Cores

Providence and the Slater System

Almy, Brown, and Slater

In 1790 astute observers could not predict that the small subregional metropolis of Providence would become the core of a cotton textile region with outliers in southern New Hampshire and upstate New York. In 1789 merchant wholesaler Moses Brown offered financial inducements to Samuel Slater, the former manager and mechanic in an English Arkwright spinning mill, because Brown recognized that profitable cotton manufacture required an experienced mill manager and skilled mechanic; Slater arrived in Providence in January 1790. The merchant firm Almy and Brown formed a partnership with Slater; Almy and Brown supplied capital and marketed goods, and Slater, for a share of profits, managed the factory and built machines. Slater designed and constructed carding and spinning machines, and cotton yarn production began in the fall of 1790 in a fulling mill; by 1793 production moved to a new factory on the Blackstone River in Pawtucket. This cotton spinning mill had no special renown; as many as six other sites on the East Coast had English mechanics supervising small cotton spinning mills. Almy and Brown's tiny capital investment suggests that the mill represented a minor sideline of the Brown merchant family. Almy, Brown, and Slater initially borrowed $10,000 from Moses

Brown, and over the ten years from 1793 to 1803 the mill's cumulative expenditures were under $100,000; in contrast, over the course of the year 1792–93 the merchant firm Brown, Benson, and Ives invested $70,000 in one voyage to the East Indies.[12] Compared to Boston, Hartford, New York, and Philadelphia merchants, who were boldly betting extensive capital on textile mills, Almy and Brown were cautious.

To compete successfully against British imports, factories produced low-cost, quality textiles and marketed them. As merchant wholesalers, Almy and Brown controlled cotton yarn marketing, but they meddled in, and undercut, Slater's factory management. They viewed the cotton mill as extending mercantile operations; it provided yarn that they sold directly or put out to weavers. Because Slater had trained as a manager and mechanic in an Arkwright cotton spinning mill, he viewed the mill as a unified package of labor organization, machines, and production processes. After experimenting with several labor forms, Slater turned to the family labor system: children and parents worked together in different jobs, labor was divided by gender, and patriarchal authority reigned. He attended to the design, construction, and operation of textile machinery, and his practice of building textile machinery in the mill where they were installed remained a model into the 1820s. Slater's marriage into the family of Orziel Wilkinson, a Pawtucket machine builder, promoted technological diffusion; this alliance gave Slater access to a network of skilled machinists, mill managers, and alternative supplies of capital, and network participants founded cotton spinning mills in Rhode Island, eastern Connecticut, and nearby Massachusetts.

Successfully competing with British imports required marketing savvy, because cotton yarn production from waterpowered spinning mills quickly saturated the local markets. Firms used wholesalers to forge long-distance sales links either directly to retailers or to other wholesalers, who in turn sold yarn to retailers; thus, the merchant wholesalers Almy and Brown had an in-house distribution arm. Starting in 1793 the firm expanded its distribution channels to country and village storekeepers in Providence's hinterland in Rhode Island, eastern Connecticut, and nearby Massachusetts and to East Coast port wholesalers, and they sent yarn on commission to unspecialized merchant firms selling at wholesale to country and village storekeepers in the vicinity of ports and at retail to local weavers or nearby farmers. Strong British competition, low levels of economic development in which barter, rather than cash, was the main currency, and uncertainty about the change from household production to the purchase of factory goods forced factories to bear all risk and retain ownership of their goods until wholesalers or retailers sold them.[13]

Slater successfully adapted Arkwright cotton spinning technology to American conditions, making him a pivotal diffuser of this technology, but early attempts to copy it failed. The Warwick Spinning Mill, established in 1794 by Providence merchant William Potter, floundered because its manager, James Greene, was neither experienced nor technically proficient and its mechanic, John Allen, had not trained in Arkwright spinning technology; thus, the mill lacked both of the essential ingredients of competitive cotton mills. In 1799 the owners sold a half-share to Almy and Brown, who promptly instructed Allen to visit their Pawtucket mill to learn about Slater's machines; Almy and Brown acquired a controlling interest in 1801. Although this expansion excluded Slater, he joined with the Wilkinson family to build another cotton mill, Samuel Slater and Company, in Pawtucket in 1799. The time lapse from the creation of the 1793 Pawtucket mill of Almy, Brown, and Slater to their 1799 separate mill expansions attests to the learning processes needed to master Arkwright spinning and to perfect Slater's production system; other entrepreneurs started cotton mills during this period, but most remained marginal. The three mills of Almy, Brown, and Slater, however, produced yarn selling competitively against British imports, and by 1801 Almy and Brown penetrated markets along the East Coast from Maine to Maryland and up the Hudson Valley to Albany. By 1804 Almy and Brown had major wholesale commission agents in New York, Philadelphia, and Baltimore, testimony that their yarn sold widely in prosperous rural hinterlands of metropolises, but they still retained a large number of retail accounts. They achieved penetration of East Coast markets through effective mill management, the use of skilled mechanics, and access to a merchant wholesaling distribution arm. Their success spurred other founders of cotton mills after 1800 to emulate them, but those mills often failed to reach beyond local markets and confronted strong competition within their local market from other American and British mills. Astute investors needed access either to mill managers and mechanics who had trained at the Almy, Brown, and Slater mills or to people receiving training from individuals with experience at those mills.[14]

Diffusion from the Providence Core

Providence's business community encompassed information networks of astute investors, because Almy and Brown and the entire Brown merchant family possessed numerous local network ties with wholesalers, leading retailers, and professionals (e.g., lawyers and physicians), and these wealthy investors' economic and social networks within Providence and its hinterland channeled cotton mill diffusion. Small mills required up to ten thousand dollars in start-up costs, and initial capital quickly increased as mill size grew; only wealthy investors amassed that much capital, but no single group in those elite networks monopolized cotton mill formation. Almy's and Brown's investment in one mill in 1799 and of Slater, in cooperation with the Wilkinson family, in another mill signaled that Providence's networks possessed multiple diffusion nodes, and diffusion operated without a coordinating propagator. Although Providence's business elite had close ties among subsets of the elite, these ties did not overlap extensively; these business leaders had diverse and selective nonoverlapping ties to the hinterland business elite of country and village storekeepers, professionals, and wealthy farmers. They formed shifting coalitions of investors, and members enticed mill managers with technical expertise and skilled mechanics, preferably with direct training under Slater or one of his disciples, to their enterprise, either as partners or hired hands. Only a few managers and mechanics, especially the Slater and Wilkinson families, accumulated sufficient capital to found mills. As supplies of good waterpower sites near Providence diminished, Providence's elite ranged farther afield into adjacent parts of Massachusetts and Connecticut. The business elite in these areas took the initiative in founding mills because local investors retained leverage over cotton mill development through their ownership of waterpower rights; thus, Providence's elite enlisted them as partners.[15]

Almy, Brown, and Slater did not rest on their laurels after their respective mill building in 1799. Samuel Slater's younger brother John arrived from England in 1803 with the latest textile technology and worked at the original (1793) Almy, Brown, and Slater mill in Pawtucket; then the investor group, now Almy, Brown, and Slaters, built a mill in Smithfield and installed machinery by 1807. With John as mill manager, expansions raised the spindle count to 5,170 before 1815, making it one of New England's three largest mills. Also in 1807, Almy and Brown formed Warwick Manufacturing Company to build a mill near their Warwick Spinning Mill; they retained majority control, but they also added several of their partners in the nearby mill, including mechanic John Allen. Shortly before 1812 they expanded production into a second building, and by 1815 the two mills together housed about 2,700 spindles.

The expansion of the Almy-Brown distribution network to include major wholesale commission agents by 1804 unequivocally demonstrated to Providence's merchants that textile investments offered lucrative opportunities, and others joined the fray, raising the ante for mill size and looking immediately to markets outside Rhode Island. In 1805 an investor group started Coventry Manufacturing Company; its organizers included Providence's John Pitman and Samuel Arnold, the latter of whom had served as an agent and bought cotton and sold yarn, natural adjuncts to his wholesale-retail operations. The firm astutely hired premier mechanics—Perez Peck and Samuel Ogden, an English mechanic—to build machinery, and production commenced in 1807 in a mill housing 1,500 spindles; within two years paid-in capital totaled $61,640, and in 1810 the firm added a six-story mill with 2,000 spindles, making it one of New England's largest firms at the time. Providence merchants Christopher and Nathaniel Olney and Ebenezer Thomas also started a cotton mill in 1805, and the Union Cotton Manufacturing Company in Providence had property valued at $40,000, among the state's largest by 1808. Along with several Providence physicians these three merchants quickly replicated operations and organized the Hope Cotton Factory Company, and production commenced in 1807.

When cotton yarn spinning started at Coventry Manufacturing Company in the same year, John Pitman, one of the investors, joined with a group led by wealthy Providence merchant Seth Wheaton—including investors from Coventry and Warwick—to organize the Providence Manufacturing Company, capitalized at $32,000. Mill production got under way in 1808, soaring to a value of $169,835 by 1814, and the Brown and Ives merchant firm joined the investment group by 1819. Wheaton had grander plans, however, and withdrew in 1808, purchasing land and a waterpower privilege on the Blackstone River immediately north of Rhode Island's border; with other wealthy Providence merchants, including Brown and Ives and the Butler family, he organized Blackstone Manufacturing Company in 1809. Capitalized at $150,000, the firm quickly completed an immense six-story stone factory with a 10,000-spindle capacity. Its initial 5,000 spindles made it one of the nation's three largest mills, and it was valued at $216,000 as of 1813.[16]

An interlocking network of mechanics, merchants, and members of the social and political elite united three firms and tied them to other firms of the Providence region. In 1807 four mechanics, including Peter Cushman and Perez Peck, a machinist for the Coventry Manufacturing Company (est. 1805), joined with Providence merchants William Potter, an investor in the Warwick Spinning Mill, and Adams and Lothrop and with Warwick's Christopher and William Rhodes to form Natick Manufacturing Company, capitalized at $32,000; the firm's three small mills in Warwick together housed 2,500 spindles by 1815. In 1809 Jonathan Adams, of Adams and Lothrop, invested in the Northbridge Cotton Manufacturing Company in his hometown of Northbridge, Massachusetts, just north of Rhode Island's border; this firm included other Providence area investors and local Northbridge investors such as Paul Whitin. William Howard, a machine maker from the Natick mill, instructed Whitin in machine building, thus transferring technical expertise learned from premier Providence machinists such as Peter Cushman and Perez Peck. The third firm in this interlocking network united leading mechanics—Peter Cushman, who was involved with Natick Manufacturing Company, and Samuel Ogden, who was involved with Coventry Manufacturing Company—with Daniel Lyman, a social and political leader. He gathered local investors, including the merchant firm S. G. Arnold and Company, headed by Samuel Arnold, who was an investor and agent for the Coventry Manufacturing Company, and founded Lyman Manufacturing Company in 1809. Mechanics Ogden and Cushman built machinery for the North Providence factory, which started production in 1810 with 1,248 spindles and doubled that by 1815, ranking it among the state's largest.[17]

Leading textile machinery builders of the Providence core launched the textile industry of the Quinebaug Valley in eastern Connecticut, and subsequent mills drew on managers and mechanics trained by the finest. In 1805 James Rhodes, the brother of Christopher and William Rhodes (investors in Warwick's Natick Manufacturing Company in 1807), took ownership of a mill privilege, property, and mills on the Quinebaug River at Putnam. This impressive industrial village—equipped with a blacksmith shop, sawmill, fulling mill, malt house, gin distillery, and gristmill—served the prosperous agricultural economy. When Smith Wilkinson arrived in 1805 on a trip to identify mill sites for the Wilkinson family, Rhodes eagerly embraced the opportunity to work with premier textile machinery builders. In 1806 they organized the Pomfret Manufacturing Company and gained complete control of the river in Putnam; the Wilkinson family acquired majority ownership, and the Rhodes brothers took the remaining shares. In 1807 production commenced in a four-story mill under Smith Wilkinson's management, and it achieved immediate success selling yarn to local farmers to weave for their own use or on account for the company. Word of the Pomfret company's achievements spread throughout the valley, and over the next five years investors organized at least eight cotton textile firms, mainly partnerships between local residents and Providence investors and machinists.[18]

The fame of Providence core textile mills spread beyond its immediate vicinity after 1799 as wholesalers in large and small metropolises along the East Coast widened cotton yarn distribution to interior towns and villages in prosperous agricultural areas. Local investors could attempt to start cotton mills on their own, but the likelihood of success increased if they hired mill managers and mechanics trained in the Providence core; thus, distant mills served as local diffusion points for the region's yarn manufacturing. Investors in rich farming areas in the Hudson Valley, southern New Hampshire, and central New York hired managers and mechanics whom Slater had trained directly or who had learned under his trainees. Job Whipple owned a flour mill at a large waterpower site in Greenwich, about thirty miles north of Albany, the subregional metropolis whose merchants controlled trade in the upper Hudson and Mohawk Valleys. Information about Providence core textile production percolated through the valley's social and economic networks: New Englanders settled this area by the mid-1780s and retained family and friendship ties to their origins, and Whipple probably heard of the Slater system because by 1801 Almy and Brown were distributing their yarn through Albany merchants. He decided to start a cotton factory and journeyed to Rhode Island in 1804 to find someone to erect the building, construct machinery, and manage the mill. Whipple offered William Mowry, a twenty-five-year-old mechanic who had been working at Slater's mill since 1799, the job along with a half-interest in the waterpower privilege; Whipple provided all fixed and working capital. The cotton factory commenced operations in 1804, and, several years after marrying Whipple's daughter in 1807, Mowry purchased his father-in-law's share. Like Slater, Mowry made the transition from mechanic to wealthy mill owner because textile investors sought his skills.[19]

By the 1790s New Ipswich, a southern New Hampshire village, housed retail merchants, professionals, and grist- and sawmill owners serving nearby prosperous farmers, and businesses shipped agricultural surpluses to Boston merchants, receiving domestic and foreign imports from them. Information about the Almy, Brown, and Slater mills followed those merchant channels; their cotton yarn reached New Ipswich by 1800, and Samuel and Nathan Appleton, sons of a wealthy New Ipswich farmer, provided one information channel. They owned a wholesale-retail firm with offices in Boston; an overseas branch that Samuel staffed in Manchester, the center of England's cotton textile industry; and retail stores in Boston and New Ipswich. The Appletons had family and friendship ties with the Lowells, Cabots, and Jacksons of Boston, future participants in the great textile mills. Out of this milieu Charles Barrett Sr., an affluent New Ipswich resident—whose son, Barrett Jr., was Samuel Appleton's partner from 1794 to 1799—invited Charles Robbins, an employee of Samuel Slater in Pawtucket, to New Ipswich in 1804 to be a partner, build cotton textile machinery, and manage the New Ipswich Cotton Factory. Family and friendship ties binding New Ipswich businesses and the Appletons' merchant firm in Boston, including Samuel Appleton's purchase of a one-quarter share in the factory in 1810, probably furnished transfer vehicles for English mechanics arriving to build new types of machinery.

The New Ipswich–Boston axis possessed another bridge to Providence core mills through Daniel Brooks, who worked at New Ipswich Cotton Factory with Charles Robbins, the Slater-trained mechanic; by 1808 Brooks had established his own cotton mill downstream from the first factory. Within a year ownership transferred to Samuel Batchelder, an affluent local retailer, and to Seth Nason and Jesse Holton from the Boston area, both of whom had trained in Providence core mills and had moved to New Ipswich. Batchelder managed the mill and by 1812 took charge of selling yarn and contracting with local weavers. Employing his mechanical skills, he invented a yarn winding machine and designed adaptations of other machines, including the power loom, and by 1820 the mill operated as Batchelder and Brown. Thus, the social, economic, and political elite of New Ipswich stood foursquare in New England's cotton textile industry, joining the Slater-trained mechanics of Rhode Island mills with the Boston merchants' great mills.[20]

The leap of a Slater-trained mechanic to Whitestown, now part of Utica, in central New York in 1808 foreshadowed one of the nation's greatest antebellum cotton textile firms, and Utica became the core of a major textile subregion whose roots can be traced back to 1793, when Benjamin Walcott Sr. worked with Slater to build the first mill for Almy and Brown. By 1802 Walcott entered a partnership to build Cumberland Mills in that town and by 1805 commenced projects with other investors: the Pawtucket Cotton and Oil Manufacturing Company grew to two mills housing 3,300 spindles by 1815, ranking it among Rhode Island's largest mill complexes, and a firm that became Smithfield Cotton Manufacturing Company formed in partnership with the leading machinist Stephen Jenks. Whitestown/Utica, about 250 miles from New York

City, might seem an odd destination for a premier mechanic trained by Slater, but Utica and environs epitomized the settings of early successful cotton mills. This subregion housed prosperous farmers and thriving villages with retailers and professionals, and the growing subregional metropolis of Utica presented tempting markets for factory cotton yarn.

Dr. Seth Capron, a physician and member of the professional elite in Whitestown, arrived from Massachusetts in 1806, having previously lived in Rhode Island; he was familiar with cotton mills and with Benjamin Walcott Sr., a leading mill builder in the Providence core. In 1808 Capron's investor group invited Walcott to erect a cotton mill and be partner in Walcott and Company, whose mill and separate machine shop were completed in 1809. Walcott returned to Rhode Island that year and sent his son to finish the machinery, take over factory management, and become a partner. Benjamin Walcott Jr. acquired superb credentials working on his father's three Rhode Island mill projects. In 1810 the firm became Oneida Manufacturing Society, reflecting broader business, and the partners raised capital to $200,000. The Oneida factory sold cotton yarn to local farm families to weave cloth for their own use and employed workers to weave cloth for sale, and in 1813 Walcott Jr. formed another partnership, Whitestown Cotton and Woolen Manufacturing Company. Although he lived far from the center of textile innovations in Rhode Island and eastern Massachusetts, his father apprized him of developments. In 1817 Walcott Jr. decided to introduce power looms and traveled to the Boston Manufacturing Company in Waltham to learn about its new looms. They proved to be unsatisfactory, but later that year he sent William Copley, one of his leading mechanics, to Pawtucket to learn about constructing Gilmour power looms. The quick installation of power looms in the Whitestown factories signaled that they were competing with other large eastern mills in cotton cloth markets. In 1818 Rhode Island's connection with central New York strengthened when the Providence merchant firm Brown and Ives, owners of the Blackstone Manufacturing Company, joined other Providence investors to build a cotton mill at Newport, about fifteen miles from Utica.[21]

Competitive Cotton Mills

The centrality of the Almy, Brown, and Slater firms, their cadre of professional managers and skilled mechanics, and alliances that the Providence merchant elite forged with their social, economic, and political peers left traces in the diffusion of competitive cotton mills (table 4.2). Through 1807 Rhode

Island dominated, with five of seventeen mills, and a dearth of mills established during the Embargo years (1808–9) suggests that investors reaching markets outside local areas viewed the economic contraction hitting ports and their immediate hinterlands as detrimental to cotton yarn markets. Similarly, Rhode Island's twelve mills started during the War of 1812 (1812–14) were consonant with the steady growth of the East's urban and rural markets and the expansion of western markets and with their restrained reaction to the absence of British competition. The influences of Providence core investors, mill managers, and mechanics pervaded successful Massachusetts mills that were established before 1820; almost half of its mills (twenty-seven of fifty-six) were within twenty-five miles of Providence. Similarly, the Pomfret Manufacturing Company, started in Putnam in 1806 by the Wilkinson machinists from Pawtucket, laid the base for the Quinebaug Valley mills of eastern Connecticut. Rhode Island mill managers and mechanics accounted for half of the twenty-four mills built in Connecticut between 1809 and 1818.[22]

When investors who lived at a distance from Providence core mill managers and mechanics hired them and even made them partners, mills became diffusion nodes of Rhode Island's production technology and management system. New Ipswich's social, political, and economic elite, who were prominent in their subregion, supplied diffusion channels after Slater employee Charles Robbins arrived in 1804; among New Hampshire's thirteen competitive mills founded before 1820, six lay within twelve miles of New Ipswich (see table 4.2).[23] Similarly, the arrival of Slater-trained mechanic William Mowry to the upper Hudson Valley town of Greenwich in 1804 probably instigated the cotton mill cluster there; investors built eleven mills between 1804 and 1815, representing over one-fourth of New York's forty-one mills started before 1820.[24] These sophisticated investors steadily expanded to serve the productive subregional farm economy and interregional economy, and they avoided the euphoria induced by the absence of British competition during the Embargo and the War of 1812.

Central New York became a leading textile subregion after the Walcotts, the prominent mill managers and mechanics whom Slater had trained, arrived in 1808–9. From the start of the Whitestown factory in 1809 through 1816, Utica's subregion gained twenty cotton mills, nearly half of New York's mills established before 1820.[25] Although thirteen factories commenced production between 1812 and 1814, the lack of British competition probably did not motivate their investment decisions; they occupied a prosperous, growing agricultural

Table 4.2. Competitive Cotton Textile Mills Established, by State, 1790–1820

Year	Maine	New Hampshire	Vermont	Massachusetts	Rhode Island	Connecticut	New York	New Jersey	Total
1790	—	—	—	—	1	—	—	—	1
1791	—	—	—	1	—	—	—	—	1
1800	—	—	—	—	—	2	—	—	2
1804	—	—	—	—	—	—	1	—	1
1805	—	—	—	1	2	—	—	—	3
1806	—	—	—	1	—	1	2	—	4
1807	—	—	—	—	2	—	—	—	2
1808	—	—	—	1	—	—	2	—	3
1809	1	—	—	1	1	1	3	—	7
1810	—	1	—	3	2	2	2	—	10
1811	—	—	—	5	1	2	1	1	10
1812	—	—	1	9	5	4	8	1	28
1813	1	—	—	13	3	3	2	1	23
1814	—	4	—	12	4	7	15	3	45
1815	—	2	—	3	2	4	3	1	15
1816	—	2	—	1	—	—	2	1	6
1817	—	2	—	—	1	—	—	—	3
1818	—	1	—	1	1	1	—	—	4
1819	—	—	1	—	2	—	—	—	3
1820	—	1	—	4	1	—	—	—	6
Total	2	13	2	56	28	27	41	8	177

Source: McLane, *Documents Relative to the Manufactures in the United States.*

Note: The number of competitive mills established in a given year is based on those still operating in 1832.

area, and, as late as 1832, leading mills near Utica in Oneida County sold up to one-third of their production to households and stores in the heart of the subregion.

Market Changes

Growing numbers of new competitive cotton mills in New England and New York forced existing mills to readjust their marketing strategies. By 1807 Almy and Brown owned four mills outright or in partnership, ranking them among the nation's largest producers of cotton yarn and cloth; sales from their four firms' 10,520 spindles dwarfed sales from the fabled Boston Manufacturing Company at Waltham up until 1819, when Waltham mills sold slightly over half of Almy and Brown's $228,000.[26] Changes in yarn and cloth distribution from their mills reveal market strategy readjustments that leading factories implemented as new mills emerged in prosperous agricultural areas. As early as 1804, their production easily surpassed Rhode Island's absorptive capacity; they rarely sold more than 3 percent of yarn and cloth there between 1804 and 1819, but Almy and Brown sold up to half of their output in New England until 1806, primarily through distributors in Boston and ports northward. After 1804 they began to distribute through wholesalers selling cotton goods on commission. In 1805 Elijah Waring established a commission firm in Philadelphia to handle cotton yarn and thread and sold to stores and weavers in Philadelphia and to retailers in surrounding towns, and similar firms emerged in Baltimore (1809) and New York (1812).

These wholesalers specialized in distribution to wholesalers and retailers in distant markets after 1810, and western markets attracted large-scale manufacturers such as Almy and Brown. They transformed distribution, sharply curtailing shares sold in New England and boosting shares sold through Philadelphia wholesalers, who dominated western trade; Philadelphia received 83 percent of the firm's output by 1819. Philadelphia's and western markets' significance to large New England mills, leaders in the drive to sell nationally, received additional validation. In 1812 Jeremiah Brown opened a wholesale agency in Philadelphia to serve Slater's mills, and from 1815 to 1819 Slater sold 70 to 87 percent of his output through Philadelphia wholesalers. In 1813 Gilman and Ammidon's wholesale firm opened in Philadelphia to serve western markets, and Providence's Brown and Ives merchant firm provided encouragement and indirect financial backing, because their Blackstone Manufacturing Company, one of the nation's largest, was producing huge quantities of cotton yarn. The relative shift to Philadelphia weavers and western markets and the substantial absolute increase in sales did not signify an abandonment of earlier markets; Almy and Brown raised sales to old markets between 1809 and 1816. Nevertheless, the increased numbers of competitive mills in prosperous agricultural areas in southern New Hampshire, eastern Massachusetts, eastern Connecticut, the upper Hudson Valley, and central New York State from 1804 to 1820 prevented Almy, Brown, and Slater from expanding their sales there; they finally retreated as their competitors usurped local and subregional markets (see table 4.2).[27]

The Switch to Power Loom Weaving

The numbers of competitive mills added annually plunged after the War of 1812 and stayed low from 1816 through the end of the decade, coinciding with a radical restructuring of the cotton textile industry (see table 4.1). Britain dumped cotton goods on the domestic market in 1815, causing temporary disruptions in cotton production, but astute investors knew that the calamity would soon pass. The resumption of British competition posed a greater threat, yet it does not fully explain the low levels of mill starts, which were even below those of 1809–11, when British competition also existed. Renewed British and other European purchases of raw cotton contributed to short-term fluctuations in raw cotton prices; simultaneously, the gross profit margins on cloth fell below any level ever seen by cotton mill investors (see table 4.1). Most mills producing cotton yarn and cloth at those gross profit levels—excluding the costs of labor, machinery, buildings, land, and waterpower—sold below production costs.

Growing numbers of British and American mills relentlessly pressured cloth prices downward, but, more important, manufacturers knew that the sales of cloth to households could never reach their potential unless factories surmounted the high costs of hand loom weaving. Between 1810 and 1815 skilled mechanics and inventors in England and the United States made significant progress designing effective power looms to weave cloth. English mills commenced installation of power looms by 1815, thus cloth import prices after the war reflected early production cost reductions, albeit modest ones, from power weaving. An eerie calm settled over cotton mills in New England and New York during the latter part of 1815 and through 1816 as many mills went bankrupt, whereas others shuttered their doors or reduced production and waited for British dumping to subside; in 1816 cloth production in New England mills fell to its lowest level in four years (see table 4.1). Rhode Islanders contributed to the shift to power loom weaving; as many as six mechanics built power looms of various designs by 1813. Therefore, when the Boston Manufacturing Company's machine shop priced looms at $125 each, significantly above alternative models, the high price and the loom's unsuitability for weaving finer yarn counts into lighter cloths encouraged mill owners to look elsewhere. Gilmour's power loom—perfected by 1817 with assistance from David Wilkinson—provided an alternative to Waltham's high-priced loom. Wilkinson, arguably the Providence core's leading machinist and independent textile machinery builder, had rights to build Gilmour looms and started production by late 1817. Within the next year Gilmour built looms that were placed in operation in several Rhode Island mills, and his looms were also installed in mills in Taunton, Massachusetts, and Utica, New York.[28]

Thus, the slow pace at which astute investors formed cotton mills between 1816 and 1820 foreshadowed the industrial transformation (see tables 4.1–2). Weaving costs per yard on power looms fell as much as 50 percent below hand loom weaving by 1823. The adoption of power loom weaving required higher levels of mechanical skills to build and maintain complicated machinery and greater management skills to coordinate a greater number of production steps. Because power looms consumed large volumes of yarn, their adoption forced mills to raise the speeds of earlier processes of spinning, warping, and dressing; mills also increased spindle counts. Firms producing coarser cloth preferred to integrate spinning and power loom weaving vertically to maintain control over yarn quality and to ensure steady supplies for power looms; this additional machinery substantially boosted the costs of labor and of building and equipping mills.[29]

By 1820 the distinctions among textile factories in New England and New York crystallized (table 4.3). In most states cotton spinning mills possessed significantly less capital investment than mills combining spinning and hand loom weavers, and capital investment in mills with power looms was up to 50 percent greater than in mills with hand looms. Numerous investors in New England and New York shifted to power loom weaving; almost 20 percent of all mills installed power looms, and the number of those looms surged from nil in 1815 to 1,308 in 1820. Astute investors slowed the pace of starting mills because vertically integrating processes from spinning through power loom weaving

Table 4.3. Cotton Textile Firms in the New England States and New York, 1820

	Cotton Spinning Firms		Cotton Spinning and Hand-Loom Weaving Firms		Cotton Spinning and Power-Loom Weaving Firms		Cotton Spinning Firms	Cotton Spinning and Weaving Firms	Number of Power Looms
	Number	Average Capital ($)	Number	Average Capital ($)	Number	Average Capital ($)	Total Spindles		
New England States									
Maine	4	$37,000	1	$22,000	1	$22,000	2,480	890	12
Vermont	4	3,375	7	13,886	3	19,000	632	3,470	38
New Hampshire	17	16,471	14	22,886	5	37,040	9,348	8,463	120
Massachusetts	14	18,650	44	29,128	11	63,873	8,294	32,992	321
Rhode Island	24	13,555	45	29,788	15	48,133	17,770	49,164	315
Connecticut	17	18,647	24	39,015	11	48,279	8,477	25,826	168
New York	6	38,833	35	51,313	15	71,741	12,330	50,236	334
Subtotal	86	18,348	170	34,082	61	54,032	59,331	171,041	1,308
United States	153	17,552	204	39,902	75	63,658	98,687	226,993	1,623

Source: Jeremy, *Transatlantic Industrial Revolution,* 98, table 5.1; 277–78, app. D, tables D.2, D.4.

required two to three times as much capital as cotton spinning mills; thus, investors initially directed their capital to existing spinning mills and to those with hand looms. As cotton spinning mills rooted in the Almy-Brown-Slater tradition emerged in and diffused from the Providence core, Philadelphia's textile industry followed a different path to form the second core of eastern cotton manufacturing.

Philadelphia's Immigrant Spinners and Weavers

From 1783 to 1812 hand spinners and weavers fled the mechanization of cotton spinning and declining prices for hand loom woven goods in England and Ireland, disembarking at Philadelphia and nearby ports. As the nation's second largest metropolis, it attracted workers needing access to contractors, raw material suppliers, and distributors. Arrivals during the 1780s established networks for a subsequent chain migration as friends and relatives communicated information about textile work in Philadelphia. During the 1790s merchant weavers and wholesalers of cloth built a textile manufacturing complex organized around a putting-out system and workshops. Female hand spinners typically worked at home producing coarse cotton yarn that skilled hand loom weavers then wove in their homes or small workshops. Growing numbers of skilled mule spinners spun higher-grade yarns by the late 1790s to 1810, whereas cotton spinning mills in Rhode Island increasingly dominated the market for coarse cotton yarn. The sharp rise in the consignment of cotton yarn from Almy and Brown's mills to Philadelphia wholesalers after 1808 signaled the pending displacement of female hand spinners.[30]

Entrepreneurs attempted to organize textile production within factories in Philadelphia before the War of 1812, but most of their efforts failed. The "Guardians of the Poor," a group of leading Philadelphia citizens who provided relief to the poor during the 1790s, organized a textile manufacturing enterprise shortly after 1800 which proved an exception. Within the factory they employed paupers to spin yarn and weave cloth, supplementing them with paid spinners and weavers, and they organized outwork networks of spinners and weavers. Until the War of 1812, when the enterprise declined to insignificance under competition from private firms, the Guardians were the largest textile employer in Philadelphia. Even as the Guardians' enterprise dominated, the firm that became Craig, Holmes and Company started manufacturing cotton textiles in the Globe Mill in 1809, and over the next decade it specialized in mule spinning fine cotton yarn, calico printing, and making

Table 4.4. Cotton Textile Industry in Philadelphia (City and County) and Rhode Island, 1810

Characteristic (Number)	Philadelphia	Rhode Island
Spinning wheels	3,643	—
Waterpowered spinning machines	8	25
Spindles	4,423	21,178
Cotton factories	8	28
Yards of cloth made in factories	65,326	734,319
Yards of web lace and fringe	716,250	20,000

Source: Coxe, "Digest of Manufactures, 1810," in *New American State Papers, Manufactures,* vol. 1, pt. 4, 269–71, 291–95.

trimmings for saddlery. With this multifaceted strategy and focus on specialty items it avoided direct competition with New England spinning mills, which were making coarse cotton yarn, and with weavers using that yarn.[31]

Sharp distinctions between Philadelphia's and Rhode Island's cotton textile industry existed by 1810 (table 4.4). Household spinners, mostly women and young girls, worked on 3,643 spinning wheels in Philadelphia, whereas hand spinners were unimportant in Rhode Island, which contained over three times as many waterpowered spinning machines as Philadelphia. Its textile industry was centered in workshops and homes, plus a few factories, whereas Rhode Island's industry centered in factories containing almost five times as many spindles and producing over eleven times as much cloth on hand looms as Philadelphia's factories. Philadelphia's shift to specialty cotton products had commenced; factories, workshops, and homes produced over thirty-five times as many yards of web lace and fringe as Rhode Island produced.

During the War of 1812 Philadelphia textile manufacturing surged, but the postwar flood of British exports forced a temporary contraction. Coincidentally, many skilled mule spinners and weavers fled to the United States following another reorganization of the British textile industry, swelling the supplies of mule spinners and weavers that local textile entrepreneurs could hire for low wages. Although Philadelphia never housed most of these immigrants, they supported a textile industry sprawling across counties of Philadelphia and Delaware in Pennsylvania and of New Castle in Delaware. Industry contours crystallized by 1820, and factories started replacing the putting-out system. The thirty-seven cotton textile factories consisted of eleven spinning mills, eleven mills combining spinning and weaving, and fifteen weaving firms; together, they employed 1,321 workers. Mule spinners and those operating spinning jennies produced most of the yarn, and few mills did power loom weaving; the

Philadelphia region accounted for only 3 percent of the nation's power looms. The average capitalization of firms differed significantly across combined cotton spinning and weaving mills ($49,473), spinning mills ($21,582), and weaving shops ($1,323). More hand loom weavers worked in combined spinning and weaving mills than in specialized weaving shops, although these shops outnumbered mills. Philadelphia dominated national commercial hand loom weaving, accounting for all fifteen specialized weaving shops. Skilled immigrant hand loom weavers, together with skilled mule spinners, underpinned Philadelphia's specialization in fine-quality yarns and fabrics.[32] This specialization avoided direct competition both with the spinning mills of Rhode Island and vicinity, which produced coarse cotton yarn, and with the newly emergent behemoths of cotton textile manufacturing in the Boston core, the third pivot of cotton textiles.

The Boston Associates and Waltham

Boston merchants' success with the "new industrial form"—large, vertically integrated cotton textile corporations combining waterpower spinning and power loom weaving—still leaves the question of why merchants in other East Coast metropolises failed to grasp the opportunities to acquire wealth based on this form of textile manufacturing.[33] Large-scale debacles over the decade of 1787–97 probably eroded their enthusiasm, whereas successes included small-scale cotton spinning mills in Rhode Island and small factories and workshops of immigrant spinners and weavers in Philadelphia. Nevertheless, neither approach attracted leading merchants because they could not effectively deploy capital in such small ventures. Wealthy merchants would have known of the large-scale successes of leading Providence merchants—Almy, Brown, and Slaters in Slatersville (est. 1807) and the Blackstone Manufacturing Company (est. 1809). The Brown merchant family distributed Rhode Island textiles to East Coast metropolises and engaged in other trade with them, thus widely conveying information about those large mills; and merchants in each metropolis had ample capital to fund huge cotton mills.[34]

The Germ of an Idea

Credit for initiating the new industrial form belongs to the Boston Associates, a loose collection of prominent merchant families. The Boston Manufacturing Company at Waltham, organized in 1813, embodied their creation, and its average stock dividends of 14 percent from 1817 to 1820 and 24 percent from

1821 to 1824 motivated them to build larger cotton mill complexes at Lowell and elsewhere in New England. This extraordinary profitability and the enormous scale of their industrial complexes seem to confirm that they broke radically from Rhode Island's approach embodied in Almy, Brown, and Slater mills and their descendants; however, that conclusion dismisses too readily the new industrial form's roots in New England's textile milieu.[35] Rhode Island and nearby parts of Connecticut and Massachusetts contained the nation's greatest concentration of managers and machinists who were experienced with Arkwright textile technology. It embodied the package of labor organization, machines, and production processes for the waterpowered spinning of coarse yarn and these processes were applicable to manufacturing coarse cloth in large, vertically integrated mills. Social networks of investors, mill managers, and mechanics raised the likelihood that new forms of textile production for the mass market would emerge in New England.

Boston's leading merchant wholesalers were New England's pivots of business information exchange, because they engaged in foreign trade and trade with other East Coast metropolises and maintained close business ties to less-specialized wholesalers in their hinterland. Many of them—including the Cabots, Jacksons, Lees, and Appletons—started as less-capitalized, smaller-scale, diversified wholesalers who sold to storekeepers in New England; thus, their networks provided information about the successes and failures in cotton textile manufacturing, technological changes, business organization and management approaches, local investors with access to land and waterpower rights, cotton textile firms past the start-up phase which were seeking capital infusions, and struggling firms that could be reinvigorated with new management and additional capital. Boston merchants imported cotton for mills in their hinterland, providing them with detailed knowledge of the creditworthiness and business operations of textile firms, and within twenty-five miles of Boston a plethora of successful cotton mills started production before 1813, providing viable business models (see table 4.2). They possessed the best information about Providence investors' huge mills: Almy and Brown had been selling cotton yarn to merchants in Boston and ports northward since the late 1790s, and Brown and Ives ranked among the five most prominent merchant firms in the China trade between 1800 and 1810, a group that included the Boston firm of James and Thomas Perkins. Through Samuel and Nathan Appleton's family and business ties to New Ipswich, New Hampshire, Boston merchants acquired firsthand information about Slater's management and production system from two New Ipswich mills.

Boston's merchant elite also acquired access to information about many mechanics' efforts to develop power looms, the pivotal machine that transformed cotton textile manufacturing. The Providence core housed many of the nation's leading machinists—most trained by Slater and Wilkinson or indirectly by their students—and the Wilkinson machine shop was among New England's most prominent; hence, the efforts of at least six Rhode Island mechanics to build different designs of power looms by 1813 became widely known. Consequently, the Boston Associates acquired a broad knowledge about cotton textile mills and experimental designs of power looms prior to Francis Lowell's departure for Britain in 1810, and they gained new information from New England sources as they progressed from the conception to the implementation of their plan. Lowell carried a formidable background: wealth, success as international merchant, mathematical facility, a Harvard education, and experience with complex, risky business ventures; thus, he comprehended British industrialists' organizational and technical efforts to master the transition to large-scale, integrated cotton textile manufacturing. Lowell met Nathan Appleton in Edinburgh, Scotland, and they apparently sketched technical and organizational requirements for a successful effort; Appleton promised his financial support. He visited factory towns, and Lowell toured cotton textile factories in Lancashire, making mental notes about techniques, organization, labor conditions, and machinery, notably power looms.[36]

Developing a Business Plan

Lowell followed standard business procedures and enlisted family, friends, and business partners to translate the germ of a business plan into reality. In 1813 his brother-in-law Patrick Jackson, his cousin Benjamin Gorham, and his partner, Uriah Cotting, joined Lowell to petition Massachusetts' legislature for corporate status for the Boston Manufacturing Company, and they received authorization to raise up to $400,000 of capital and to manufacture cotton, woolen, and linen goods. The corporate form provided a long-term structure: owners entered and exited through the sale of shares without disrupting ongoing management, and the legal structure allowed bylaws that penalized stockholders. A critical bylaw permitted the corporation to set and collect capital assessments, thus establishing the means to expand. By 1816 the Boston Associates had codified their business plan's eight elements: corporate management structure, extensive liquid capital, accounting systems, vertically integrated production, a focused effort on technology and products, labor organization, a marketing strategy, and a protective tariff. Corporate management had three tiers: leading stockholders determined policy about investment and management, reserving final authority to themselves but delegating extensive power to the board of directors; the board worked closely with the treasurer, who was elected to serve as chief administrative officer and to supervise operations (e.g., collecting capital assessments, construction, accounts, and purchases); and a plant manager supervised day-to-day technical aspects, especially the production and maintenance activities of the machine shop. This organization suited a future multiplant, multilocational (formed as separate firms) structure of cotton textile production. Top-level managers stayed in Boston, accessible to information about the regional and national economy and to local capital sources, whereas day-to-day plant managers located at factory sites. This structure represented a refinement of, rather than a radical departure from, approaches of leading Providence core firms; however, they would not implement this structure as fully nor gain similar levels of efficiency as the Boston Associates' firms.[37]

Investors in the Boston Manufacturing Company followed typical mercantile procedures for large-scale, uncertain ventures; they spread risk and provided extensive liquid capital—the plan's second element. The total capital subscription of $100,000 came from twelve investors: Jackson ($20,000), Lowell ($15,000), and ten others ($5,000 to $10,000 each). Nathan Appleton contributed only $5,000, a tiny fraction of his net worth of $200,000; he supported the experiment but preferred to see results before committing more capital. Compared to financing Asian voyages, these textile investments appear modest: on one voyage in 1810 the total investment upon departure from Boston probably reached $40,000, and that same year Lowell agreed to invest $20,000 to $25,000 in a shipment from India which Jackson organized, an amount greater than Lowell's original investment in the Boston Manufacturing Company. The Boston Associates astutely kept capital highly liquid and maintained substantial reserves to cover contingencies, policies that Appleton recommended based on his analysis that textile firms often failed to allow sufficient working capital. Initial capitalization of $100,000 was not new: merchants contributed much greater capital to the 1787-97 textile debacles, and Brown and Ives contributed $50,000 to the $150,000 initial capitalization for the Blackstone Manufacturing Company in 1809. Investors transferred careful mercantile record keeping practices to the Boston Manufacturing Company, thus establishing accounting systems—their plan's third element. They allocated direct costs to departments handling activities; they divided labor costs between cloth and machinery accounts; and to the extent that it was feasible they distributed overhead costs including executive salaries across accounts. Although lacking the details of modern accounting systems, investors possessed sound measures of costs and net profits.[38]

The plan's fourth element—vertically integrated production—was settled quickly in 1813. The firm scrapped production of diverse textiles and only weaving cloth; instead, it vertically integrated, using waterpower to spin cotton yarn and weave cotton cloth. They targeted markets for coarse, unbleached cloth, the product most consumers demanded and households' first purchases when they ceased home weaving, but the Boston Associates did not innovate vertical integration. Although in 1813 most cotton mills only spun yarn, some mills also housed hand loom weavers, and Silas Shepard built fifteen to twenty power looms for installation in Taunton's Green Mill in 1812. The Boston Manufacturing Company's focused efforts on using new technology and products—the plan's fifth element—placed the firm on the path that Almy and Brown blazed when they hired Samuel Slater. The Boston Manufacturing Company sought the best technical talent to design and equip its factory and chose Paul Moody, who possessed wide experience with textile machine building and factory design, and under his leadership the firm addressed the entire productive system. In early 1814 he established a machine shop at Waltham to build the mill's machinery, a practice that most cotton firms followed.

Initially, Moody perfected the power loom, completing this work with Lowell's help in late 1814, based partly on ideas borrowed from mechanics who had sold them power looms. Once this loom reached operational status, Moody turned to improving designs for other machines; nevertheless, the Boston Manufacturing Company often relied on outside machinery builders, especially in New England, to equip its mill. Moody's technical improvements solidified the Boston Manufacturing Company's leadership in the decade following 1814. Lowell and Moody targeted the lowest technological level—a simple power loom that weaved coarse yarn into heavy sheeting—thus increasing the likelihood of attaining technological success, creating a product that would reach the largest market, and placing the firm against the weakest competitors, unskilled, unproductive hand loom weavers of coarse cloth.

Labor organization—the plan's sixth element—diverged from the dominant family labor system Providence core firms used; instead, the Associates chose the boardinghouse system, a derivative of their managed approach. They paid premium wages to attract bright, educated young women from farms and villages to take unskilled positions monitoring waterpowered spinning and weaving machines. The factory built boardinghouses and hired matrons to supervise women, and, as the needs materialized, they added a bank, library, church, and store to their factory village. Speculations that investors devised the boardinghouse system to avoid the evils of British industrialism and maintain social harmony rest on comments they made decades later and on interpretations of their societal views. Their decision to pay premium wages to attract high-quality laborers probably represented cold calculations that high productivity offset high wages, leading to a low labor cost per yard of cloth.[39]

The Boston Manufacturing Company developed a marketing strategy—the plan's seventh part—which funneled the entire output to one sales agent for introducing goods into national wholesale distribution. It chose Benjamin C. Ward and Company, a wholesale partnership founded by Nathan Appleton in 1815, which took charge of most sales over the next few years, receiving a sales commission; soaring production provided incentives to market textiles effectively, and in 1819 the firm's directors authorized Ward and Company to serve as its exclusive sales agency. By then Appleton's wholesale house provided standard commission services for manufacturing clients: buying raw materials and selling goods, arranging shipping and insurance, and advancing funds before sales. The Boston Manufacturing Company followed the path that leading mills in Providence and vicinity had blazed earlier through the use of commission merchants in New York, Philadelphia, and Baltimore. Skilled in international trade, the founders knew that sales growth required blocking international competitors; they added a tariff as the plan's final element. Lowell successfully lobbied Congress to pass a tariff in 1816 levying a 25 percent ad valorem duty on cotton goods. This protected precisely the cheap cotton goods that the Waltham mill produced, but the tariff did not represent a unique approach to protection; many firms in cotton textiles and other industries sought protection.[40]

The eight core elements of the Boston Manufacturing Company's business plan powered enormous success. Sales jumped from $23,628 in 1816—the first year with a full complement of power looms—to $124,748 in 1818 and to $260,658 in 1820; the firm's directors responded by dramatically expanding. In

1817 they started a separate machine shop and commenced sales, especially of power looms, to other textile factories, and they built a second mill; by 1821 capital totaled $600,000, spindle capacity reached 8,000, and annual cloth output soared to 1.8 million yards, vaulting the Boston Manufacturing Company to the top of the cotton textile industry. The Waltham complex generated high rates of return—dividends reached 13–17 percent between 1817 and 1820—on rapidly rising capitalization; thus, in 1820 investors began plans to enlarge production on a broad scale, thus establishing the Lowell project.

To claim that the Boston Manufacturing Company radically departed from previous cotton textile firms reads history backwards from large mill projects at Lowell and other sites to their roots in Boston and Waltham in 1813. Individual components of the business plan had parallels in other leading cotton textile mills in New England and New York or represented modifications (e.g., the boardinghouse system) which had alternatives (e.g., family labor system), and, after the Boston Manufacturing Company, the Associates' firms were not technological leaders. Instead, their business plan constituted an innovative leap, but they did not recognize their efforts as an official plan. Their replication of it on a larger scale at Lowell and subsequent, almost mechanical replication (with fine-tuning) on a grand scale at other sites suggests that they developed a sophisticated management approach to capture national markets, which set them apart.[41]

Laggards in the New York City Region

The New York City social, political, and economic elite's impressive failures in cotton textile manufacturing probably dampened enthusiasm. They formed the New York Manufacturing Society in 1789, commenced manufacturing cotton yarn and cloth, employed Samuel Slater for three weeks, and then lost him to Providence's Moses Brown. Slater's low opinion of this factory proved prescient—it expired by 1793. They repeated their failure to hire experienced mill managers with technical expertise and skilled mechanics familiar with the Arkwright spinning system and soon supported one of the largest industrial failures—the Society for the Establishment of Useful Manufactures' (SUM) effort to build a factory city at the Great Falls of the Passaic in nearby Paterson, New Jersey, from 1791 to 1796. Merchants, including John Livingston, a member of the patrician family dominating New York politics, started another cotton mill in New York City—Dickson, Livingston and Company, but they invested too much capital in buildings and equipment, in contrast to Almy and

Brown's cautious approach. They commenced Arkwright spinning and hand loom weaving around 1793; however, immigrant English mechanics were insufficiently skilled, and investors moved their machinery to New Haven, Connecticut, by 1795. The new firm, John R. Livingston and Company, also started with a large factory, but the mill never reached profitability and languished within a decade. Thus, only fourteen, mostly small, cotton mills operated within about seventy miles of New York City by 1810.[42]

Competitive cotton mills formed two subcenters near New York City by 1810–15 (see table 4.2). The Hudson Valley from New York City to Kingston (about one hundred miles north) constituted one subcenter; Dutchess County alone housed five mills, including the prominent Matteawan Manufacturing Company at Fishkill (est. 1814). Its formation signaled serious efforts by New York merchants to again engage in cotton manufacturing—Schenck family members headed the project; Abraham managed the mill and Peter, a ship owner and commission merchant, served as the selling agent in New York City. The factory adopted Gilmour power looms, and in 1820 it housed forty-four looms making about 200,000 yards per year; nevertheless, its output represented only 11 percent of the Boston Manufacturing Company's production. Other Hudson Valley counties housed a total of five competitive factories, including the famous Ramapo Works (est. 1814) at Hempstead, immediately north of New York City; it started as a nail and metalworking firm, added cotton manufacturing, and adopted Gilmour power looms early. The firm's machinists started building looms over the course of the years 1817 and 1818 and soon had installed fifteen; the Ramapo Works operated seventy-three power looms by 1820.

Paterson emerged as the second, and primary, subcenter near New York City. The huge waterpower potential of the Great Falls of the Passaic remained little used following the debacle of the SUM in the early 1790s; this delayed exploitation for cotton manufacturing confirms the laggard position of New York City's environs. Roswell Colt, the son of Peter Colt, former superintendent of the SUM, provided an impetus for change when he purchased the site and waterpower rights in 1814 and invested forty thousand dollars to improve the dam, basin, and canals. At that time Essex County, including Paterson, housed twenty cotton mills with total spindles of 32,500, and it sold much of their cotton yarn to Philadelphia's hand loom weavers. Colt's improvements encouraged textile expansion in Paterson; four competitive mills began operating between 1813 and 1816, whereas only two began elsewhere in northern

New Jersey (see table 4.2). By 1818 Paterson's five cotton factories housed 20,000 spindles, ranking them among the nation's larger mills, though they were still smaller than the big New England mills. After 1820 its cotton textile firms grew significantly, but the Paterson and lower Hudson Valley mills never equaled the cotton textile cores of Providence, Philadelphia, or Boston.[43]

The Stage Is Set

Commerce, agriculture, and manufacturing grew slowly in the period from 1790 to 1820. Agricultural technology stagnated, manufacturing employed little machinery outside textiles and processing (e.g., flour milling and sawmilling), and few transportation improvements, other than roads, were completed. Small annual rises in per capita incomes and the tiny urban share of the population seemed to confirm that this growth was merely a prelude to subsequent changes, yet these indicators dismiss too readily the significance of the East's transformation by 1820. Rich agricultural areas surrounded metropolises, and distant from them other prosperous farming areas had emerged; this flourishing economy of urban and rural dwellers supported thriving manufacturing firms in metropolises, cities, villages, and rural areas. The continued growth and transformation of urban and rural economies set the stage for rapid urban industrial growth in the East by the 1840s.

Part 2 / The Late Antebellum, 1820–1860

CHAPTER FIVE

Tightening Ties That Bound the East

> A connection of the Atlantic and Western waters by a canal, leading from the seat of the National Government to the river Ohio, regarded as a local object, is one of the highest importance to the states immediately interested therein, and considered in a national view, is of inestimable consequence to the future union, security and happiness of the United States.
>
> —PROCEEDINGS OF THE CHESAPEAKE AND OHIO CANAL CONVENTION, WASHINGTON, 1823

The continued population surge across the Appalachian barrier posed opportunities and threats to eastern businesses. Lucrative profits awaited merchant wholesalers controlling trade in agricultural output destined for eastern markets and trade in manufactures for midwestern markets, but the mercantile elite in each metropolis—Boston, New York, Philadelphia, and Baltimore—feared that competitors in other metropolises would capture most of the benefits. Migration posed an additional dilemma for eastern industrialists, because midwestern craft shops and factories might capture local and regional markets, blocking eastern producers. Eastern firms manufacturing high-value goods relative to weight—such as boots and shoes, cotton textiles, clocks, and buttons—had little to fear if they remained technological leaders and low-cost producers. Firms that manufactured low-value goods relative to weight, however, such as furniture, iron products, and farm machinery or firms possessing no technological advantages had difficulty capturing midwestern markets; their profits came from the expansion of the eastern market. Market shift (measured by population) assumed alarming proportions: during the thirty years from 1790 to 1820 population growth in the South and Midwest reduced the East's share of the national population from 60 percent to 50 percent, and its share fell to 36 percent by 1860 (see table 1.2).

The Economy Gathers Steam

The number of hands laboring on farms and in factories and stores soared; population more than tripled to 31.5 million between 1820 and 1860, representing a compound annual growth of 3 percent and identical to that during the period from 1790 to 1820. Swelling immigrant arrivals fueled this growth, with immigrants constituting ever larger shares of total population; the cumulative numbers of new arrivals (ignoring births and deaths) during the 1820s accounted for 1 percent of the population in 1830, but during the 1840s cumulative arrivals reached 7 percent of the 1850 population. If the nation had not received these immigrants, growth would have slowed because women's fertility rate had declined, and skilled industrial mechanics and laborers among these arrivals contributed to manufacturing growth, just as the textile group and others had done before 1820. Measured by the change in real gross domestic product, the economy grew by 4.1 percent annually, and growth accelerated; by 1860 GDP was over five times larger than in 1820, enhancing the opportunities for market integration (see table 2.1).

As in the period before 1820, exports exerted little impact on the buoyant economic growth; as a share of gross national product, exports never exceeded 7.3 percent. The composition of exports changed little, and the primary sector accounted for about 80 percent; between 1840 and 1860 primary sector exports accounted for no more than 13 percent of total primary production, and manufactured exports as a share of industrial output fell from 6 to 4 percent. Export demand stimulated the growth of specific crops (e.g., cotton) or industries (e.g., clocks), but these stimuli typically had cyclical rather than long-term secular impacts. Tariff rates rose from about 40 percent around 1820 to a peak near 60 percent around 1830 and then declined to about 20 percent in 1860. In the absence of the late 1850s tariff, cotton textiles would have been about one-third smaller and other manufactures perhaps 15 to 25 percent smaller, and the Midwest would have settled somewhat faster. Thus, agricultural growth and regional industrialization in the East rested mostly on developmental processes within the nation and the East, not on foreign sources, but the tariff supported higher levels of industrialization.[1]

Rising Real Incomes

Economic expansion outpaced population growth between 1820 and 1860, boosting incomes; GDP per capita rose 61 percent, amounting to compound annual growth of 1.2 percent and almost double the 0.7 percent rate for the period from 1793 to 1820. The contributions to rising GDP per capita of intersectoral labor shifts out of farming into other sectors declined from 55 percent (1800–1820) to 25 percent (1820–40) to 15 percent (1840–60) as productivity gains in nonfarm sectors such as distribution, commerce, mining, and manufacturing increasingly spurred growth. The cumulative rise in GDP per capita noticeably impacted individuals' lifestyles: they purchased more food and better housing and increased absolute and relative amounts of money spent on goods and services above these basic necessities. As a share of GDP per capita, items above necessities rose from 27 to 44 percent between 1820 and 1860, and expenditure growth on them far exceeded the growth of GDP per capita (see table 2.1); this expenditure shift powered industrial expansion. Rising real wages for selected occupational groups independently confirm the national gains in GDP per capita translated into the growth in real income in the East. Between 1821 and 1860 annual real wages increased by 1.2 percent for laborers (e.g., common laborers and teamsters), 1.4 percent for artisans (e.g., blacksmiths, carpenters, machinists, masons, and painters), and 1.3 percent for clerks; this translated into growth and the diversification of demand for manufactures (fig. 5.1). Other measures support these conclusions; the real wages of unskilled nonagricultural laborers in the nation rose by 1.7 percent annually, and the wages of manufacturing workers in the Brandywine increased by 1.1 percent annually (see fig. 2.2).[2]

The Business Cycle

The 1820–60 economic expansion can be divided into three periods: growth (1820–36), contraction (1837–43), and renewed expansion (1844–60). The first period commenced with moderate growth and declining prices between 1820 and 1830; then prices turned sharply upward, reaching a peak in 1836 (see fig. 5.1). The 1837 financial crisis is often identified as the onset of a depression lasting—with a brief interruption in 1839—until 1843, and its severity approached the Great Depression of the 1930s, yet this interpretation overstates the case. During the 1837 financial crisis banks suspended specie payments (gold and silver backing of currency), and a mild price deflation quickly ensued. Nevertheless, the goods sector of the economy suffered little; domestic trade fell only 10 percent from a peak in 1836 to a trough in 1838 (fig. 5.2). Capital flows from Britain resumed, and the federal government supported rising state government expenditures through the transfer of revenue from public land sales to states, thus boosting demand. These developments quickly

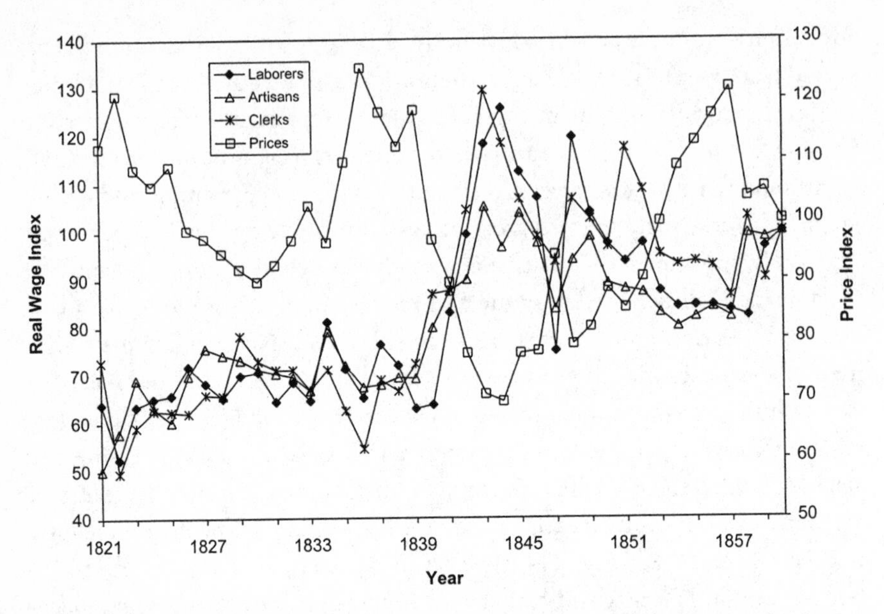

Fig. 5.1. Real Wage Indices for Laborers, Artisans, and Clerks in the East, and Price Index for the East, 1821–1860 (1860 = 100). *Source:* Margo, *Wages and Labor Markets in the United States, 1820–1860*, 70–73, tables 3A.8–3A.11.

ended the recession, and domestic trade jumped 7 percent (1838–39). Another brief financial crisis hit in 1839, but not as severely as in 1837, because many banks maintained specie payments.

The 1839–43 economic downturn—a subperiod of the second-period contraction—did not produce major distress. Real gross national product rose 22 percent between 1839 and 1844, just one year into the recovery, and domestic trade fell by only 11 percent from 1839 to 1840 and then remained stable through 1843 (see fig. 5.2). The greater plunge in prices than in goods production translated into rising real GNP; a breakdown of the real values of flows of goods to consumers reveals the economy's underlying resiliency. From 1839 to 1844 the dominant perishables sector, encompassing agricultural production, grew 22 percent, matching GNP expansion, whereas semidurable and durable manufactures surged 94 percent; thus, manufacturing commenced a major upturn. Because agriculture still dominated, most people fed themselves while agricultural trade languished; unemployment probably remained concentrated in large metropolises accounting for a small share of the population. The third-period economic upturn commencing in 1844 carried through to

1860, with several modest contractions. Prices recovered from deflation (see fig. 5.1), and real GNP tacked on successive annual growth rates of 4.2 percent (1844–49), 6.5 percent (1849–54), and 3.9 percent (1854-59); these GNP rate changes corresponded somewhat with wavelike fluctuations in domestic trade (see fig. 5.2).

Through these business cycle perturbations the economy transformed from agriculture to manufacturing; during the two decades from 1839 to 1859 agriculture's share of commodity output fell from 72 to 56 percent, and manufacturing's share rose from 17 to 32 percent (table 5.1). Nevertheless, agricultural value added almost doubled and was 74 percent larger than manufacturing by 1859; thus, agriculture continued to support industrial expansion. The real value of industrial production catapulted almost fivefold, representing a compound annual rate of 7.5 percent and significantly exceeding the 5 percent annual growth from 1859 to 1899, considered the apogee of industrial growth. The 1839–59 industrial growth coincided with falling tariffs on manufactures,

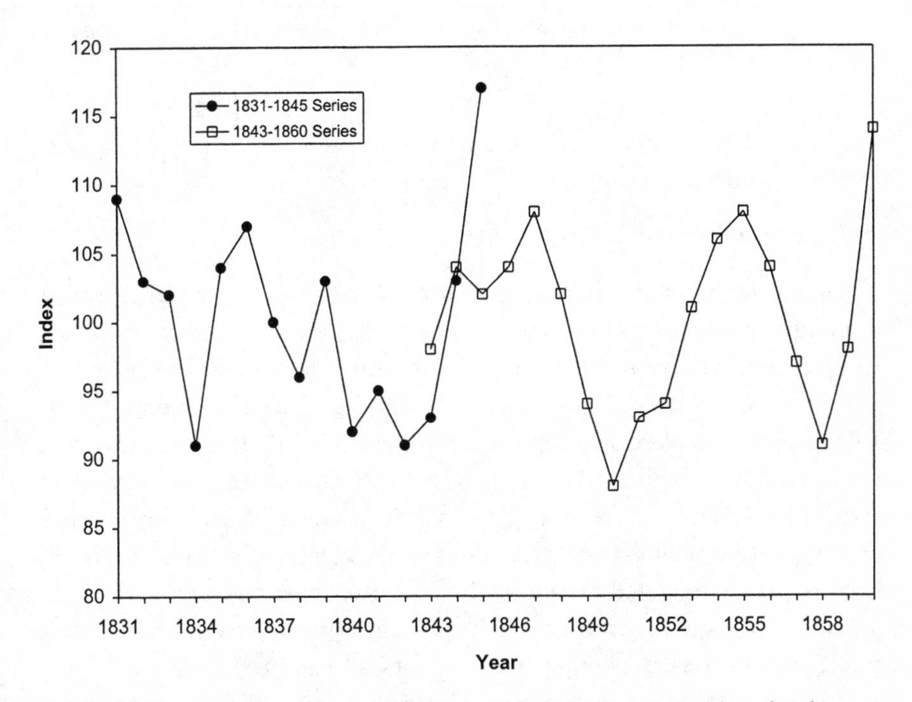

Fig. 5.2. Indices of Volume of Domestic Trade, 1831–1860. *Source:* Smith and Cole, *Fluctuations in American Business, 1790–1860,* 73, 104, tables 18, 28.

Table 5.1. Commodity Output Measured by Real Value Added, 1839–1859 (1879 Prices)

Year or Period	Agriculture	Manufacturing	Construction	Mining	Total
		Value Added (millions of dollars)			
1839	787	190	110	7	1,094
1844	944	290	126	14	1,374
1849	989	488	163	17	1,657
1854	1,316	677	298	26	2,317
1859	1,492	859	302	33	2,686
		Composition (percentage)			
1839	72	17	10	1	100
1844	69	21	9	1	100
1849	60	29	10	1	100
1854	57	29	13	1	100
1859	56	32	11	1	100
		Period Growth (percentage)			
1839–44	20	53	15	100	26
1844–49	5	68	29	21	21
1849–54	33	39	83	53	40
1854–59	13	27	1	27	16

Source: Gallman, "Commodity Output, 1839–1899," in *Trends in the American Economy in the Nineteenth Century,* 24:43, table A-1 (variant A).

suggesting that firms improved productivity and competed effectively against foreign manufacturers.[3]

Capital Investment Spurs the Economy

The rising shares of national production allocated to investment propelled economic growth; investment's share averaged 6.2 to 7.0 percent between 1805 and 1840, but then it rose to over 10 percent during the 1840s and to 12 percent during the 1850s. Annual growth of real total capital stock accelerated from 1805 to 1860, and after 1840 capital deepening occurred as growth in capital stock exceeded growth in GDP by 15 percent in the 1840s and 32 percent in the 1850s. In a reversal of pre-1815, per capita investment in land clearing and breaking rose at small rates between 1815 and 1860 (except the 1840s), probably because people migrated to new lands across the South and to the Midwest, and per capita investment in animals continued growing from pre-1815 to a peak from 1815 to 1840, but then it fell (see tables 2.1 and 2.2).

Minimal gains in per capita farm investment did not reflect a rural decline; instead, increasingly productive agriculture propelled rural-to-urban migration. The expanding economy required larger inventories to sustain production; nevertheless, the gains in per capita inventories approximated those in GDP per capita between 1815 and 1860. Thus, the economy's distribution sectors (e.g., wholesaling, warehousing, and transportation) raised productivity sufficiently to allow producers to build inventories at the pace sustaining higher per capita production. Although capital investment in farm structures rose, escalating urban industrial growth propelled acceleration in per capita growth in structures. Growth rates of per capita equipment investment also accelerated, and investors expanded capital stock per capita to almost triple the growth in GDP per capita after 1815, providing additional confirmation that capital was deepening (see tables 2.1, 2.2, and 2.4).

Consonant with agriculture's declining share of commodity output, shares of capital stock in clearing and breaking fell to 22 percent by 1860, less than half the 1774 level (56 percent), and shares of capital stock in animals fell from the peak of 15 percent from 1815 to 1840 to 9 percent by 1860. Shares of capital stock in inventories remained about 18 percent between 1815 and 1860, underscoring that the distribution sectors effectively supported growing output and rising incomes. In contrast, shares of capital stock in structures, which remained around 27 percent between 1805 and 1840, jumped to 42 percent by 1860 as urban industrial growth accelerated, and equipment's share of capital stock rose from about 5 percent between 1774 and 1815 to 9 percent by 1860 as the economy required more transportation vehicles, machinery, and tools (see tables 2.3 and 5.1). Nevertheless, equipment's share paled next to its 28 percent share by 1900, as accumulated antebellum advances in machinery and machine tools swept through manufacturing. Highly integrated capital markets in the East—at least since 1835 and probably earlier—funneled capital to all economic sectors and regions at market prices reflecting competitive interest rates; thus, the economy expanded without undue hindrance from capital shortages, and this investment boosted productivity.[4]

Productivity Gains

Agricultural productivity per worker accelerated from 1800 to 1860 (see table 2.7), demonstrating that farmers intensified their responses to market opportunities, including growing urban populations and more workers in resource extraction (e.g., mining and lumbering), trade, transportation, and manufacturing. Productivity advances were also rooted in competition among farmers; nearby or distant farmers boosting productivity pressured prices downward, and thus farmers failing to raise productivity faced lower returns. Manufacturers encountered the same predicament; rising demand encouraged greater production, and they needed to raise productivity if competitors' productivity gains lowered prices. By 1820, however, only a few industries, such as cotton textiles, possessed machinery suitable to expand production. Equipment's share of domestic capital stock remained minuscule, and the per capita increase in equipment investment stayed low before 1850; therefore, substantial productivity gains came from sources other than machinery investments (see tables 2.2 and 2.3).

If production economies of scale existed, greater factory size offered one avenue to boost productivity; between 1820 and 1850 average factory employment increased across a range of industries—boots and shoes, cotton textiles, hats and caps, iron and iron products, and paper—suggesting that owners took that route. By 1850 factories with more than fifteen workers contributed substantial shares of value added in production in boots and shoes (58 percent); coaches, wagons, and harnesses (52 percent); furniture (63 percent); hats (93 percent); paper (68 percent); and cotton textiles (100 percent); nevertheless, the small sizes of most 1850 factories and the modest change in size over the preceding thirty years cautions against assuming that larger factories provided productivity advantages. Factories achieved greater productivity than artisan shops, but sources of scale economies rapidly dissipated as factory size increased. Nonmechanized factories with over fifteen employees were not significantly more productive than those with six to fifteen employees; therefore, sustained increases in factory productivity did not derive from systematic factory enlargement. Nonetheless, manufacturing labor productivity soared by 2.1 percent annually between 1820 and 1850 and by 3.2 percent from 1850 to 1860. Boosts in capital investment per worker contributed little to those gains, because structures and inventories absorbed much of the capital, not machinery (see table 2.3), and this applied across the gamut of industries—including labor-intensive, less-mechanized, capital-intensive, and mechanized ones.[5]

Distinctions between urban and rural manufacturing suggest clues to explain the growing productivity; urban manufacturers faced larger, more competitive markets than rural firms. Although urban firms in labor-intensive and less-mechanized industries accumulated experience with increasing productivity by 1820, high transportation and marketing costs protected their rural counterparts from competition, and limited local markets provided little incentive to raise productivity. After 1820 transportation improvements threw rural firms into competitive battles to survive, and rising agricultural prosperity provided market incentives to find productivity enhancements, which might explain higher labor productivity growth in labor-intensive and less-mechanized industries in rural areas than in urban areas between 1820 and 1860. Yet firms in capital-intensive and mechanized industries achieved higher rates of productivity advance in urban areas between 1820 and 1850, possibly because these firms served larger market areas that still remained beyond rural firms' reach. Nevertheless, rural firms boosted productivity faster than urban firms during the 1850s, substantially narrowing the productivity gaps between them. Rural firms' gains and narrowed productivity gaps with urban firms comport with the convergence of farm and rural manufacturing wages on urban wages in the East from 1820 to 1860.

The fact that most eastern manufacturers achieved large productivity gains in small firms with little machinery and rural firms became more competitive suggests an intriguing hypothesis: local and subregional rural markets in areas of rising agricultural prosperity, as well as urban markets, undergirded eastern industrialization after 1820, just as they had earlier. To claim that rural markets powered industrialization challenges the view that the East's failing rural sector succumbed to midwestern competition, thus providing a reservoir of manufacturing workers supporting industrialization; these industrial workers also could have been released from thriving agriculture that was achieving productivity advances. Small nonmechanized factories in rural areas offered manufactured goods to local and subregional customers, and they boosted productivity through simple technical advances in tools, changing labor organization (introducing a division of labor), and modifications to managing production; therefore, they increasingly drove artisan shops out of business.[6]

The Soaring Urban Population

The sevenfold explosion of eastern urban dwellers between 1820 and 1860 demonstrates agriculture's success in freeing laborers for manufacturing and other sectors. The four large metropolises became vast centers of consumption and production, and New York surpassed one million people by 1860; other urban places grew faster, thus metropolises' share of the East's urban population dropped from 69 to 53 percent. Rising shares of population in urban places (from 11 to 35 percent) points to a dramatic structural change in the eastern economy, and this swept along every state, albeit at differential rates. Massachusetts, Rhode Island, and eastern New York witnessed sharp increases in the percentages of their population that were urban during the 1820s, and most of the East—except Vermont, western New York, and western Pennsylvania—urbanized rapidly by the 1830s. Consistent with rapid industrial growth between 1840 and 1860, these decades had the greatest absolute increases in the percentage of the population in urban places in most states, and, except for New Hampshire and Vermont, these changes exceeded the increases made from 1860 to 1890; thus the East's pivotal structural transformation occurred in the antebellum (see tables 1.2, 2.4, and 2.5).[7] Transportation improvements spurred industrialization directly through widening markets for distributing manufactures, and they indirectly widened industrial markets by lowering the costs of shipping agricultural goods, which raised farm incomes, boosted rural demand for manufactures, and lowered food costs for urban dwellers—allowing them to shift consumption expenditures toward manufactures. Transport improvements required large amounts of capital, for which the mercantile elite in regional and subregional metropolises, with major stakes in promoting economic growth, were important sources.

Metropolitan Leadership

The elite in New York, Boston, Philadelphia, and Baltimore possessed the greatest financial stake in ensuring that the East continued an upward trajectory of economic growth and development after 1820. They controlled intra- and interregional exchanges of commodities (e.g., wheat, flour, butter, coal, lumber, and textiles) and capital (e.g., credit, cash transfers, and bonds), and their capital underwrote investments in large-scale projects of raw material extraction (e.g., mines and lumbering), manufacturing (e.g., steam engine works), and transportation (e.g., canals and railroads). The Midwest beckoned as its population swelled by 4.9 percent annually from 1820 to 1860, over twice the East's growth rate; by 1840 the Midwest's 4.1 million people exceeded the East's entire population in 1810 (see table 1.2). Costly canals and railroads were weapons in the battles for the intra- and interregional supremacy over economic exchange; the mercantile elite encouraged transportation improvements, but they needed to avoid locking up substantial capital in infrastructure, such as canals, railroads, and ships, because these investments reduced capital in order to support riskier activities that promised higher returns, such as financing and buying and selling commodities.

Interrelated paradoxes emerged from the construction of interregional canals and railroads to the Midwest. When long-distance lines reached there from an eastern metropolis, competitor merchants in other eastern metropolises accessed these lines; little extra transshipment and ton-mile costs were added along the short, integrated lines connecting East Coast metropolises. Competitor merchant agglomerations in the Midwest (e.g., Cincinnati, St. Louis, and Chicago) controlled intraregional exchange, thus forcing eastern merchants to deal with midwestern merchants in interregional exchange. Consequently, merchants in eastern metropolises used their access to market information and capital to battle one another to control interregional exchange with midwestern merchants; battles over interregional transport improvements were less important. Merchants with greater capital specialized in buying and selling larger volumes of commodities over wider territories, and they exchanged with merchants in other metropolises in the Midwest and East who were operating at less-specialized and less-capitalized levels.[8] If this eastern mercantile hierarchy arose before the interregional transportation improvements to the Midwest had been completed or before commodities moved over these routes in significant amounts, transport lines would have a limited impact on the differential capacity of merchants in eastern metropolises to control the exchange of commodities and capital.

Transport Projects to the Midwest

The New York legislature's efforts to encourage the construction of a canal across the state started in 1792 but bore little fruit until 1817, when it passed a bill authorizing that a canal be built connecting the Hudson River and Lake Erie. The passage of this bill and the immediate start of construction propelled the state and New York City into the lead in constructing interregional transport lines to the Midwest, setting off agitated debates in other states about appropriate responses. New York completed the Erie Canal in 1825, and the following year Pennsylvania's legislature finally achieved a consensus, after two years of debates between railroad and canal advocates, and authorized construction of the Mainline Canal, a railroad and canal hybrid connecting Philadelphia and Pittsburgh. Construction got under way in 1826 and finished in 1834. Baltimore considered the Chesapeake and Ohio Canal but chose the railroad, presumably because it was cheaper to build and provided better service. Maryland authorized the incorporation of the Baltimore and Ohio Railroad in 1827; construction started from Baltimore in 1828, reaching the Wheeling, West Virginia, terminus on the Ohio River in 1853. Faced by the Erie

Canal's completion in 1825, the Massachusetts legislature and Boston business leaders lengthily debated construction, first, of a canal to connect to the Erie at Albany and, then, of a railroad, but they failed to achieve consensus. Finally, private investors completed the Boston and Worcester Railroad between these cities in 1835 and extended it to Albany in 1841.[9]

A Battle Won before It Starts

If interregional transportation lines gave eastern mercantile elite chances to capture midwestern trade, the completion dates of the Erie Canal (1825), Pennsylvania Mainline (1834), Baltimore and Ohio Railroad (1853), and Boston and Albany Railroad (1841) approximate the probabilities of success. New York merchants won by 1825; they had had nine years to dominate midwestern trade before Philadelphia merchants could fully use the Mainline. Midwestern exports did not enter the Erie above token amounts until the mid-1830s, however, about the time that the Pennsylvania Mainline opened, and tonnage on both canals fell far below tonnage moving on the Ohio-Mississippi River system to New Orleans (fig. 5.3).[10] Therefore, interregional lines between East and Midwest could not have impacted the eastern mercantile elite's capacity to capture midwestern trade until after 1835.

A metropolis' population size indirectly measures its mercantile elite's importance in interregional (and intraregional) trade. The wholesaling-trading complex generates initial multiplier effects in banking, insurance, stevedoring, warehousing, and transportation (e.g., shipping and wagons) and secondary multiplier effects in retailing, professionals (e.g., lawyers and physicians), and local government. Entrepôt (e.g., processing) and commerce-serving (e.g., shipbuilding and publishing) manufactures expand along with trade, setting off secondary multipliers, and continued growth stimulates new construction and more multiplier effects. Based on population, New York merchants may have achieved dominance in the East by 1820, before completion of the Erie Canal (see table 2.5); measures of foreign and domestic trade confirm this. Between 1803 and 1820 New York merchants exported more domestic goods than other merchants in East Coast metropolises, and by 1820 they exported 76 percent more than their nearest rivals in Baltimore. By 1825, when the Erie Canal opened over its entire length, New York merchants exported 83 percent more than merchants in the three other ports combined.

New York merchants led in imports beginning in the late 1790s, and the port

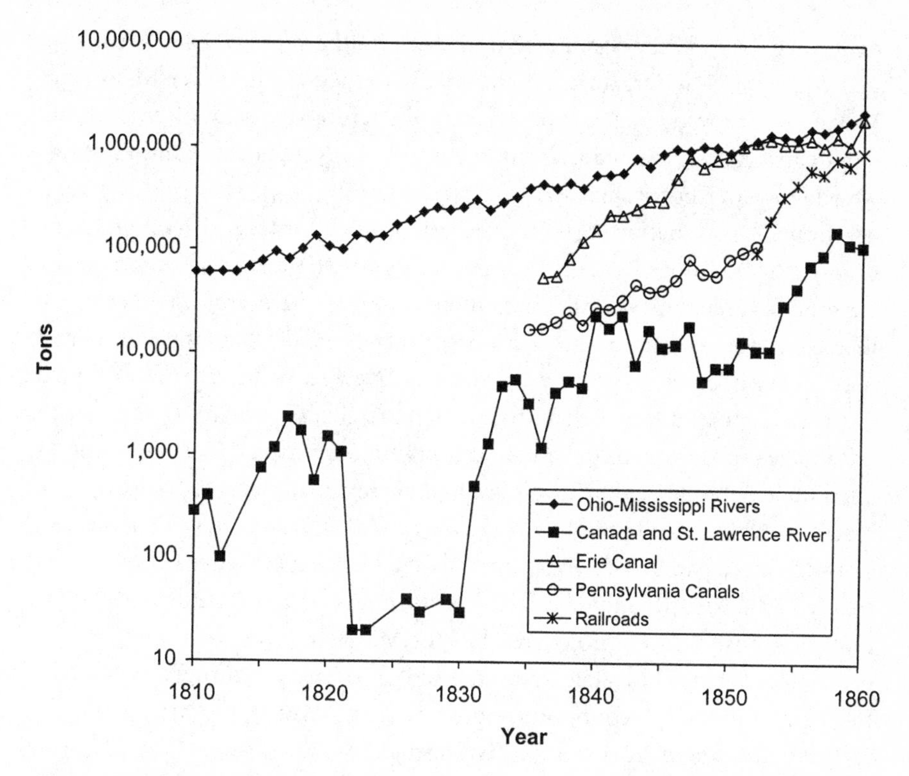

Fig. 5.3. Freight Tonnage from the Midwest by Rivers, Canals, and Railroads, 1810–1860. *Source:* Haites, Mak, and Walton, *Western River Transportation,* 124–28, app. A, tables A-1-A-3.

collected more import duties than Boston, Philadelphia, and Baltimore from 1796 to 1812. When the War of 1812 ended, British merchants dumped vast quantities of manufactures on the New York market, which, in 1815, collected twice as much from duties on imports as Philadelphia, the second-rank port. New York merchants' dominance of imports widened; by 1825, at the Erie's completion, they accounted for 58 percent of the value of imports to the four leading ports and controlled over three times as much trade as Boston and Philadelphia merchants. The start of specialized shipping firms (packet lines) in 1817 reinforced New York merchants' dominance over the export and import trade, and their capacity to support this premium service and other ports' failure to offer more than token competition underscored their overwhelming lead in foreign trade.[11]

By 1820 New York merchants also led in interregional trade among the four eastern ports, and they dominated trade with southern ports. Their leadership in trade with New Orleans, the entrepôt for trade of the lower Mississippi Valley and for midwestern exports using the Ohio-Mississippi River system, demonstrates that they had captured control of most exports from the Midwest before the Erie Canal was completed (see fig. 5.3). New York's pivotal position in information flow in the East by 1820 buttressed its merchants' dominance of domestic trade: they received domestic news and domestic circulation of foreign news earlier than merchants elsewhere, and, after New York received news, the next in line were merchants in major cities of its hinterland and those in other large ports. Therefore, New York merchants exploited business opportunities earlier than other merchants, providing them with advantages in dominating trade and accumulating capital.[12] They strengthened their national lead by dominating their regional market's trade; by 1820 New York's hinterland population was the largest in the country (see table 2.6). Thus, its merchants acquired commanding leadership as the nation's most highly capitalized, highly specialized controllers of trade before any interregional transportation project reached the Midwest. Promoters of every project assumed that midwestern trade would flow over the transportation artery, thus contributing to economic growth and development; the timing and extent of that trade mirrors its capacity to benefit the East.

The Midwest's Impact on the East

Exports of agricultural products from the Midwest first flowed over interregional canals in significant amounts after 1835, and the exit route through Canada never carried much (see fig. 5.3). Pennsylvania's canals carried little tonnage compared to amounts exiting over the Erie Canal, and the gap between them widened considerably; by the late 1840s Pennsylvania's canals carried just one-tenth of the Erie's tonnage. Until the mid-1840s freight shipments down the Ohio and Mississippi Rivers to New Orleans dwarfed those over interregional canals, but most arrivals before the mid-1840s came from Kentucky and Tennessee, not from states north of the Ohio. New Orleans and vicinity and the Gulf Coast ports were the main markets for foodstuffs, and this trade continued until the Civil War. The Midwest north of the Ohio River started sending large quantities of foodstuffs to New Orleans after 1840, and these entered foreign markets or moved to eastern consumers. The South consumed few midwestern foodstuffs because southern farms, including those

Table 5.2. Traffic Clearings on the Erie Canal System, 1838–1852

	Average Annual Tonnage (thousands)			
Period	Midwest Clearings	Instate Clearings	Total Clearings	Arrived at Tidewater
1838–40	151	985	1,136	424
1841–43	265	903	1,168	549
1844–46	420	1,325	1,745	956
1847–49	827	1,717	2,544	1,294
1850–52	1,187	1,866	3,053	1,508

Source: Ransom, "Interregional Canals and Economic Specialization in the Antebellum United States," 21, table 7.

in Kentucky and Tennessee, produced sufficient quantities to meet demand. Interregional exports of manufactured goods (principally textiles) to the Midwest from New York City over the Erie and from Philadelphia over the Mainline remained small before the mid-1840s. These goods reached an annual average of 50,000 tons (31,000 for the Erie; 19,000 for the Mainline) between 1840 and 1844, an unimpressive total considering that wagons carried 30,000 tons annually to Pittsburgh over the Pennsylvania Turnpike from 1818 to 1824. Wagons carried an indeterminate, but sizable, amount over turnpikes after the canals opened and continued to do so into the 1840s.[13] Therefore, interregional canals exerted little impact on eastern agricultural specialization and did not begin to meet promoters' expectations of purchasing agricultural goods from midwestern farmers or selling manufactured goods to them until the mid-1830s, and substantial trade waited another decade.

Traffic patterns on the Erie Canal system—including the major lateral canals—demonstrate that interregional canals primarily served intraregional markets in the East, not trade with the Midwest, until the early 1850s (see table 5.2). The tonnage being traded within New York State (i.e., instate clearings) dwarfed the amounts being traded with the Midwest through the early 1840s, and by the early 1850s New York's internal trade exceeded its traffic with the Midwest by almost 60 percent. Shipments destined for tidewater (via New York City) grew dramatically after 1843, but even that trade of heavy resource products to urban markets or for export constituted only half of the total tonnage on the Erie Canal system until the early 1850s. The Pennsylvania Mainline and lateral canals carried little trade with the Midwest (see fig. 5.3). Most of the goods remained in the state or were exchanged with other eastern markets. Midwestern exports over the Erie and Pennsylvania Canal systems point to the mid-1840s, when the economy began to recover from recession, as the start of extensive exchange between Midwest and East. Other evidence on interregional trade flows involving the East confirms these conclusions: by 1839 domestic and foreign exports from the East accounted for only 10 percent of its income, and this rose to only 15 percent by 1860.[14] Prior to the mid-1840s transportation improvements within the East primarily impacted eastern regions. Paradoxically, many of the canals and railroads lost money or had low returns on investment, implying that they had neither simple nor obvious impacts on eastern economic development.

The "Spirit of Improvement"

The "spirit of improvement," a movement motivating efforts to construct turnpikes, canals, and railroads, promoted the exercise of public spirit and the capture of immediate economic gains, which participants in the transportation debates translated into distinctions between developmental and exploitative projects. Frontier settlement gave transportation advocates justification to label many projects "developmental": residents neither generated sufficient traffic from existing economic activities to pay for projects nor accumulated enough capital to fund them, but, after they were completed, increased settlement and development created economic gains that justified the costs of the projects. Because private investors could not readily appropriate deferred, diffuse gains—such as rising land values, increased tax revenues, and wider economic opportunities—transportation improvements required government support as public enterprises or as private ones with public support. Several thousand projects—including turnpikes, canals, and railroads—received state and local government support; purely public enterprises were infrequent. In contrast, investors built exploitative projects to serve existing trade routes or to link commercial centers, when their calculations showed that revenue from traffic ensured profitable returns on their invested capital.

Although developmental and exploitative justifications seem incontrovertible, they mask dilemmas. Project losses may exceed diffuse economic gains, thus negating developmental justifications, and both rationales assumed that existing patterns of regional development were the only possible ones, whereas alternative transport modes might generate different patterns and equivalent economic growth rates. Advocates of developmental and exploitative projects might underestimate competition from alternative transport modes, such as when wagon teamsters forced canal companies to keep tolls low, reducing canals' profit margins or causing losses. Analyses of projects' social savings—the extent that the costs of transport services fell below the costs for the best alternative transport mode—permit an assessment of transport projects' impact on eastern economic development.[15]

Waterway and Wagon Alternatives to Canals and Railroads

The navigation of sailing vessels on inland rivers, bays, and protected coastal waterways in the East posed formidable competition for canals to transport low-value, bulky commodities. The high costs of steamboats meant that they primarily carried passengers and high-value freight, and steamboat operations focused on New York City, the pivot of steam navigation including New York harbor, nearby New Jersey, Long Island Sound, and the Hudson River to Albany; steamboat service also ran on the Delaware River to Philadelphia and Trenton and on the Chesapeake Bay. Arks, flatboats, and rafts on inland rivers continued generating strong competition for canals to transport low-value, bulky products to metropolitan markets. Fierce wagon competition posed more formidable difficulties for canals and railroads, because wagons captured high-value freight, the most lucrative to transport. The flexibility of wagons over the fixed routes of canals and railroads, low fixed costs, and requirements that wagons transport all goods to and from canals and railroads meant that wagons dominated transportation within the inner hinterlands of East Coast metropolises—the areas of heaviest demand for transport services. Before 1860 canals and railroads successfully carried low-value, bulky freight only if they ran for long distances through prosperous agricultural areas or rich resource areas (e.g., mining and lumbering).[16]

The Impact of Canals on the East

Canal construction incurred heavy costs; twenty-one out of twenty-nine canals each cost over $1 million to build, including the famous Erie ($7.1 million), Pennsylvania Mainline ($7.2 million), and Chesapeake and Ohio ($10.1 million) and the lesser-known Genesee Valley ($5.7 million). Per-mile building costs averaged $29,000, about twenty-five times typical turnpike costs and triple those of the best turnpikes with stone and gravel bases, and labor-intensive work and lock construction in hilly terrain boosted these costs. From

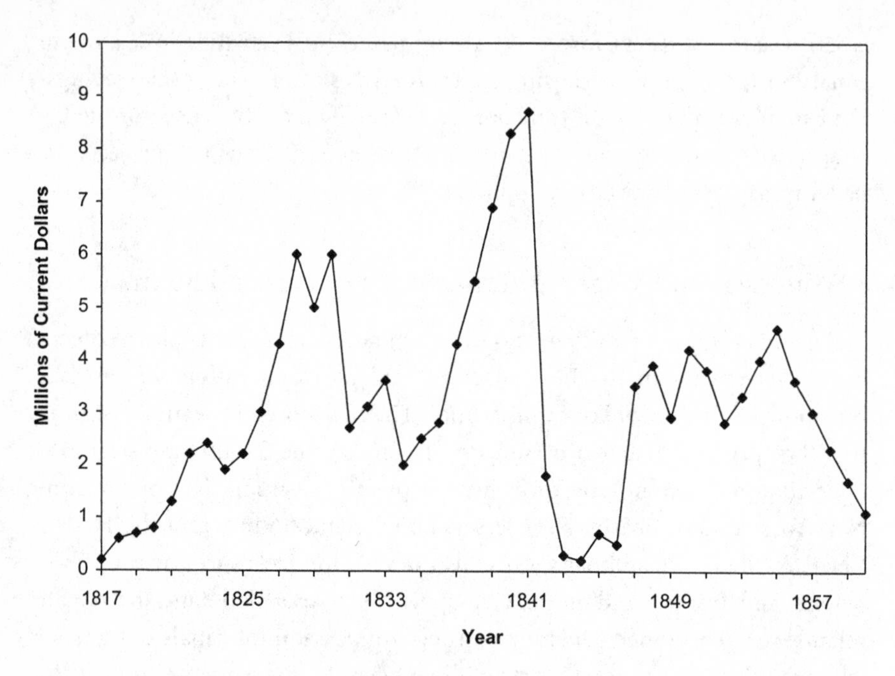

Fig. 5.4. Annual Canal Investment in the East, 1817–1860. *Sources:* Cranmer, "Canal Investment, 1815–1860," 24:555–56, table 3; Segal, "Cycles of Canal Construction," 208–10, table 2.

1817 to 1860 eastern construction tracked three waves corresponding to national waves, because the East accounted for 72 percent of the $188 million invested in canals (fig. 5.4).

The first wave matched the long economic expansion that occurred between 1820 and 1836, and expenditures on the Erie and Champlain Canals in New York accounted for 60 percent of the cumulative $12.3 million spent on canals through 1825, when both were mostly completed. By 1834, at the end of the first wave, construction finished on all future heavily utilized canals; the total cost of $48 million was 36 percent of the East's investment through 1860. These canals were ready to support almost three decades of economic development, and, if eastern canal building had ceased after the first wave and future expenditures had been kept to improving existing canals, the public and private sectors would have avoided huge losses. The second wave crested in 1841, and the third wave commenced several years after the economic recovery began in 1844, but expenditures never reached previous levels; much of the money, except for the Erie's enlargement, funded the completion of projects started during the second wave. Delays in completing the Erie expansion meant that New York State failed to gain benefits when they would have contributed the most to its economy, and by the 1850s railroads started offering competition to canals (see fig. 5.3). Investments in the Genesee Valley and Chesapeake and Ohio canals—both noteworthy debacles—brought a fitting end to eastern canal investments that should have ceased, except for improvements, in 1834.[17] In every eastern region lowly wagons exerted withering competition; few canals survived, as social savings evaporated.

Wagons Defeat the New England Canals

Southern New England's canals—Middlesex, Blackstone, and Farmington (with an extension)—traversed prosperous farmland and were financial debacles, thus demonstrating the wagon's capacity to dominate transport of most goods, except low-value, bulky goods (e.g., wood and stone for construction and cordwood for fuel), within about one hundred miles of regional and subregional metropolises (map 5.1). Construction of the Middlesex Canal from Boston to the future industrial city of Lowell (twenty-seven miles northwest) began in 1794 and finished in 1803. Its initial cost of $530,000 ($20,000 per mile) ranked low compared to later multimillion dollar canals, but its total represented a small down payment to the final bill. By 1814 a seventy-eight-mile navigation route opened for canal boats between Boston and Concord, New Hampshire. Boston's Brahmans furnished leadership as investors, managers, and engineers in the enterprise known as the Proprietors of the Middlesex Canal. These judges, politicians, and officers of Harvard College (but few merchants) grasped the opportunity to make a profit, yet decisions over the canal's long history imply that public spirit dominated. They provided superb management and upgraded and maintained the canal in excellent condition until its demise during the 1850s. By 1818 owners' cumulative investment totaled $1.2 million, including their original capital and a return of 6 percent compounded which they had not received. That total compounded at 6 percent until 1853, minus payments to investors of about $0.5 million, raised total losses over $8 million—assuming that no shares changed hands; their original investment was a gift to canal users, and they had foregone a fortune.[18]

From 1815, marking the opening of navigation on the upper Merrimack River to Concord, until the mid-1830s, the Middlesex Canal and river extension into New Hampshire captured most of the downward transport to Boston of low-value, bulky commodities—stone, cordwood, boards, shingles, and

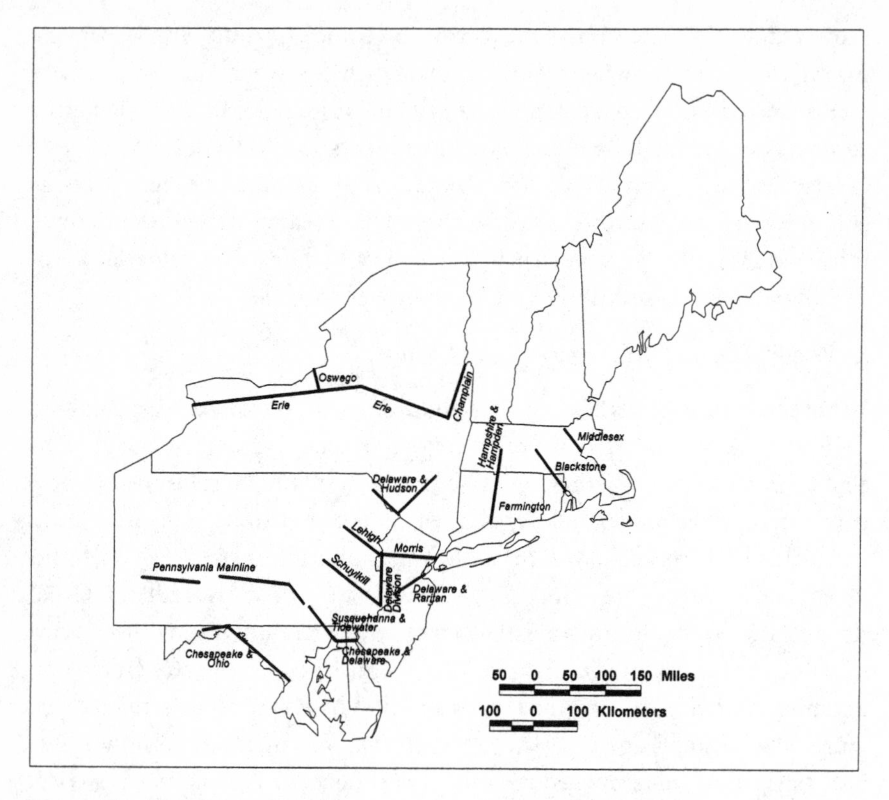

Map 5.1. Leading Eastern Canals

empty barrels—produced in the Merrimack basin. It also dominated the upward shipment of salt, lime, manure, and plaster and by 1825 of bituminous coal and by 1830 anthracite coal. With the opening of textile mills in Lowell in 1823, the canal competed effectively to transport raw cotton. The Middlesex faced fierce competition from wagon owners, however, on carrying agricultural products—including barreled beef, cider, butter, cheese, and ashes (for fertilizer)—and manufactures (e.g., textiles) to Boston and groceries, manufactures, and fish inland; these shipments always constituted tiny shares of canal tonnage. The successive opening of railroads squeezed freight from the canal, and toll receipts fell from a peak of $45,500 in 1833 to $1,200 by 1851. Throughout its operation the canal lowered toll charges, first, in competition with wagon teamsters and, later, in competition with teamsters and railroads. The freight rates of private companies running boats on the Middlesex Canal and the Merrimack River between Boston and Concord testify to the harsh competition canals faced from wagons; rates including canal tolls and warehouse costs along the waterway and excluding truckage such as delivery to any place in Boston declined from ten to fifteen cents per ton-mile around 1816 to five to six cents during the early 1830s. Adding wagon charges at origin and destination to canal and river freight rates raised the total near the rates of fourteen to eighteen cents per ton-mile which teamsters and other hired haulers charged.

Canal management recognized its enemy; in 1819 a company report lamented: "a great proportion of the trade between Town and Country is still done on *wheels.*" In 1822 James Baldwin, a canal superintendent, analyzed the competition that wagons posed; he stressed that wagon teams took goods directly between origins and destinations, without the expense of transshipment between the wagon and the canal at both ends, and all-wagon transport permitted smoother control over flows of goods. In contrast, canal shipment and transshipment raised coordination and time costs at the origin and destination, and, because transshipment required the frequent handling of goods, this increased damage and raised insurance costs. Baldwin's incisive analysis portrayed the canal company as helpless before the logic of wagon transport in Boston's inner hinterland.[19]

The Blackstone Canal opened in 1828, connecting Providence, Rhode Island, and Worcester, Massachusetts. Providence merchants, led by the Brown and Ives families and other local wealthy families, oversubscribed the first $400,000 of capital, and in 1831 the canal company raised an additional $150,000. Costs per mile ($13,000) placed it among the cheapest canals, but the total investment of about $700,000 by the mid-1830s, when its failure as a paying concern became evident, translated into large opportunity costs for investors. During its operating years, from 1828 to the early 1840s, the canal carried cordwood and coal to Providence and cotton, gypsum, salt, flour, molasses, and oil inland. Manufactures for interior consumers, textiles from numerous Blackstone Valley mills, and farm products forwarded to Providence remained conspicuously absent from canal traffic; the forty-five-mile waterway never competed with wagons for that trade. The failure of the Blackstone Canal demonstrated that even astute merchants such as Brown and Ives misunderstood transport economics and the developmental impacts of canals on port hinterlands.

Leading promoters of the canal from New Haven to the Massachusetts border, the Farmington Canal, and its extension to Northampton on the Connecticut River, the Hampshire and Hampden Canal, resembled the proprietors of the Middlesex Canal. Prominent project leaders included Simeon Baldwin, a noted Connecticut jurist, and James Hillhouse, the treasurer of Yale College. Similar to the Blackstone, financing commenced in euphoria; major investors resided in towns along the canal, and wealthy New York City residents subscribed $200,000. An additional $300,000 of capital came from the purchases of canal stock by two banks, whose new state charters required support of the canal. Construction started in 1825, and the canal finally reached Northampton in 1835; an unmitigated failure, the canal attracted little traffic, and by the mid-1840s it ceased operations. At an initial construction cost of $830,000, the debacle exceeded the costs of both the Middlesex and Blackstone Canals. Similar to the other canals, talented social, political, and economic elite failed to recognize that prosperous farmers and manufacturers along the canal's route used wagons to transport products cheaply to markets and to waterway shipping points; no producers of large supplies of low-value, bulky products existed near the canal.[20] Investors in this canal and the Blackstone decided to build while the Erie Canal neared completion and its success seemed assured; nevertheless, the Erie, seemingly the epitome of success, raised a conundrum (see map 5.1).

Did New York State Need the Erie Canal?

Modern analysts agree with contemporaries that the Erie—the definitive "developmental" canal—exerted powerful economic impacts on New York State, but this claim assumes that the state would not have developed economically, at least to the same extent, in the absence of the canal. This assumption has weak support. With the settlement of western New York after 1800, farmers forged increasing ties to Montreal merchants, thus threatening those in New York City. That merchant competition infused discussions of proposals for a canal that either used Lake Ontario as a passage or followed a land route south of the lake, and early completion of the Erie's middle section between the Mohawk and Seneca Rivers entailed an attempt to undercut Montreal. Even if western New York farmers did not need the Erie to reach market outlets, they could support the canal—if the entire state shared costs—and acquire a cheap benefit. Moreover, they gained access to a second low-cost transport route—besides via Montreal—to New York City, forcing both cities' merchants to compete for western New York's business. The support of counties in central New York from Utica to Seneca Falls, like counties west of Seneca Falls, exuded a self-serving interest, because they accessed Lake Ontario ports by wagon or could access them through the construction of a short canal; the thirty-eight-mile Oswego Canal was built from 1825 to 1828 (see map 5.1). Before 1820 residents of central and western New York had legitimate concerns about Montreal's commercial liquidity and its merchants' capacity to market flour in New York City and overseas. Nevertheless, as wheat production soared in the Genesee country of western New York, Montreal merchants would have rectified marketing inadequacies and the British government would have enhanced exchange earlier through Canada; open trade existed between the United States and Canada by the early 1830s. Astute farmers near Utica and downstream along the Mohawk River shifted into cheese before the Erie was finished, because they could send that high-value good by wagon to Albany; they gained little from the Erie, unless others covered much of the cost. Among the Erie's opponents, the New York counties along Pennsylvania's border recognized wagon transport to the canal negated its benefits compared to the costs of shipping farm products down the Susquehanna River to Baltimore and Philadelphia. To gain their support, leading promoter DeWitt Clinton promised lateral canals—a financial time bomb—and a road across southern New York. Farmers in the Hudson Valley and on Long Island opposed the canal because they gained nothing from it and, instead, would meet increased competition for New York City's market. Wealthy landowners in the Hudson Valley and New York City elite speculating in land near the canal route in central and western New York supported the Erie. Thus, its construction represented less a project driven by necessity than a political compromise permitting farmers within about one hundred miles of the canal to access a low-cost channel underwritten by the state's credit. Their transport benefits exceeded their costs as taxpayers, but many farmers had alternative access to markets under different scenarios.[21]

The Erie's early success seems to support the claim it was indeed a developmental canal, because it stimulated the settlement and growth of central and western New York; even recognizing, however, that some growth anticipated the canal's completion in 1825, this interpretation dismisses too readily the rapid retardation of population growth. West of Utica the regional population soared 70 percent, to 256,047, during the 1810s, and then growth slowed to 55 and 26 percent, respectively, during the 1820s and 1830s, just when the developmental effects should have boosted growth rates, and population change in Oneida (including Utica) and Onondaga (Syracuse), key counties along the canal route, followed similar patterns. The textile mill built by some of the nation's leading machinists and mill builders near Utica in 1809, eight years before construction started on the Erie, testified that central New York already was on a high developmental trajectory. The initial traffic surge on the canal's middle section between Syracuse and Utica between 1820 and 1825 probably represented the diversion of traffic from the Lake Ontario–St. Lawrence route to Montreal and from wagons either traveling to Albany or transshipping to boats at Utica for the Mohawk River route to Schenectady and then by wagon to Albany.

From 1820 to 1840 canal shipments on the Erie mostly remained within the state: cordwood, flour, and wheat moved to the tidewater; and merchandise manufactures, especially textiles, and imported foods such as sugar moved inland. The Champlain Canal carried simple traffic—lumber and timber—to the tidewater; few goods moved inland (see map 5.1). These canals met conditions for success: they carried huge volumes of low-value, bulky commodities to distant markets, and canals could compete with wagons to transport them. Other states tried to emulate the Erie's successful model, but their efforts mostly failed because decision makers did not understand the critical conditions; they can be excused, because leading participants in the Erie project also failed to grasp them. Benjamin Wright, the chief engineer for the Erie's middle section, surveyed the routes of the Blackstone and Farmington canals during 1822 to 1824, and in 1828 DeWitt Clinton, a leading advocate for the Erie after 1815, toured the route of the Farmington; these eminent personages pronounced the imminent financial debacles suitable projects that offered reasonable returns on investment.[22]

Pennsylvania Builds an Uncompetitive Canal

The start of construction on the Erie in 1817 goaded Pennsylvanians into action, but protracted discussion ensued until 1824, when the legislature finally appointed commissioners to survey the canal routes. As the Erie neared completion, justifications for building a canal between Philadelphia and Pittsburgh crystallized: proven success of the Erie ensured that Pennsylvania's canal would succeed, threats from New York City and Baltimore to Philadelphia's commercial competitiveness required immediate action, and the canal would benefit all sectors—farmers, merchants, and manufacturers. Debate raged over building a railroad instead of a canal, calling into question its financial feasibility, but canal proponents drowned the opposition's logic by appealing to broad developmental benefits, even if the canal proved unprofitable. This assumption's fallacy loomed: if the canal experienced losses and had little developmental impact, the state faced disastrous losses.

The legislature's decision in 1827 to build the Mainline and a set of lateral canals resulted from sectional disagreements over the project and represented crass political "logrolling." Opposition by residents in southeastern Pennsylvania—the state's richest farmland and within one hundred miles from Philadelphia and Baltimore by wagon—should have warned proponents that they held dubious assumptions about instate canal traffic. The inclusion of lateral canals represented a blatant attempt to attract voters of the Susquehanna Basin north of Harrisburg; although farmers could use the river to transport goods, the entire state subsidized canals. By 1834 the completed Mainline stretched 395 miles and evolved into a hybrid monstrosity of railroad, portage railroad, and canal sections costing $12.1 million ($31,000 per mile), over 50 percent greater than the Erie's per-mile cost; lateral canals added $4 million more (see map 5.1). The Mainline confronted insurmountable challenges to compete with the Erie and still repay total debt, and freight companies faced much higher costs because rates covered the maintenance of five transshipment depots (i.e., the costs of warehouses, supervisors, and laborers), whereas Erie freight companies needed two to three depots.[23]

Evaluation of the Mainline system requires answers to two questions: did it contribute to Philadelphia merchants' competitive position vis-à-vis those in New York City and, secondarily, Baltimore; and did its contributions to Pennsylvania's economic development exceed the financial losses? When the Mainline opened the spigot to allow traffic to flow from Pittsburgh to Philadelphia, it trickled through, and midwestern tonnage remained a small fraction of the Erie's (see fig. 5.3). Midwestern shippers without access to the Erie sent products down the Ohio-Mississippi Rivers to New Orleans and then by ship to eastern ports. The expansion of Philadelphia's intraregional trade over the Mainline eventually compared favorably, though still smaller, to the Erie's impact on intraregional trade involving New York City between 1845 and 1849, just prior to the Mainline's demise (table 5.3). Aggregate tonnage figures, however, overstate its benefits: coal constituted the majority of tonnage sent to Philadelphia during the 1840s, and by the end of that decade it accounted for two-thirds of tonnage; most of this trade originated on lateral canals. Nevertheless, coal carried by the Mainline and lateral canals accounted for less than 10 percent of Pennsylvania's coal production, because specialized coal canals carried most of the coal to markets.

Table 5.3. Intraregional Trade on the Erie and Pennsylvania Mainline Canals, 1835–1849

	Average Annual Tonnage (thousands)			
	Erie Canal		Pennsylvania Mainline Canal	
	To New York City	From New York City	To Philadelphia	From Philadelphia
1835–39	294	90	65	40
1840–44	350	103	169	46
1845–49	583	174	412	127

Source: Ransom, "Interregional Canals and Economic Specialization in the Antebellum United States," 15, table 2.

Pennsylvania's canal system compiled horrendous losses, as heavy interest costs on debt dragged the system toward insolvency; by 1840 Governor Porter acknowledged its failure. The legislature approved extensive taxes to cover annual revenue shortfalls in 1844, and a movement started to dispose of the canals, but that effort did not finish until 1857. At that date expenses totaled $101.5 million, and citizens paid taxes to cover a loss of $57.7 million. Pennsylvania faced the worst case: state canals carried little traffic and thus did not benefit Philadelphia vis-à-vis New York City and Baltimore; canals had limited developmental impact; and they incurred appalling financial losses that threatened the state's solvency. Philadelphia's merchants vastly overestimated how soon midwestern commodities flowed directly east in large quantities; they thought this would occur by the late 1820s, but it lagged until the mid-1840s. Coal, the only product to move over state canals in large quantities, had alternative outlets over specialized coal canals which easily accommodated shipments at minimal marginal cost. And canal proponents underestimated the competition from wagons for transporting commodities from southeastern Pennsylvania's prosperous farms to markets.[24] Unfortunately, Philadelphia's merchant competitors in Baltimore made the same mistake.

Maryland Squanders Its Public Revenues

By the early 1820s Baltimore merchants secured control over a productive agricultural hinterland extending to western Maryland and north about seventy-five miles into Pennsylvania. Baltimore's radial network of seven turnpikes provided efficient wagon transport for retailers, farmers, and other resource producers to ship commodities to Baltimore and to obtain manufactures and other items. Grandiose dreams of Baltimore merchants to capture the trade of the West and Susquehanna Valley to New York's border overcame their initial skepticism of the value of a canal to the West, and they supported the Maryland legislature's decision to build the Chesapeake and Ohio Canal (see map 5.1). That project, along with the Susquehanna and Tidewater Canal, wasted public tax revenue and provided few gains to hinterland trade which merchants already dominated through wagon transport. Proponents of the Chesapeake and Ohio Canal from Georgetown to Cumberland, Maryland, justified it, as with other eastern canals, saying it would reduce transport costs to one-tenth that of roads. This diverted attention from more appropriate contrasts between the canal's cost—for example, in capital investment and the interest on borrowed money—and the Potomac Company's works. It made river improvements—such as channel clearing and the construction of short canals around the falls of the Potomac—from 1784 to the early 1820s and provided economical downriver transport for bulk commodities. Canal proponents also failed to consider that farmers calculated wagon transport cost to the canal as part of total cost and then compared that with wagon transport directly to Georgetown or Baltimore.

Construction commenced at Georgetown as a grand public works project in 1828 and continued, without any rational economic reassessment, until reaching Cumberland in 1850. The directors constructed it at over twice the dimensions (i.e., width and depth) as other major canals and to the highest engineering specifications possible, producing an overpriced, deluxe canal. Maryland took over the project in 1836, contributing about half of the final cost of over $10 million; per-mile cost ($55,000) ranked among the nation's highest. It needed to generate tolls of over $600,000 annually (6 percent interest) to pay interest charges on debt, excluding operating expenses and retiring debt, but toll revenue averaged only $139,626 annually between 1851 and 1860.[25] It never climbed out of financial disaster, and its developmental impact on the Potomac River Valley and broader gains to Maryland and Virginia remained the only potential benefit; on that account it failed miserably.

Wheat farmers, the leading commercial crop producers, should have received the most benefits. From 1842 to 1860 the canal typically carried 150,000 to 200,000 barrels of flour annually, but that did not substantially exceed the peak of 118,222 barrels carried on the Potomac River in 1811. The Potomac Company invested less than one-tenth as much in river improvements as the canal company's investment, while providing similar service, and modest additional investment could have provided the capacity to handle equivalent shipments as the canal. The canal set toll rates below the costs of operations, interest, and the amortization of debt; rates on wheat and flour stayed at two cents or less per ton-mile during the 1840s. In 1842 the canal carried the equivalent of 199,648 barrels of flour, but this amounted to only 39 percent of flour sold outside the Potomac Valley. Most of the flour probably moved by wagon; thus, the canal had little developmental impact on wheat farming. After completion of the canal to Cumberland in 1850, coal producers west of the city accessed the canal, and it quickly transmogrified into a coal canal. Nonetheless, it had dubious developmental consequences, because it charged a heavily subsidized rate of a quarter-cent per ton-mile. Between 1853 and 1860 the highest annual coal tonnage constituted only 24 percent of maximum coal tonnage shipped over the Lehigh Canal.[26] This profitable, privately owned canal in eastern Pennsylvania could have served the coal demands of the Chesapeake region by itself; coal extraction in western Maryland survived because the canal offered subsidized rates. Maryland citizens did not stop with this debacle.

Baltimore merchants were leading advocates of a canal along the west bank of the Susquehanna River—bypassing its inadequate channel—to connect Pennsylvania's state canal works at Columbia on the Susquehanna with Havre de Grace on the Chesapeake Bay. The canal was their weapon to capture the trade of the Susquehanna Basin, and agricultural counties in Pennsylvania along Maryland's border and lumber and coal businesses in the upper basin offered support. Construction started on the forty-five-mile Susquehanna and Tidewater Canal in 1836 and finished in 1839; per-mile cost ($72,000) made it the second highest in the East, raising a formidable barrier to profitability (see map 5.1). Tolls exceeded estimated interest costs in only two years between 1840 and 1860, and accumulated deficits probably reached as high as $5 million by 1860.

These horrendous losses might be excused if the canal had offered an avenue for high volumes of goods with no other way to market, but it provided doubtful benefits. Shippers took advantage of low canal tolls to ship wheat instead of flour, and the canal transmogrified into a coal carrier as shippers enjoyed heavily subsidized rates at the canal owners' expense. Before 1860 peak coal traffic reached about 230,000 tons annually, an amount easily transported by profitable anthracite canals in eastern Pennsylvania. Baltimore merchants also advocated the canal as an avenue to send merchandise north into the Susquehanna River Basin, thus capturing business from Philadelphia merchants; tonnage, however, remained trivial.[27] Baltimore merchants and their Maryland allies consistently underestimated their grip on Maryland's trade and that of the border agricultural area in Pennsylvania. These farmers possessed convenient wagon access to Baltimore, and wagons provided efficient return shipments of manufactured goods. The Upper Susquehanna Basin always remained problematic as a trade area because Philadelphia captured the most lucrative trade through transportation projects northwest of the city, and New York City forged transport tentacles from the Hudson River to the basin. Therefore, canal efforts remained useless for Baltimore's inner hinterland and uncompetitive for the distant river basin.

Canals along the East Coast Corridor

Metropolitan business leaders recognized that they needed to lower transport costs for low-value, bulky commodities moving in large volumes along the transportation corridor between Boston and Baltimore. The Boston–New York leg used the Atlantic Ocean and Long Island Sound, whereas the New York–Philadelphia leg required circuitous trips around southern New Jersey to the Delaware River; the Philadelphia-Baltimore leg needed an even longer detour around the Delaware peninsula to Chesapeake Bay. Construction of the Chesapeake and Delaware Canal commenced after 1823 (see map 5.1); Pennsylvania's government and leading Philadelphia citizens contributed most of the initial capital, followed by the federal government, Delaware, and Maryland. Philadelphia merchants believed they gained because the canal would allow them to compete better against Baltimore merchants for Susquehanna River trade reaching Havre de Grace on the Chesapeake Bay. Completed in 1829, the fourteen-mile canal had the dubious honor of being the East's costliest per mile ($196,000), and it accumulated enormous unpaid interest because revenues stayed low until the late 1840s. Most canal commodities consisted of wheat bound for Brandywine flour mills near Wilmington, Delaware, and flour and lumber; by the mid-1850s coal constituted about one-third of total tonnage, but coastal ships could easily have handled these commodities.

The New York–Philadelphia leg had greater traffic potential than the Philadelphia-Baltimore leg, and in 1830 New Jersey's legislature passed charters to build the Delaware and Raritan Canal between New Brunswick on the New York Harbor and Bordentown on the Delaware River and to build the Camden and Amboy Railroad parallel to the canal (see map 5.1). After the canal opened, in 1834, tonnage and toll revenue rose rapidly, but the sixty-five-mile canal cost $44,000 per mile, one of the East's costlier canals; this resulted from high land costs in this long-settled area. Annual toll revenue did not match the interest costs of $170,400 (6 percent interest) until the mid-1850s, when surging trade caused tolls to soar. It carried bulk commodities such as lumber, but coal from the anthracite fields of northeastern Pennsylvania dominated; tonnage reached about 1.3 million by the late 1850s. Merchandise, including manufactures, also surged, and tonnage rose from 100,890 in 1850 to 227,394 in 1860. The Delaware and Raritan Canal was a linchpin in supplying coal to New York and New England and a channel for eastern interregional trade after 1850, but it was built a decade too early.[28] Success of the anthracite canals underscored that canals needed to carry huge volumes of low-value, bulky goods to achieve profitability.

Exceptional Anthracite Canals

By 1820 many of the technical problems of using anthracite coal were solved, but coal transported by wagon from anthracite fields could not compete with coal from Richmond, Virginia, or from England; the solution—coal canals. The Lehigh (on which coal traffic began in 1820), Schuylkill (1826), and Delaware and Hudson (1829) reached into the anthracite fields and formed the core of the canal network. Philadelphia provided the first market in the early 1820s, and transshipment to coastal vessels there reached other East Coast markets. Completion of the Delaware and Hudson, with direct ties to Hudson Valley and New York City markets, and of the final links—the Delaware Division (1830), the Morris (1831), and the Delaware and Raritan (1834)—supplied direct, multiple links to Philadelphia and New York City, the largest markets (see map 5.1); coal dealers had low-cost access by coastal vessels to all East Coast markets from them. Coal tonnage on the Lehigh, Schuylkill, and Delaware and Hudson rose rapidly; collectively, shipments climbed from 175,984 tons in 1830 to 3,015,340 in 1859. The per-mile cost ($18,000) to construct these privately owned canals placed them near the bottom in cost, and huge coal volumes generated soaring toll revenue, making them highly profitable.[29] They possessed the capacity to supply the entire East Coast market; thus, the transmogrification of the Mainline, Chesapeake and Ohio, and Susquehanna and Tidewater into coal canals did not offer a retroactive justification for these overpriced canals.

A Developmental Mirage

The Erie, Champlain, and anthracite coal canals constituted the chief exceptions to the sorry squandering of public and private resources on eastern canals, and there is no evidence that planning at the national level would have succeeded either.[30] Successful canals shared common attributes: builders kept construction costs low, canals exploited regions with large supplies of bulky, low-value commodities, and wagons could not compete. The siren claim that canals generated developmental benefits not reflected either in canal traffic or toll revenue lured governments and private individuals into most canal failures, but few, if any, examples exist of consequential development from money-losing canals. Southern New England canals penetrated rich agricultural and industrial areas where wagons hauled goods cheaply over short distances to ports and other distribution points. Other canals—the Mainline's eastern portions and the Susquehanna and Tidewater—offered subsidized, low-cost transport for areas with alternative transportation (e.g., rivers) or alternative production sites and transportation (e.g., anthracite coal canals) supplied markets. Some money-losing canals—the Chesapeake and Ohio and the lateral canals of the Erie and the Mainline—encouraged the development of marginal agricultural areas in Appalachian valleys whose coal and lumber production, existing only because canals offered heavily subsidized rates, competed with other areas—such as anthracite coal mining in northeastern Pennsylvania and lumbering in the Adirondacks and Maine—supplying ample amounts. Successful canals exploited existing regions of settlement and resource extraction and permitted rapid production increases. By the time railroad construction commenced during the mid-1830s, the developmental rationale retained little substance, because few areas of the East fit the appellation of frontier settlement; therefore, railroads faced competition from existing transport by waterway (e.g., river, coastal, and canal) and road (e.g., wagons and stagecoaches).

The Impact of Railroads

Railroads did not acquire technology to transport large volumes of low-value, bulky commodities long distances competitively until after 1860. Large fixed costs and moderately high variable costs, coupled with speed and high-quality service, generated competitive advantages by moving passengers and high-value, low-bulk freight; railroads needed large volumes to cover costs, however, and the demand for the long-distance shipment of passengers and high-value freight remained small until the late antebellum. Inner hinterlands of major East Coast ports and trade areas of large interior cities housed high-density rural and urban populations and produced large quantities of agricultural and industrial goods, fuel, and construction materials. If railroads attracted substantial volumes, they might achieve profitability, but stagecoaches and wagons offered formidable competition for passengers and freight. These vehicles possessed greater flexibility than railroads—which followed fixed tracks—and passengers and freight traveling short distances on railroads in the inner hinterlands still utilized stagecoaches and wagons at origins and destinations, adding two transshipment points for every trip. Therefore, railroads contributed minimal direct economic benefits until the late 1850s, and their social savings—aggregate savings in transport cost compared to the best alternative mode—remained a small share of the economy.[31]

A Minor Impact on Industry

Because railroads require substantial inputs—including the construction of rights-of-way, tracks, equipment, and machinery—they provide backward linkages to industry. Preparation of rights-of-way constituted about 40 percent of total costs, but expenditures mainly covered the employment of unskilled hand labor. Iron rails and equipment amounted to about one-fifth of costs before 1840 and about one-third afterward; however, the domestic production of rails remained trivial before 1849 and accounted for only 15 to 20 percent of rail consumption during the early 1850s. Railcar components of equipment mostly involved wood craftwork, and locomotives amounted to under 10 percent of machinery production as of 1860; therefore, backward linkages of railroads generated minimal stimuli to antebellum industrial growth. Railroads offered forward linkages to industry by providing transport services; railroads could stimulate factory growth along rail lines and cause factories without service to decline, and evidence for this influence should appear in New England, the center of eastern industrial growth and early railroad expansion. Many cotton textile mills appeared long before railroads reached their sites, however, and in Connecticut metal manufactures west of the Connecticut Valley expanded years before railroad construction commenced; some industrial areas did not obtain railroads until the late 1840s. Massachusetts' shoe industry remained concentrated within forty miles of Boston, easily accessible by wagon transportation, and, although railroads dominated shoe shipments by the late 1850s, their high-value, low-bulk character made rail transport suitable, not essential. Thus, before the late 1850s railroads exerted little impact on New England's industrial growth.[32]

Why Build Railroads?

If antebellum eastern railroads generated minimal direct economic benefits and exerted little impact on industrialization, why build railroads? Governments (states, cities, and towns) claimed railroads as a public good, and, because railroads (and canals) provided benefits mostly to people along their lines, public funding even in the face of limited financial returns suggests that these expenditures became embroiled in political deals as much as in appeals to the broader body politic. Astute, wealthy investors—including leading industrialists, merchant wholesalers, commodity dealers, and financiers in Boston, New York, Philadelphia, and Baltimore—rarely committed much capital to railroads before the 1850s; their reticence spoke volumes about limited railroad profitability. Commonly, rich investors contributed $20,000 or less in many railroads—small fractions of their wealth—and often quickly sold shares. The New York City and Hudson Valley elite who subscribed to $2,321,000 of stock in the Mohawk and Hudson Railroad in 1826 held only $669,000 in 1831, the year it opened for service; none held more than $18,000 of stock. Except for Erastus Corning, no leading financiers such as Edwin D. Morgan, Cornelius Vanderbilt, and Russell Sage who participated in the New York Central Railroad after 1850 invested much capital in the early railroads eventually making up the line. Even the fabled Boston Associates kept capital contributions modest compared to textile investments, and they reduced their participation in railroads after flurries of promotional activity; their heavy funding of the Boston and Lowell Railroad and the firm control they exerted over it was an exception, rather than the rule. Boston's wealthy, such as the China traders, spread investments across eastern railroads, but they kept financial risks limited.[33]

Antebellum railroad funding followed a script—it spread risks among numerous investors; few harbored delusions that they would earn more than 6 percent, the equivalent of returns on safe government securities; and many expected limited returns or total losses. Repeated railroad investments made sense only if viewed as investor strategies to risk small individual capitals but to pool them into large sums and hope for indirect benefits to business. With few exceptions, private investors in the large metropolises came from merchant wholesaling, commodity brokerage, banking, insurance, and manufacturing, and they invested in radial railroads binding hinterland places to the metropolises. Outside these large urban areas investors occasionally came from the farming ranks, hoping for rising land values, but most lived in cities and towns

Table 5.4. Railroad Mileage and Passenger and Freight Revenues, 1839–1860

	Mileage			Revenues ($1,000)					
				1839		1849		1855–56	
	1840	1850	1860	Passenger	Freight	Passenger	Freight	Passenger	Freight
Maine	11	217	473	$10	$10	$149	$32	$761	$721
New Hampshire	34	439	658	0	0	486	547	672	1,216
Vermont	0	296	562	0	0	25	32	568	949
Massachusetts	277	1,077	1,284	696	338	3,379	2,590	4,741	3,977
Connecticut	128	442	604	20	10	503	471	1,923	1,145
Rhode Island	49	59	107	74	20	118	57	157	98
New England	499	2,530	3,688	800	378	4,660	3,729	8,822	8,106
New York	463	1,387	2,677	977	61	3,232	2,473	8,560	9,690
New Jersey	192	234	539	661	272	1,402	622	2,118	992
Pennsylvania	667	1,082	2,662	739	618	1,362	2,882	4,055	10,835
Delaware	36	36	127	—	—	—	—	—	—
Maryland	242	278	370	430	283	645	1,066	1,013	3,835
Middle Atlantic	1,600	3,017	6,375	2,807	1,234	6,641	7,043	15,746	25,352
East	2,099	5,547	10,063	$3,607	$1,612	$11,301	$10,772	$24,568	$33,458

Sources: Baer, *Canals and Railroads of the Mid-Atlantic States, 1800–1860;* Fishlow, *American Railroads and the Transformation of the Ante-Bellum Economy,* 322–23, 326–28, tables 41–43; Taylor, "Comment," in *Trends in the American Economy in the Nineteenth Century,* 24:526–27, table 1.

along railroad lines. Typically, they had interests in small-scale wholesaling, banks, factories, newspapers, large retail stores, and hotels or worked as professionals. Sometimes individual investors contributed personal capital, but often such funding came from firms, thus dispersing the risks among partners. Railroads provided high-quality service, including greater speed, less bothersome damage to goods, and passenger comfort; although these benefits had minimal direct economic value, passengers and high-value freight switched to railroads from stagecoaches and wagons whenever feasible. Railroad rates fell somewhat below road transport rates; thus, passengers and shippers gained the benefits of railroads without incurring extra costs, even if their savings constituted small shares of total costs.[34]

Early railroad construction followed this model closely (table 5.4). In 1840, after more than a decade of construction, New England mileage remained concentrated in Massachusetts, Connecticut, and Rhode Island—high-density areas of agriculture, industry, and urban places. In the Middle Atlantic the high-density corridor through New Jersey (from New York City to Philadelphia) contained disproportionate amounts of mileage compared to New York and Pennsylvania. Most Pennsylvania railroads bound the rich southeastern area to Philadelphia and linked that metropolis with Baltimore, and most New York railroads linked cities and towns in the rich agricultural heartland between Albany and Rochester. The jump in railroad mileage during the 1840s, about 1.6 times the length built before 1840, represented a substantial slippage in astute construction. Mileage in southern New England rose appropriately, but the 952 miles built in northern New England were mainly Boston investors' vain attempts to capture trade in low-density agricultural and lumber areas and to entice western trade from New York City using Canadian routes. In New York the New York and Erie Railroad debacle accounted for about 40 percent of the mileage built.[35] During the 1850s railroads built almost as much mileage as they had in the preceding two decades, and network intensification coincided with efforts to extend links to the Midwest and the large absolute industrial expansion.

The dominance of passenger over freight revenues before 1840 demonstrated that travelers quickly voted for faster speeds and greater comfort of railways (see table 5.4), yet freight shipments required new infrastructures such as warehouses and transshipment centers. Passenger revenues more than tripled during the 1840s, whereas freight revenues soared almost sevenfold, but the former still exceeded the latter by a small margin as of 1849; eastern railroads shifted to freight by the early 1850s, while passenger revenues continued rising, because railways were preferred for long-distance travel. Railroads in the densely populated southern New England and New Jersey corridor between New York City and Philadelphia remained heavily oriented to passenger travel, whereas railroads in Pennsylvania, New York, and Maryland became industrial freight lines; New York railroads also had large passenger flows. The doubling of passenger revenues and tripling of freight revenues from 1849 to the mid-1850s, coinciding with huge railroad construction, implies that eastern railroads had by then embarked on a new business era. Eastern growth powered part of this surging traffic, but a substantial jump in commodity flows from the Midwest over the Erie Canal during the late 1830s alerted members of the eastern business elite to impending transformation of trade between the East and the Midwest (see fig. 5.3); they were galvanized to action. From the mid-1840s to the early 1850s they pushed to completion—through construction or organizationally—four long-distance railroads: the New York Central, New York and Erie, Baltimore and Ohio, and Pennsylvania.

The Pennsylvania Mainline Canal captured little of this swelling Midwest trade, and, thus, Philadelphia merchants and financiers confronted a dilemma (see fig. 5.3). The Baltimore and Ohio reached Cumberland, Maryland, in 1842, and then became embroiled in political battles, exacerbated by financial problems, over the last section to the Ohio River. Resolving these issues would give Baltimore a railroad with which to tap midwestern trade, and railroad expansion there would provide lucrative business. Promotion efforts for a Pennsylvania railroad to reach between Harrisburg and Pittsburgh, thus completing the cross-state link with Philadelphia, commenced in earnest; Philadelphia's elite led this effort, and the incorporation act of the Pennsylvania Railroad passed the legislature in 1846. By the end of 1852 the Harrisburg-Pittsburgh line had opened, providing service over connecting railroads to Philadelphia, and the organization of all Philadelphia-Pittsburgh lines under one ownership was achieved in 1861.

New York City merchants and financiers controlled midwestern trade pouring over the Erie Canal, but they observed soaring freight tonnage on railroads in the East and competitor railroads heading toward the Midwest border: the Baltimore and Ohio progressed west of Cumberland and reached Wheeling on the Ohio River in 1853; under hard-driving J. Edgar Thomson, the Pennsylvania relentlessly headed for an 1852 completion; and the New York and Erie neared its completion date of 1851. There were, however, nine railroads end to end under separate management crossing New York's rich heartland between

Albany/Troy and Buffalo, and the region was served by the Erie Canal. The provisions of incorporation acts for many of those railroads restricted freight shipments to protect the canal's revenue, but these restrictions terminated in 1851, and the railroads consolidated as the New York Central in 1853, valued at $30.8 million.[36]

Why did New York City merchants and financiers and other business elite in upstate New York, such as Erastus Corning, a wealthy merchant and industrialist from Albany and the first president of the New York Central, fail to press earlier for elimination of cumbersome restrictions on railroad freight and to consolidate the railroads? Railroads could not compete with the Erie for freight until volumes reached levels high enough to underwrite the costly construction of railroad infrastructure, such as warehouses and cargo-handling facilities, needed to compete with the Erie for low-value commodities.[37] Large volumes of instate shipment of bulky commodities offered few enticements compared to the lucrative returns from carrying passengers. Railroads could pay canal surcharges on instate high-value freight shipments, and wagons offered fierce competition over short distances, but, to make profits on long-distance freight, railroads needed vast volumes. Swelling volumes of agricultural commodities flowing from the Midwest by the late 1840s and early 1850s and return flows of high-value manufactures and immigrants finally offered incentives to consolidate the railroads. These business elite would have gained little if they had expended political capital earlier to alter protections for the Erie, and costly railroad freight infrastructure would not have paid its way.

Intra-East Development

Between 1820 and 1860 the nation experienced rising real income and large-scale economic growth, and these changes also unfolded in the East. Agriculture expanded, but manufacturing grew faster, resulting in the economy's structural shift from agriculture to manufacturing; most factories remained small, yet they achieved rapid productivity gains. Interregional trade between the East and Midwest stayed insubstantial until the mid-1840s, and transportation improvements based on canals and railroads served intra- and interregional markets within the East, but wagons and stagecoaches provided fierce competition. Thus, intra- and interregional development processes within the East powered much of its growth until at least the early 1850s, and these growth processes dramatically transformed regional industrial systems.

CHAPTER SIX

Agriculture Augments Regional Industrial Systems

> Articles manufactured in the interior are frequently transported to New York for a market, and carried directly back by retailers to be vended in the neighborhood of the manufactory. . . . [P]roductions of large establishments in the interior have to bear the burden of transportation to the city of New York for a market, as wholesale operations are not carried on to any considerable extent in the interior. —ALVIN BRONSON, 1832

As an Albany resident, Alvin Bronson observed the close ties of factories in Albany, Troy, and other Hudson Valley towns with New York City, and after he had surveyed factories in New York State in 1832, he came to a striking conclusion. Small factories still sold goods at retail prices locally, sometimes through barter, but large factories were integrated with New York City wholesale markets. This produced a seemingly bizarre consequence: factory products were transported to New York and sold to wholesalers, who in turn resold the goods to retailers in the factory's vicinity—goods made a round trip.[1] Greater intra- and interregional exchanges of factory goods signaled new levels in the transformation of agriculture and manufacturing in the East, as thriving agriculture buttressed regional industrial systems.

Agriculture Transforms and Flourishes

The Rural Population, Labor, and Land

Eastern agriculture could not fully absorb the natural increases in the farm population after 1820; thus, some of the surplus population migrated to new farmlands in the Midwest, yet even with out-migration the number of rural dwellers rose between 1820 and 1860 (see table 1.2).[2] The rates of rural increase slowed markedly by the 1840s, coinciding with a jump in the percentage living in urban areas; however, the absence of rural population decline, even as the eastern economy expanded, suggests that rural economies stayed robust. This vibrancy was rooted in agriculture, but thriving farms also demanded processing, transportation, and services. Because farm laborers could choose between work in factories or on farms, their numbers indirectly measure agriculture's health.

Between 1820 and 1840 farm laborers increased in New England—slowly in the industrial states of southern New England (Massachusetts, Connecticut, and Rhode Island) and rapidly in northern New England (Maine, New Hampshire, and Vermont)—but then their numbers slipped. Yet farm labor counts in every New England state in 1860 remained near their numbers in 1820, before large-scale industrialization commenced (see fig. 2.4). New York and Pennsylvania witnessed huge increases in farm labor forces from 1820 to 1840, then New York's numbers leveled off, whereas Pennsylvania's continued rising. New Jersey maintained a slow, steady rise, and farm labor forces in Delaware and Maryland stayed within a band during the period from 1820 to 1860. These trends provide no evidence of declining eastern agriculture; in most states slippage after 1840 amounted to less than 10 percent by 1860, and the East's total number rose from 1820 to 1860. Nevertheless, some eastern agriculture experienced stress; isolated, hilly areas with poor soil became less competitive as farms with better access to markets and good soil raised productivity. By the mid-1840s increased volumes of midwestern agricultural products arriving over the Erie Canal exacerbated these competitive conditions, but eastern farmers had alternatives—they could shift to the production of higher-value goods for urban markets.

Farmworkers' productivity climbed at much faster rates after 1820; except for the anomaly of the 1840s, annual productivity accelerated in each decade from its low point during the 1810s. This growth released labor from agriculture and supported the jump in the share of the East's population in urban places from 11 percent in 1820 to 35 percent in 1860 (tables 1.2 and 2.7). National rates of productivity increase included substantial additions of midwestern and southern farms with low productivity during the land-clearing stage; thus, productivity figures understate the gains in eastern agricultural areas that boosted production for urban markets. Improvements in many agricultural implements—plows, sowing machines, harrowers, cultivators, scythes and grain cradles, and threshers—between 1815 and 1830 supported the transition to higher productivity. Backed by state agricultural societies, almanacs widely discussed these improvements, and they expanded beyond planting directions

Table 6.1. Farm Comparisons for the East, Midwest, and Nation, 1850 and 1860

	Number of Farms (thousands)		Acreage per Farm		Value of Farm			
					Property per Farm($)		Property per Acre ($)	
	1850	1860	1850	1860	1850	1860	1850	1860
New England	168	184	110	109	2,221	2,589	20	24
Maine	47	56	97	103	1,173	1,413	12	14
New Hampshire	29	31	116	123	1,890	2,285	16	19
Vermont	30	32	139	136	2,129	2,988	15	22
Massachusetts	34	36	99	94	3,202	3,462	32	37
Connecticut	22	25	106	100	3,240	3,607	31	36
Rhode Island	5	5	103	96	3,170	3,616	31	38
Middle Atlantic	350	413	121	113	3,397	4,414	28	39
New York	171	197	112	107	3,250	4,078	29	38
New Jersey	24	28	115	108	5,030	6,520	44	60
Pennsylvania	128	156	117	109	3,197	4,234	27	39
Delaware	6	7	158	151	3,114	4,720	20	31
Maryland	22	25	212	190	3,988	5,726	19	30
Midwest	368	587	136	124	1,824	2,958	13	24
Ohio	144	180	125	114	2,495	3,770	20	33
Illinois	76	143	158	146	1,261	2,854	8	20
United States	1,449	2,044	203	199	$2,258	$3,251	$11	$16

Source: U.S. Bureau of the Census, *Historical Statistics of the United States,* ser. K17-81.

Note: Midwest is defined as the East North Central census region (Ohio, Indiana, Illinois, Michigan, and Wisconsin).

to discussing agricultural information. The appearance of new farm papers in 1819 such as the *American Farmer* and the *Plough Boy* attested to farmers' interest in agricultural implement advances and other ways to improve farming. Growing urban markets encouraged agricultural implement inventors to intensify their efforts and for farmers to take an interest in agricultural advances.[3]

While industrial growth surged during the late antebellum period, people started new farms in the East which farms compared favorably in size to its midwestern counterparts (table 6.1). Except for northern New England, eastern farms remained more valuable enterprises than midwestern farms. The average values of farm property (i.e., land and buildings) per farm in southern New England states, the rapidly industrializing area, were comparable to Ohio's values, the richest midwestern farm state in 1860, and in the Middle Atlantic states values significantly exceeded midwestern farm values in both 1850 and 1860; New Jersey farms, situated between the nation's greatest urban markets, New York and Philadelphia, occupied the pinnacle. The value of farm property per acre indirectly measures farm intensity, and Ohio represented the future Midwest by 1850; its farms had greater values per acre than northern New England farms, and they widened the gap over the next decade. Delaware and Maryland farms kept pace with the increasing intensity of Ohio farming, but more significant was the extraordinary lead in intensive agriculture of farms in Massachusetts, Connecticut, Rhode Island, New York, and Pennsylvania over typical midwestern farms, including Ohio, and over the nation's farms in 1850 and 1860. These eastern states possessed similar values per acre, implying that farmers responded to common demands for agricultural products from swelling urban industrial populations, and New Jersey occupied a unique position; property value per acre far exceeded that of other eastern farms. Thus, increasing wealth of eastern farm populations certainly provided a growing, diversifying market for manufactures between 1820 and 1860.

Farmers Respond to Opportunities

Massachusetts

The transformation of Massachusetts agriculture after 1820 built on earlier changes; the cooperative, noncash local exchange of labor time and materials continued declining, reducing it to modest levels. This slide did not signify lesser social obligations and bonds, which remained intact within politics, churches, families, and friendship groups. Farmers increased labor productivity without farm machinery and with similar amounts of land per farm and stable labor force, achieving gains of 0.6 percent annually from 1820 to the early 1840s, but productivity stagnated into the mid-1850s (see figs. 2.1 and 2.4). Successful farmers devised sophisticated strategies to boost their productivity and to respond to competitors from elsewhere: they increased the use of hired hands, especially as their sons moved to establish farms elsewhere or shifted into nonfarm occupations; they improved more land, cutting woodlands and draining marshes to convert the land to pasturage; and they cultivated higher-value crops, stabled and stall-fed cattle, innovated feed supplements, enhanced breeding, and improved hay quality through greater planting of English hay (e.g., clover and timothy).

When the Erie Canal opened in 1825 and competition increased from New York farmers, Massachusetts farmers already were more competitive, and they enhanced their productivity by the mid-1840s, when midwestern farm products started coming east in large volumes over the Erie. Massachusetts farmers, along with other eastern farmers, continued following Von Thunen principles and adjusted their production intensity to their distance from urban markets, thus creating more valuable farms by the 1850s (see table 6.1). Between 1820 and 1840 these trends impacted farmers at varying distances from Boston (see map 2.1): Brookline, which bordered the city; Concord, seventeen miles away; Shrewsbury, thirty-six miles away; and the Connecticut Valley, eighty to one hundred miles away.[4]

The Period from 1820 to 1840

Swelling urban populations constituted lucrative markets for farmers, and Boston beckoned as the largest; over these two decades its population more than doubled to 118,857 (see table 2.5). Other cities in eastern Massachusetts housed large markets, such as the textile city of Lowell, which transformed from farmland in 1820 to having 20,796 urban residents in 1840, then the nation's eighteenth largest city; and the state's total urban dwellers soared by 160,000, reaching 279,000. Agriculture in Brookline, near Boston's mercantile core, also changed: grain production plunged and farm labor numbers jumped 50 percent, to 325, as production intensity increased. Farms increasingly supplied feed, fresh meat for consumers, and horses for urban transport; hay production, cattle, and swine doubled, and the number of horses tripled. Farmers' shifts into higher-value vegetables and fruit overwhelmed these changes, however, which by 1840 made up 60 percent of farm production value. They achieved this growth with greater fertilization and innovative practices such as the "Boston hotbox," in which products were grown under glass sheets.[5]

Concord farmers leveraged their proximity to Boston and its satellites to increase their specialization in oats, hay, and wood; they cut down almost 40 percent of woodland, land under tillage rose 10 percent, and acreage in English grasses jumped by 36 percent. Oats and hay went to urban draft animals and horses, and wood supplied fuel for urban dwellers; farmers also employed growing hay output to raise beef cattle and cows, shipped cattle to Boston's Brighton auction market, and produced butter. Concord farmers (Middlesex County) comprised a subset of farmers in eastern Massachusetts which included Suffolk County (Boston) and surrounding counties (Essex, Middlesex, and Norfolk); by 1840 these counties were highly specialized producers for Boston and other local urban markets (table 6.2). Suffolk County, whose population was largely Boston residents, contained few farmers, and city expansion soon engulfed them; they provided fresh vegetables, fluid milk, and nursery products for city residents. Farmers in Essex, Middlesex, and Norfolk Counties produced huge quantities of fresh vegetables, nursery and florist products, dairy products (e.g., fluid milk and butter), and fruit—placing them at a high ranking nationally: as a share of national value, Middlesex County produced 12 percent of nurseries and florists and Norfolk County 6 percent of vegetables.

Shrewsbury's Ward family exemplified the typical responses of large farms, farther from Boston, to opportunities arising after 1820; they hired common carriers to ship products to Boston and other urban markets in eastern Massachusetts. Regular trips allowed farmers to sell high-priced fresh products in urban markets; thus, the Ward's shifted their dairy mix toward perishable butter, which commanded a higher price than cheese, a product with a long shelf life. They competed with distant dairy farmers on cheaper land who sent cheese to Boston but could not compete on butter, except in cold weather, and, increasingly, they hired drovers to take greater numbers of cattle to Brighton market to obtain higher beef prices, rather than slaughtering them locally before shipping them.[6]

Connecticut Valley farmers accessed growing local urban markets: Springfield, a wholesale-retail center, almost tripled, to 10,985, and Northampton, a

Table 6.2. Agricultural Production and Specialization in Eastern Massachusetts, 1840

County	Population Size	Number of Farmers	Vegetables		Nurseries		Dairy		Fruit	
			VP ($)	LQ	VP ($)	LQ	VP ($)	LQ	VP ($)	LQ
Suffolk (Boston)	95,773	348	$8,338	34.2	$2,500	44.9	$16,435	5.2	$1,277	1.9
Essex	94,987	7,607	10,216	1.9	21,791	17.9	193,808	2.8	63,642	4.3
Middlesex	106,611	14,170	77,949	7.9	70,750	31.2	375,900	2.9	86,092	3.1
Norfolk	53,140	6,035	152,213	36.0	12,880	13.3	268,600	4.9	54,226	4.6

Source: U.S. Bureau of the Census, *Compendium of the Sixth Census, 1840.*

Note:

VP: value of production, expressed in dollars.

LQ (location quotient): value of county production per farmer divided by value of national production per farmer.

Value of national production per farmer: $0.70 Vegetables (market gardeners); $0.16 Nurseries (nurseries and florists); $9.08 Dairy; $1.95 Fruit (orchards).

retail center, rose 31 percent, to 3,750. Nevertheless, urban and rural nonfarm markets in the valley paled compared to those in eastern Massachusetts; thus, valley farmers continued focusing on products that could withstand long-distance shipment. Lowland farmers increased their output of hay crops and grain, primarily corn, for animal feed and improved pasturage to expand their production of beef cattle (e.g., in Northampton) and dairy products (e.g., in Amherst), and those in Hadley and Hatfield tried broomcorn. A symbiosis emerged in which lowland farmers moved their cattle to hill pasturage during the summers, and hill farmers thus acquired manure for their hay fields. Valley farmers drove their cattle to Brighton market in Boston or to other urban markets, where by 1830 they competed with cattle from Massachusetts, New York, and Ohio.

Farmers' incomes rose, and they dramatically reduced home textile production (e.g., spinning and weaving), shifted to purchasing factory textiles at local stores, and improved home interior decorations.[7] They made sophisticated adjustments to competition from farmers outside New England, staying competitive in local and subregional markets against outsiders. Because midwestern agricultural products did not begin to flow in large amounts over the Erie Canal until the mid-1840s, major rivals came from other eastern states, especially New York and Pennsylvania; both were major wheat and cattle producers, and their farmers' products also withstood long-distance shipment (see fig. 5.3). Valley farmers possessed advantages: they occupied superb agricultural land, and their fattened cattle had shorter distances to travel to Boston, and therefore could retain more weight, than cattle from outside New England.

The Period from 1840 to 1860

After 1820 Massachusetts' farmers remained uncompetitive in wheat production, because they occupied expensive farmland; production served local needs, and they failed to supply enough wheat to satisfy them. By 1840 this deficit assumed immense proportions even in the prime wheat area of the Connecticut Valley (in Franklin, Hampden, and Hampshire Counties); it raised about 10 percent of local consumption requirements, and the state's farmers raised about 4 percent. New York farmers' surplus wheat could meet less than 8 percent of Massachusetts' deficit, whereas Pennsylvania farmers' wheat surpluses totaled 4.7 million bushels; Massachusetts needed 75 percent of this surplus to meet its deficit. By 1840 wheat moved over the Erie Canal from the

Midwest, and Ohio became a supplier; Massachusetts' deficit would have taken 39 percent of Ohio's huge surplus of 9.1 million bushels. The flow of midwestern agricultural products coming over the Erie by the mid-1840s, supplemented by railroads after the early 1850s, challenged Massachusetts' farmers to adjust (see fig. 5.3); besides rising volumes of wheat, distant farmers increasingly sent live cattle and processed beef and pork. Massachusetts' farmers faced opportunities in their midst as urban populations rocketed by 163 percent; the increment of 454,000 equaled the state's population in the first decade of the nineteenth century. Although eastern Massachusetts continued to gain the largest number of urban dwellers, industrialization offered swelling local markets everywhere.[8] Farmers capitalized on their proximity to these markets and shifted more toward the production of fresh, high-value, bulky commodities such as garden vegetables, fruit, butter, milk, and fresh meat.

In the early 1840s Brookline started becoming a residential suburb of Boston. Farm numbers and farm size declined, grain production almost ceased, and hay and dairy output stayed about the same, but that of garden produce and fruit rose; after 1870 Brookline's farmers found greater remuneration selling high-value land for urban uses. Concord farmers were too far from Boston to be engulfed by suburbanization: they switched heavily into producing butter, eggs, fruits, garden vegetables, and fluid milk, and milk production jumped after the railroad reached Concord in 1844. Shrewsbury's Ward family expanded beef production for Boston, taking advantage of its proximity and offering high-quality cattle fed with hay and corn, but with growing midwestern competition beef production offered limited long-term possibilities. By the 1840s the family looked more toward nearby Worcester, a burgeoning commercial-industrial city, and it shifted back and forth between milk and butter, depending on prices; its local market production grew so much the family turned to specialized marketing firms to handle its sales.[9]

Connecticut Valley farmers copied their peers farther east, clearing more land in response to growing local urban markets and gaining short-term advantages from selling lumber for construction and fuel. Hill farmers reduced the woodlands by 43 percent, and under competition from farms outside the region they sharply cut the sizes of their sheep herds. Valley farmers boosted hay and feed corn production to support the rising dairy output and increasingly focused on fresh meats, milk, butter, and fresh vegetables for local markets. By the early 1850s larger farms on prime flat land of the valley, like those in New York State and the Midwest, were using more equipment, such as seed drills, horserakers, and mowing machines.

Farmers benefited from a new crop offering export opportunities, when valley farmers in Connecticut innovated specialized cigar wrapper tobacco. By 1840 they produced 472,000 pounds, and this number soared to 6 million by 1860; Massachusetts valley farmers started production around 1850 and by 1860 produced 3.2 million pounds. They entered sophisticated export markets with cigar factories in Connecticut and major wholesale houses in Hartford, New York City, and Baltimore, and traveling buyers acquired tobacco for large-city cigar manufacturing; thus, farmers possessed a lucrative, specialized crop for which they lacked competition from farmers elsewhere in the East or Midwest.[10] Massachusetts farmers responded effectively to competition from other eastern farmers through the mid-1840s and after that time to competition from midwestern farmers occupying cheap, fertile land. They shifted out of wheat, beef cattle, and, to a lesser extent, cheese, which they conceded to distant competitors, and into fresh, high-value commodities, such as milk, butter, vegetables, and fruit, for urban markets. Yet farmers on hilly, poor soil, distant from urban centers, were uncompetitive; some remained as semisubsistence farmers, but most of them shifted into other occupations or moved elsewhere. New York State possessed much greater quantities of good farmland, yet its farmers faced similar competitive pressures.

New York

The Period from 1820 to 1840

Urban markets north of New York City developed mostly along the Hudson-Mohawk Valleys and Erie Canal, and cities added 90,000 people (see map 2.2). By 1840 they totaled 80 percent of New York City's size in 1820, and the Albany-Schenectady-Troy conurbation's 59,836 residents in 1840 exceeded Boston's size in 1820 (see tables 2.5 and 6.3). The New York City region, an enormous market of one million people by 1840, housed almost 6 percent of the nation's population; it attracted vast quantities of agricultural goods from elsewhere, and local farmers developed finely honed specializations requiring the purchase of agricultural goods from other farmers in the region and outside it (table 6.4). Sizable numbers of farmworkers operated in the city's constituent counties of New York (2,773 employed), Kings (3,234), Queens (6,138), and Richmond (844). The region's wheat deficit reached immense proportions, 3.8 million

Table 6.3. Population Sizes of Selected Cities in New York, 1820–1860

City	1820	1830	1840	1850	1860
New York City	152,056	242,278	391,114	696,115	1,174,779
Albany	12,630	24,209	33,721	50,763	62,367
Auburn	—	4,486	5,626	9,548	10,986
Buffalo	2,095	8,668	18,213	42,261	81,129
Elmira	2,945	2,892	4,791	8,166	8,682
Rochester	2,063	9,207	20,191	36,403	48,204
Schenectady	3,939	4,268	6,781	8,921	9,579
Syracuse	250	2,565	6,259	22,271	28,119
Troy	5,264	11,556	19,334	28,785	39,235
Utica	2,972	8,323	12,782	17,565	22,529
Total selected cities	183,964	315,887	512,553	920,798	1,485,609
Total excluding New York City	31,908	73,609	121,439	224,683	310,830

Sources: Miller, *City and Hinterland,* 50, 53; U.S. Bureau of the Census, *Twelfth Census, 1900,* vol. 1, pt. 1: *Population,* 430–33, table 6; *Seventeenth Census, 1950,* vol. 1: *Number of Inhabitants,* 146–47, table 23.

bushels by 1840, with New York City alone accounting for 1.8 million bushels. Although much of the shortage might have been met from elsewhere in New York State, it is unlikely; for example, prime-quality flour from the Genesee Valley sold throughout the East. The region's wheat deficit would have required 81 percent of Pennsylvania's surplus or 42 percent of Ohio's.

The agglomeration of urban residents and prosperous farmers possessed a market capacity to reshape agriculture throughout the East and into the Midwest. Regional farmers specialized in fresh, high-value, bulky products for urban consumers and prosperous farmers, just as expected in a Von Thunen land-use system (see table 6.4). Every county in New York City specialized in fresh vegetables, although Kings and Queens Counties supplied most of the surplus; they probably supplied much of the region and were extraordinary national specialists. Farmers in Bergen and Monmouth Counties in New Jersey, adjacent to Manhattan (New York County), also were highly specialized in vegetables and produced substantial volumes for the region. Urban populations demanded ornamental plants and decorations, and New York and Queens Counties specialized in these areas, providing most of the surplus production. Numerous counties specialized in dairy products, but New York State's counties accounted for much of the surplus. New York County faced a huge deficit because it produced few dairy products, and Hudson County, New Jersey, was the only other county with a deficit. Although fluid milk, butter, and cheese data cannot be distinguished, innermost counties—Kings,

Queens, and Bergen—probably specialized in fluid milk, whereas other counties specialized in butter and produced fluid milk for local consumption; cheese for commercial sale was unimportant because farms in upstate New York specialized. Many counties specialized in fruit: in New York City, only Queens both specialized and supplied substantial surplus fruit, but nearby Westchester County had extraordinary specialization and produced mammoth amounts sold throughout the region. It produced surpluses in vegetables, nurseries, dairy, and fruit; nevertheless, estimates based on national figures understate the region's consumption, because its urban dwellers and prosperous farmers had greater propensity to consume these high-value products. A modest amount of these products may have exited the region for other East Coast cities.[11]

The Albany-Schenectady-Troy conurbation reproduced the impact, albeit at smaller scale, of the New York City region on agriculture; urban residents comprised about one-third of the four-county population (table 6.5).[12] Although local farmers raised 127,576 bushels of wheat in 1840, the conurbation's deficit reached 797,044 bushels; this equaled 43 percent of New York City's deficit and 21 percent of the New York City region's, which constituted the nation's greatest concentration of demand for wheat. Thus, the Albany-Schenectady-Troy conurbation materially impacted the organization of wheat production and trade in New York State and the East and loomed as an impetus to midwestern wheat production. Its urban residents lured farmers to develop finely honed specializations: farmers in Albany and Rensselaer Counties specialized in vegetables, with output nine to eleven times that of national per-farmer production; every county specialized in dairy and fruit production, and Rensselaer farmers generated enormous dairy surpluses, while Albany County farmers contributed most of the conurbation's nursery and florist output.

Smaller cities along the Erie Canal exerted much less of an impact on local farm specialization, implying that cities had to reach a threshold size before local agriculture was significantly reorganized. In the upper Mohawk Valley, Utica housed only 15 percent of Oneida County's population, and it had a small deficit in vegetable production; although some farmers certainly produced vegetables for Utica's market, specialized vegetable production did not reach equivalent levels as in the Albany-Schenectady-Troy conurbation (see table 6.5). Oneida specialized in fruit, but dairy products were its premier specialty, and farmers generated huge surpluses; nearby Herkimer County also specialized in dairy products and generated an equivalent surplus. These counties'

Table 6.4. Agricultural Surpluses, Deficits, and Specializations in the New York City Region, 1840

County	Population Size	Vegetables		Nurseries		Dairy		Fruit	
		SD ($)	LQ	SD ($)	LQ	SD ($)	LQ	SD ($)	LQ
New York City									
New York	312,710	19,734	34.33	13,705	55.56	−593,639	0.89	−130,538	0.15
Kings	47,613	76,908	37.13	1,784	6.67	151,432	8.35	−11,789	1.30
Queens	30,324	100,172	24.37	14,479	15.82	82,674	2.56	25,768	3.22
Richmond	10,965	15,374	28.81	−384	0	4,905	3.46	1,543	3.74
Subtotal New York City	401,612	212,188	29.96	29,584	21.00	−354,628	3.70	−115,017	2.12
New York State									
Dutchess	52,398	−5,190	0.24	−1,834	0	540,610	4.42	27,039	1.57
Orange	50,739	−5,711	0.15	4,224	2.10	569,910	4.13	16,681	1.09
Putnam	12,825	−1,784	0.06	−449	0	123,967	5.25	11,736	2.81
Rockland	11,975	−1,796	0	−419	0	−10,664	0.56	21,850	5.42
Suffolk	32,469	−4,820	0.01	−1,056	0.06	84,573	2.06	14,923	1.84
Ulster	45,822	−6,873	0	−754	0.68	143,114	3.31	18,064	2.46
Westchester	48,686	−6,323	0.17	−1,504	0.15	261,076	4.74	184,445	12.66
Subtotal New York State	254,914	−32,497	0.13	−1,792	0.70	1,712,585	3.83	294,736	3.24
Total New York State	656,526	179,691	5.19	27,792	4.14	1,357,958	3.81	179,719	3.05

Connecticut									
Fairfield	49,917	−7,338	0.03	−1,647	0.10	31,986	2.23	15,303	2.89
Litchfield	40,448	−6,067	0	−1,416	0	323,365	5.30	2,805	1.21
Subtotal Connecticut	90,365	−13,405	0.01	−3,063	0.04	355,351	3.96	18,108	1.94
New Jersey									
Bergen	13,223	30,511	18.65	−463	0	39,213	2.89	52,249	11.91
Essex	44,621	−2,093	2.06	1,144	5.30	16,437	3.60	8,857	4.43
Hudson	9,483	4,663	10.63	1,145	11.29	−15,363	0.45	−1,623	1.48
Middlesex	21,893	−334	1.53	−561	0.46	19,992	2.52	30,292	7.34
Monmouth	32,909	27,381	8.46	1,348	2.86	9,534	1.50	51,772	6.17
Morris	25,844	−3,877	0	−905	0	32,299	2.84	22,617	5.33
Passaic	16,734	423	3.67	−586	0	18,699	4.98	3,149	4.57
Somerset	17,455	−2,341	0.12	−611	0	67,917	3.38	6,912	2.19
Sussex	21,770	−3,266	0	−762	0	186,631	6.04	23,897	4.05
Subtotal New Jersey	203,932	51,066	4.39	−250	1.62	375,359	3.22	198,122	5.47
Total Outside of New York City	549,211	5,164	1.19	−5,104	0.84	2,443,295	3.70	510,965	3.62
Grand Total	950,823	217,353	4.36	24,479	3.06	2,088,668	3.70	395,948	3.46

Source: U.S. Bureau of the Census, *Compendium of the Sixth Census, 1840.*

Note:

SD: Surplus is positive; deficit is negative. Production of areal unit minus consumption of areal unit, expressed in dollars. Consumption equals population of areal unit multiplied by national output per capita.

LQ (location quotient): value of areal unit production/farmer divided by value of national production/farmer.

National Output		
Per capita	Per farmer	
$0.15	$0.70	Vegetables (market gardeners)
$0.035	$0.16	Nurseries (nurseries and florists)
$1.97	$9.08	Dairy
$0.42	$1.95	Fruit (orchards)

Table 6.5. Agricultural Surpluses, Deficits, and Specializations in New York State, 1840

City/County	Population Size	Vegetables		Nurseries		Dairy		Fruit	
		SD ($)	LQ	SD ($)	LQ	SD ($)	LQ	SD ($)	LQ
Albany-Schenectady-Troy Cities	59,836								
Albany	68,593	52,214	10.79	3,299	4.31	−9,785	1.67	4,203	2.05
Schenectady	17,387	−1,058	0.66	−609	0	50,807	2.81	5,868	2.02
Rensselaer	60,259	39,332	8.81	−1,559	0.44	154,006	3.83	19,744	2.95
Saratoga	40,553	−5,805	0.04	−1,419	0	77,514	1.72	14,827	1.62
Total Counties	186,792	84,683	5.45	−288	1.32	272,541	2.39	44,642	2.14
Utica City	12,782								
Oneida	85,310	−3,230	0.84	−2,986	0	669,330	5.66	42,676	2.47
Herkimer	37,477	−5,222	0.05	−1,312	0	597,521	5.87	13,906	1.21
Rochester City	20,191								
Monroe	64,902	−3,420	0.90	5,703	4.96	44,887	1.89	42,302	3.55
Buffalo City	18,213								
Erie	62,465	4,095	1.75	−186	1.13	−14,395	1.09	−1,264	1.16

Source: U.S. Bureau of the Census, *Compendium of the Sixth Census, 1840.*

Note:

SD: Surplus is positive; deficit is negative. Production of areal unit minus consumption of areal unit, expressed in dollars. Consumption equals population of areal unit multiplied by national output per capita.

LQ (location quotient): value of areal unit production per farmer divided by value of national production per farmer.

National Output		
Per capita	Per farmer	
$0.15	$0.70	Vegetables (market gardeners)
$0.035	$0.16	Nurseries (nurseries and florists)
$1.97	$9.08	Dairy
$0.42	$1.95	Fruit (orchards)

farmers had built a formidable cheese production complex over the previous two decades, specializing in high-quality cheeses for East Coast urban markets. By 1815 New York City brokers appeared locally, and by 1835 specialized commission houses from that city, and secondarily Philadelphia, had local agents contracting for cheese. Although Oneida and Herkimer farmers developed little specialization for local markets as of 1840, they were a huge import market for wheat. Their production of 322,882 bushels left them with a deficit of 284,914 bushels, amounting to 18 percent of the deficit of New York County, the core of the national metropolis.[13] These cheese counties contributed to the reorganization of eastern agriculture by drawing in huge quantities of surplus wheat raised elsewhere, suggesting, in simplified form, that the New York City region and other East Coast urban markets drew in immense quantities of agricultural goods from eastern farmers; in turn, these eastern subregions, which were specialized suppliers, imported agricultural products from other rural areas of the East and, increasingly, from the Midwest by 1840.

Rochester's consumer market also exerted a limited impact on agricultural specialization of its surrounding county of Monroe as of 1840 (see table 6.5). It had a small deficit in vegetables and modest specialization in dairy products, probably milk and butter for Rochester's market; nevertheless, surplus dairy products remained tiny compared to Oneida and Herkimer Counties' surpluses, indicating that Monroe farmers exported little cheese and butter to distant markets. Specialization in nursery and florist products foreshadowed their future importance as farmers increasingly served East Coast and midwestern urban markets over the next several decades. Monroe farmers, along with those elsewhere in the Genesee Valley south of Rochester, dramatically expanded the acreage of cleared land between 1822 and 1835, tripling the amount in Monroe and doubling it in Livingston and Genesee (see map 2.2).

Specialized wheat production soared, and Rochester became a flour milling center; by 1831 the city's mills manufactured 240,000 barrels of flour, and by 1845 production more than doubled again. Considerable wheat arrived from Ohio and other midwestern states after the mid-1830s, but most milled wheat continued to come from western New York until the 1850s. By 1840 Genesee Valley counties (Monroe, Genesee, Livingston, and Orleans) ranked among the East's greatest wheat-producing regions; their surpluses totaled 2.6 million bushels, enough to supply 68 percent of the New York City region's wheat deficit. Buffalo, like Utica and Rochester, remained a small city with a limited impact on agriculture in Erie County (see table 6.5). Some specialized vegetable farmers emerged, but most other urban market goods probably came from modest surpluses of individual farmers. The production of 207,492 bushels of wheat still left a deficit of 101,710 bushels, however, which would soar when Buffalo's population surged in the 1840s.[14]

Thus, New York farmers specialized in order to serve swelling urban and industrial markets by 1840, and farmers became markets for some agricultural goods, adding incentives to specialize even beyond urban market products. Farmers along the lower Hudson Valley purchased cattle that drovers had brought from western New York and the Midwest and fattened them for the New York City region, but that enterprise faced a limited future because midwestern meatpacking centers soon shifted into high-volume export production. New York farmers raised 28 percent of the nation's potatoes in 1840, even though they constituted just 12 percent of the farmers, and within the state Jefferson, Oneida, Otsego, Saratoga, and St. Lawrence Counties each raised over one million bushels, accounting for 6 percent of the nation's crop; nevertheless, this specialization declined by the mid-1840s. A barley corridor of nine counties lined the Erie Canal between Albany and Syracuse, together producing 1.5 million bushels—61 percent of the state crop and 37 percent of the nation's. Hop growing was concentrated even more; Madison and Otsego Counties produced 62 percent of the New York crop and 22 percent of the nation's. Albany firms managed the nation's barley market in 1840, and, together with Troy, they housed twelve massive breweries making 2.7 million gallons of beer, over twice the volume of New York City's fifteen breweries. Philadelphia's nineteen breweries produced 11.3 million gallons, however, dwarfing the Albany and Troy as well as New York breweries; Philadelphia made 49 percent of the nation's commercial beer, but Pennsylvania produced little barley and hops. Philadelphia breweries probably purchased these staples from New York, because its farmers raised 61 percent of the nation's barley and 36 percent of the hops.[15]

The Period from 1840 to 1860

During the two decades after 1840 New York farmers continued transforming agriculture; they supplied swelling urban industrial populations and responded to competition from farmers elsewhere in the East and Midwest. New York City's market soared as the city added 783,665 people, over three times as many as during the previous twenty years (see table 2.5). At 1.2 million strong in 1860 (ignoring nearby cities), this market generated sweeping impacts on eastern agriculture. Fluid milk sheds spread outward as radial railroads provided fast transportation for this highly perishable product; by the 1840s Orange County farmers, fifty miles northwest of New York City, started shifting from butter to more profitable milk and shipped it via the Erie Railroad. Milk production spread north along the tracks of the Hudson River Railroad, bringing the counties on the east bank of the Hudson into the milk shed, and vegetable and perishable fruit production in the New York region expanded dramatically.[16] At a smaller scale cities of the Albany-Schenectady-Troy region almost doubled, to 111,181 people; thus, by 1860 this conurbation's population (ignoring county rural populations) was almost three-fourths the size of that of New York City in 1820, when it offered such a large market that the region's agriculture was reshaped (see table 6.3).

Utica almost doubled in size to 22,529 people by 1860, offering greater opportunities to produce fluid milk and butter for local markets. Farmers in Oneida and Herkimer Counties continued improving milk production to make higher-quality cheese: they dramatically boosted the amounts of improved land, raised hay quality, and introduced specialized milking barns. Cheese farmers ranked among the state's wealthiest farmers; Oneida's farms were 19 percent more valuable than the average state farm in 1850 and 30 percent more valuable by 1860. Syracuse blossomed as a canal city after 1820 but remained too small to make a significant impact on local agriculture until after 1840; by 1860 its population of 28,119—larger than Utica's—stimulated specialization in fluid milk, vegetables, and fruit production on nearby farms of Onondaga County. To the south farmers shifted into cheese and butter production; they exported cheese to other markets in the East, and fruit farmers, especially apple growers, entered those markets. Genesee Valley wheat farmers prospered for a short time after 1840, yet they succumbed to midwestern wheat grown on naturally fertile, cheaper land. Although they added expensive fertilizer to maintain yields, wheat production peaked around 1849 and plummeted from 3.4 million bushels in 1855 to 1.2 million in 1859; farmers' competitive solution combined specialization and diversification. Rochester's population more than doubled, and Buffalo's exploded fourfold; thus, farmers faced growing markets for fluid milk, fresh vegetables, and fruit by 1860. Farmers in western New York looked to butter, cheese, fruit, and, increasingly, corn, as export crops, and those in the Rochester vicinity specialized in nursery and florist products for eastern and midwestern markets (see table 6.3).

From 1820 to 1860 New York farmers continually adapted to burgeoning opportunities in the New York City region and, secondarily, in the Albany-Schenectady-Troy conurbation, and after 1840 cities along the Erie Canal became submarkets. Farmers near them specialized in vegetables, nursery products, fluid milk and butter, and fresh fruit, and some products—including cheese, barley, hops, nursery products, and wheat—entered other East Coast markets. Agricultural specialization transformed farmers into markets for food grown elsewhere in New York State and the Midwest. Outside of these leading agricultural areas farmers mostly met their consumption needs; nevertheless, by the late 1850s over half of New York's farmers produced surpluses for markets.[17] Philadelphia also had catalytic impacts on eastern agriculture, but the scope remained far less than New York City's, and interior Pennsylvania experienced little large-city growth.

Philadelphia and Its Region

Philadelphia region farmers faced lucrative opportunities as the metropolis' population swelled to over a half-million by 1860 (see table 2.5). Income for Pennsylvania farmers indirectly indicates that those in the Philadelphia region fared well; by 1840 the state's farm income per worker had reached $255, below New Jersey's rich farmers (who earned $276) but 24 percent above the level of New York ($206), whose farm incomes were comparable to New England's richest—Massachusetts ($199) and Connecticut ($208). Shipments of farm products (mostly wheat) and foods (mostly flour) on the Schuylkill, Lehigh, and Mainline canal systems toward Philadelphia from 1826 to 1840 measure gains by Philadelphia region farmers. The real value of farm products that were shipped rose at a compound annual rate of 32 percent, but this figure underestimates shipment values because these goods also came by wagon.

The Von Thunen land-use structure existing around Philadelphia by 1820 was elaborated by 1840: farmers in Philadelphia and nearby counties specialized in vegetables, fruit, fluid milk, and nursery products for the huge city market; grain (wheat) and hog farming occupied the next ring, premier wheat areas of Lancaster and York Counties; and distant counties in the Susquehanna Basin specialized in butter, cheese, livestock raising, and wheat. The Philadelphia region diverged from the New York City and Boston regions, which imported wheat from distant suppliers, whereas the Philadelphia region still produced surpluses. Philadelphia County farmers raised only 67,000 bushels of wheat for the county's 258,037 people, a deficit over consumption of 1.2

million bushels, and, if the rest of the state's eastern district (according to the census definition)—one of the nation's greatest surplus wheat areas—met that entire shortage, it would have required 54 percent of the surplus; some of this probably went to nearby New Jersey and to the New York City region. Nevertheless, the 345,106 people added to Philadelphia between 1840 and 1860 eliminated that surplus (see table 2.5). Philadelphia's region did not develop as much urban market specialization in the interior as happened in eastern Massachusetts and along the Erie Canal; by 1860 only Reading, with 23,162 people, had a population that exceeded 20,000.[18] The Baltimore region developed even less urban market specialization.

Baltimore and Its Region

Baltimore's impact on agriculture in its region seemingly resembled Boston's: both metropolises had similar population sizes during 1820–60 (see table 2.5). Nevertheless, Baltimore's population constituted 91 percent of Maryland's urban population in 1860; rural and village dwellers surrounded Baltimore, whereas budding industrial cities ringed Boston. By 1840, when Baltimore City housed 102,313 people, farmers in Baltimore County specialized in vegetables, nursery and florist products, and, to a limited extent, dairy products, and in adjacent Anne Arundel County farmers specialized in vegetables. Frederick County farmers, fifty miles northwest of Baltimore, developed low-level specialization in dairy products and fruit; some of that probably reached the Baltimore market, although the city of Frederick generated a small demand, and Washington city's 23,364 people probably attracted some products. The counties of Prince Georges and Montgomery in Maryland and Fairfax in Virginia did not specialize in urban market foods, consistent with Washington's small size. Maryland farmers, especially in Frederick County, raised large quantities of wheat, and Baltimore County's deficit required only one-third of the excess wheat produced elsewhere in Maryland. Much of this wheat came to Baltimore as flour on the Baltimore and Ohio Railroad, and, after subtracting local flour consumption, merchant firms exported most of the surplus to foreign markets in the Caribbean and South America.[19] After 1840 Baltimore's growing population stimulated greater urban market specialization, especially of vegetables and fluid milk, but farmers north of Baltimore in Maryland also used the dense regional railroad network to send foods to the larger Philadelphia market.

Farm and City Cooperate

Eastern farmers responded directly to opportunities presented by the burgeoning urban markets. Farmers near large cities increasingly specialized in urban market foods, whereas distant farmers boosted production of foods that could withstand long-distance shipment, and they continually experimented with crops to find cash specialties. Farmers occupying poorer-quality land, distant from markets, produced modest surpluses above subsistence which they exchanged for goods. Rich farmers, especially outside the immediate inner ring of intensive production for local urban markets, often produced diverse foods to supply their consumption needs even as they sold growing amounts of specialized foods. Boston, New York, Philadelphia, and Baltimore and, secondarily, industrial cities in eastern Massachusetts and large cities such as the Albany-Schenectady-Troy conurbation spurred agricultural changes, and the rising urban share of the eastern population indicated an increased commercial market orientation of farmers (see table 1.2).

The 1840s accelerated pressures on farmers to specialize and meet food demands of booming regional and subregional metropolises, industrial cities, and large central places, and even rich farmers began eliminating some diversified production; midwestern wheat and, increasingly, meat started displacing eastern production on a large scale. Eastern farmers, except those in the most fertile areas, no longer competed effectively over these products with midwestern farmers, who occupied fertile, low-cost land. At the start of the 1850s contemporary observers promoted commercial attitudes toward farming, downplaying the advantages of "self-sufficiency."[20] The transformation of eastern agriculture consisted of declining farming in hilly, poor-soil areas distant from markets and intensifying production through greater investment in fertilizer, improved grasses, drainage, and clearing remaining woodlots. Investments made economic sense only if farmers shifted further into products with high returns per acre, either urban market products (e.g., vegetables, nursery, dairy, and fruit) or extremely high-value crops such as tobacco, in the Connecticut Valley; therefore, eastern farmers increasingly became markets for specialized food production, though still far less than urban consumers.

This agricultural transformation—coupled with growing inflows of cheap midwestern food after 1840—released ever-larger numbers of farm laborers for nonfarm jobs, especially for industrial and commercial employment in eastern cities. By 1840 New England and Middle Atlantic nonagricultural incomes per worker were over twice as high as agricultural incomes, contributing to industrial demands, and farmers' rising prosperity increased their capacity to purchase manufactures. In 1860 eastern per capita farm income for owner-occupiers ($174) and renters ($78) significantly exceeded farm income in the Midwest for owner-occupiers ($113) and renters ($61); thus, prime rural markets for manufactures remained in the East throughout the period from 1820 to 1860.[21]

Financing Industry

The greater commercialization of rural economies and the elaboration of central place hierarchies contributed to capital accumulation and multiplied the nodes of capital—the rural wealthy, farm services (e.g., feed stores and elevators), retailers, bankers, transport agencies, professionals (e.g., lawyers and physicians)—supporting broad-based industrial expansion. During the years between 1820 and 1840 capital accumulation jumped in rural areas near large metropolises as farm economies intensified their urban market production, and these rural areas and central places helped underwrite early metropolitan industrial satellites. Simultaneously, rich agricultural areas distant from metropolises—the Connecticut Valley, along the Erie Canal, and the lower Susquehanna Valley—witnessed extensive central place development at lower levels and at nodal centers. These areas supported the rise of small workshops and nonmechanized factories, but poorer agricultural areas in northern New England, along the Pennsylvania–New York border, and rugged areas of New York and Pennsylvania remained outside the circle of capital accumulation; they gained few manufacturing enterprises, except for processing (e.g., lumber and smelting) financed by metropolitan merchants. As the central place hierarchy matured after 1840, central place businesses in areas near large metropolises and emerging large cities continued boosting capital accumulation, supporting firms in industrial satellites. Distant, rich agricultural areas shifting into higher-value products as wheat and cattle entered from the Midwest also enhanced capital accumulation, especially in large central places; their businesses underwrote some early, small factories.

The Connecticut Valley of Massachusetts exemplifies this transfer of capital to industry in a rich agricultural area distant from metropolises; by the 1820s the increased productivity of valley farms spurred the transformation as farm households decreased home production of textiles and of other goods and boosted purchases of store goods. Employing surplus farm labor, retail merchants intensified control of an outwork system built on consumer goods manufactures such as buttons and palm leaf hats; merchants supplied the raw materials to farm families and, in turn, distributed the goods they produced to external markets. By the mid-1830s button merchants began concentrating production in central shops in towns such as Williamsburg, whereas palm leaf hat manufacture started centralizing in shops around 1840. Until the late 1830s most other manufacturing in the valley north of Chicopee consisted of craft workers and mechanics who were expanding their shops to meet the growing rural demand for manufactures; some shops concentrated in larger central places such as Northampton and Amherst, but many dispersed across the valley. The 1837 financial crisis, coupled with the business failure of many small workshops, exposed the difficulties that those shops encountered, however, in shifting to larger-scale production. Major Northampton retail merchants with the necessary capital to underwrite industrial transformation eagerly filled this breech. They underwrote factories clustering in small villages—Haydenville, Leeds, and Florence—along the Mill River between Williamsburg and Northampton from the mid-1840s to 1860 and manufactured cotton textiles, brass hardware, buttons, and sewing silk; factory villages also emerged elsewhere in the valley such as Easthampton and Amherst.[22]

Although large retail and service businesses in subregional and regional metropolises also underwrote the industrial expansion and firms grew through the reinvestment of retained earnings, the more significant suppliers of fixed capital included the largest, highly capitalized, specialized wholesalers of regional metropolises (namely, Boston, New York, Philadelphia, and Baltimore) and less-specialized wholesalers in those cities and in subregional metropolises such as Hartford, Providence, Albany, and Utica. Providing the extensive working capital that manufacturers required remained an equally important activity of merchant wholesalers, brokers, and commodity dealers. After 1820 commercial banks continued as efficient accumulators of community capital for investment in manufacturing. In New England and the Middle Atlantic the number of banks soared between 1820 and 1837; New England bank capital surged, whereas the Middle Atlantic's stagnated from speculative losses in infrastructure until the early 1830s (see fig. 3.1). By the late 1850s New England housed about 500 banks and the Middle Atlantic about 450, whereas the Midwest housed under 300 banks. Bank money per capita available for credit rose significantly in most eastern states during the period from 1830 to 1860, indicating that business enterprises were accessing ever-larger pools of capital; banks recirculated money at ever-higher rates as bank credit per capita climbed significantly. Per capita bank money and bank credit in Massachusetts, Connecticut, Rhode Island, and New York exceeded other states by 1850 (earlier in some cases), and gaps widened significantly by 1860.These New England states leveraged their prosperous agriculture and commerce into a capital credit machine that funded these sectors and thus supported industrial expansion; New York's climb reflects similar processes, as does the growth of New York City as the nation's financial center. During 1830–60 most eastern states exceeded the average per capita bank money and bank credit in the Midwest, often by huge margins, and this reflected the East's leadership in the transformation of agriculture and industry.[23]

Banks were prominent in regional and subregional metropolises, but they also proliferated in larger central places and nodal centers along the transportation networks throughout the East's prosperous agricultural areas; thus, these areas possessed efficient vehicles to accumulate surplus capital and to transfer it to other sectors. The nexus of prosperous agriculture, a central place hierarchy, and the growth of local manufacturing constituted a powerful adjunct to eastern industrialization by supporting manufacturing growth outside industrial agglomerations in metropolises and their satellites. Eastern urban and rural capital markets were highly integrated by the years 1835 to 1860; thus, bank interest rate differentials stayed low and closely tracked New York City rates.[24] Consequently, manufacturers across the East faced similar risk-adjusted interest rates, allowing industrial growth to blossom in numerous locations without hindrance from excessive capital costs relative to other places. Nevertheless, industrialists needed to keep abreast of the technological changes sweeping the East; their failure to do so led them to lose their competitive advantage.

A Technological Advantage

Inventive activity, measured by the number of patents, continued to follow the business cycle after 1820 (see fig. 3.2). A sharp uptick in inventions from the early 1820s to 1836 reflected entrepreneurs' efforts to capture gains during robust economic growth, and inventions remained at a high level rather than plunging during the 1837–43 contraction, indicating that inventors perceived economic opportunities. The resumption of soaring numbers of inventions to new, higher levels from the mid-1840s to 1860 matched a strong economic expansion. Inventiveness across aggregated sectors (agriculture, construction, transportation, and manufacturing) and separately for manufacturing, the largest patent sector, exhibited similar patterns (see table 3.1). After rising to a high level of inventiveness between 1805 and 1811, annual inventive rates stabilized through the period from 1843 to 1846; regional ranking according to inventiveness retained the order that had been established earlier even when inventive rates soared from the 1840s to 1850s: southern New England and New York ranked first and second, respectively, and both were significantly higher than other regions; and northern New England, Pennsylvania, and the southern mid-Atlantic shared the lower tier.

After 1820 residents of large cities retained the high inventive rates established earlier, and the absolute gap between them and the much lower rates of small cities and rural areas continued. In spite of the higher inventiveness of large-city residents, people living elsewhere accounted for most inventions up to 1836; subsequently, large-city residents and those in small cities and rural areas had similar shares of total patents. During the period from 1805 to 1811 high inventiveness was concentrated in the earliest prosperous agricultural areas of eastern Massachusetts, focused on Boston; the Berkshires of western Massachusetts, western Connecticut, and the Hudson Valley, focused on New York City; and southeastern Pennsylvania, focused on Philadelphia. By the years from 1830 to 1836 additions to these regions included other prosperous agricultural areas that had been swept into the urban market and diverse specialized agriculture in the Connecticut Valley of Massachusetts, Worcester County, southeastern Connecticut, the northern half of Rhode Island (in the Providence region), southern Vermont and New Hampshire, along the Erie Canal and south central New York State, and the Delaware and Susquehanna Valleys (oriented to New York City and Philadelphia). This spread of inventiveness away from regional metropolises and environs tracked the expansion of local and subregional farm markets and central place populations; hence, inventiveness in construction, transportation, and agriculture increased. And this growth carried workshops and small nonmechanized factories into inventive processes as they competed with one another and with craft shops for nearby growing markets.[25]

This interpretation of inventiveness after 1820 matches inventors' characteristics, just as it did earlier. From 1823 to 1846 those patenting just one invention over their careers accounted for about 60 percent of patents, and those with two career patents contributed an additional 17 percent, confirming that widespread invention continued to take place across areas of agricultural prosperity; those patenting extensively (six or more career patents) accounted for only 8 percent of patents. The share of patents held by the elite in commerce and the professional classes in urban areas declined from over 40 percent before 1820 to 19 percent between 1836 and 1846. Artisans (e.g., carpenters; shoemakers; and makers of watches, jewelry, and instruments) maintained about 30 percent of patents, most likely because their specialties still operated outside factory competition, or if they invented in sectors adaptable to factory production, such as low-cost shoes, they benefited from selling or licensing their patents. From 1805–22 to 1836–46 the share of patents held by machinists, toolmakers, and dealers of metal products (e.g., stove manufacturers and blacksmiths) rose from 22 percent to 38 percent, reflecting the growing significance of producer durables to expanding output across the gamut of industries, even though capital equipment, including machine tools and machinery, remained a small share of manufacturing. Although elite and artisan craft workers still accounted for almost half (48 percent) of all patents as of the period from 1836 to 1846, patenting had shifted to those meeting the varied demands for manufactures from urban and rural populations.[26]

Famous inventors never accounted for more than 2 percent of patents during the antebellum; their characteristics define an extreme bound for innumerable patentees whose histories remain obscure. Like the distribution of patents generally, southern New England and New York together accounted for the largest number of patents by great inventors (about two-thirds). This elite's share of college attendees reached 24 percent, far beyond that of the population as a whole, yet, like most people, 48 percent of them never attended school or went for several years. At first their occupations—particularly engineers (33 percent), manufacturers (23 percent), and merchant/professionals (23 percent)—were oriented to the urban industrial, not the rural, economy. Great inventors had experience: only 31 percent of them started inventing before age thirty, the patent careers of 65 percent of them extended more than a decade, and 64 percent of them worked in occupations linked to their patent. Inventions derived from the workaday world of inventors, typically constituting solutions to production problems accompanying the rising demands for goods and services. Inventors did everything possible to exploit their inventions for commercial advantage, gaining income from 76 percent of their patents through manufacturing products and licensing them to other industrialists.[27]

In regional industrial systems centered on Boston, New York, Philadelphia, and Baltimore the continued agricultural transformation and urban growth after 1820 generated broad-based manufacturing demand and supported capital accumulation to fund infrastructure and industrial expansion. The profit lure encouraged inventors to find new products and better ways of making them, and this industrial transformation ended household manufactures, as entrepreneurs with access to rich markets grasped new opportunities. These industrialists in regional and, to a lesser extent, subregional metropolises surged to commanding positions, but their counterparts in distant rich agricultural areas also captured markets because large-scale production economies could not yet be achieved in many manufactures.

CHAPTER SEVEN

Metropolises Lead the Regional Industrial Expansion

> The branches of [manufactures] carried on in the city of Boston are so numerous. . . . The manufacture of . . . types and stereotype plates was commenced about ten years ago. Steam engines have not been built here till within five or six years. Cork cutting, hook and eye making, chain cable making, suspender making, pocket book making, and lithographic engraving, have all been introduced within the same period of time. The iron foundry, and various other branches of the work done in iron, for machinery of various kinds, has been much extended within the last six or eight years. . . . The principal articles manufactured in this county [Suffolk] are consumed in this State and other parts of New England.
>
> —JOHN TYLER, 1832

Because New York State housed the greatest economy in the East, its shift out of household textile production was a prototype of industrial transformation. Total volume of household output plunged from 16.5 million yards to 0.9 million between 1825 and 1855, but declines unfolded erratically and unevenly across the state (table 7.1). Rising agricultural prosperity and falling prices of factory textiles during the economic expansion after 1820 caused the 50 percent drop in total volume of household textiles over the decade from 1825 to 1835, and the smaller decline of 19 percent in the following decade overlapped with the 1837–43 business contraction; with the subsequent upturn the plunge was renewed, and volume became trivial by 1855. Nonetheless, the 33 percent drop in per capita household textiles from 1835 to 1845 demonstrates that the shift out of household manufactures continued relentlessly even during economic difficulties. Population shifts from rural to urban areas in New York boosted the number of people ending household manufacturing, inexorably reducing per capita production (see table 2.4).

The change in county volumes of household textile manufacturing matched statewide trends; nevertheless, differences in per capita production across

Table 7.1. Household Textile Manufactures in New York State, 1825–1855

	Total Yards of Textile Goods				Yards per Capita			
Selected Counties	1825	1835	1845	1855	1825	1835	1845	1855
New York City								
New York	1,172,859	868,500	904,081	0	7.1	3.2	2.4	0
Kings	13,070	8,689	132	0	0.9	0.3	0	0
Queens	99,099	13,193	80,219	0	4.9	0.5	2.5	0
Richmond	6,524	165	0	0	1.1	0	0	0
Hudson Valley								
Orange	442,111	80,098	34,056	2,948	10.6	1.8	1.1	0
Dutchess	390,143	116,086	37,909	3,193	8.4	2.3	0.7	0.1
Albany	227,665	142,568	83,472	17,960	5.3	2.4	1.1	0.2
Rensselaer	402,864	229,140	95,688	4,949	9.1	4.1	1.5	0.1
Erie Canal								
Montgomery	370,709	233,620	88,137	12,287	9.1	4.8	3.0	0.4
Herkimer	349,884	163,982	102,834	14,110	10.6	4.5	2.7	0.4
Oneida	478,937	241,211	409,857	27,989	12.7	3.1	4.8	0.3
Onondaga	391,237	230,810	159,992	15,589	8.1	3.8	2.3	0.2
Wayne	213,317	163,190	147,555	13,064	8.0	4.3	3.5	0.3
Erie	198,676	177,211	183,709	26,654	8.2	3.1	2.3	0.2
Other								
Allegany	170,415	220,047	213,982	30,963	9.4	6.2	5.3	0.7
Broome	133,004	110,619	111,089	27,251	9.6	5.5	4.3	0.7
Lewis	96,837	78,819	71,297	15,801	8.3	4.9	3.5	0.6
New York State	16,469,422	8,773,813	7,089,984	929,241	10.2	4.0	2.7	0.3

Source: Tryon, *Household Manufactures in the United States, 1640–1860,* 304–5, table 16.
Note: Corrections made to selected computations from the source.

counties persisted (see table 7.1). In 1825 per capita output in counties of New York City fell substantially below the state average, but continued moderate levels in New York and Queens Counties signified merchant-managed home manufacturing, rather than autonomous household production. Increased urban market agricultural production did not differentially alter the household allocation of labor by 1825. Counties in the Hudson Valley and along the Erie Canal, which benefited the most from access to urban markets, retained per capita levels of textile production matching both counties distant from the canal and the statewide average; Albany county's lower level resulted because it housed the bustling subregional metropolis of Albany. By 1835, however, a clear differentiation existed: farm households in most Hudson Valley counties sharply reduced textile output as they intensified their production of fruits, dairy products, oats, and hay for cities in the valley and for the New York City market; and household textile production declined sharply in Erie Canal counties as they increased their commercial agricultural production of cheese, wheat, and barley. Nevertheless, per capita rates of textile output along the Erie—corresponding to lower levels of commercial agricultural production—remained above Hudson Valley rates; counties distant from the canal had the least commercial agriculture and highest per capita rates of textile production. From 1835 to 1845 this county differentiation persisted, even as per capita rates continued declining; by 1845 Hudson Valley counties produced few household textiles. During the following decade per capita rates of textile output collapsed statewide, consonant with the economic upturn, accelerating a shift to urban market agricultural production and urbanization.

Because most areas of New York state faced similar textile prices, they benefited equally from falling prices; therefore, inverse relations between agricultural prosperity and per capita production of household textiles across subregions of New York unequivocally demonstrate that rising farm incomes spurred industrial growth. New York State's experience mirrored the broader declines of household manufactures in the East between 1840 and 1860 (table 7.2). Declining industrial prices, intensified agricultural production for urban markets, and swiftly rising shares of population in urban areas caused the total value of household manufactures to plunge 56 percent during the 1840s and 36 percent during the 1850s (see table 1.2). In contrast, national total value declined slowly because higher household manufactures persisted in the South, and new farms in the Midwest—with greater rates of household manufactures—compensated for rapid shifts of established farms to commer-

Table 7.2. Total and per Capita Value of Household Manufactures in the East, 1840–1860 (in Dollars)

	Total Value			Value per Capita		
	1840	1850	1860	1840	1850	1860
New England	2,526,532	1,598,844	1,107,836	1.13	0.59	0.35
Maine	804,397	513,599	490,786	1.60	0.88	0.78
New Hampshire	538,303	393,455	251,052	1.89	1.24	0.77
Vermont	674,548	267,710	63,334	2.31	0.85	0.20
Massachusetts	231,942	205,333	245,886	0.31	0.21	0.20
Connecticut	226,162	192,252	48,954	0.73	0.52	0.11
Rhode Island	51,180	26,495	7,824	0.47	0.18	0.04
Middle Atlantic	6,379,431	2,292,195	1,374,808	1.26	0.35	0.17
New York	4,636,547	1,280,333	717,898	1.91	0.41	0.18
New Jersey	201,625	112,781	27,588	0.54	0.23	0.04
Pennsylvania	1,303,093	749,132	544,728	0.76	0.32	0.19
Delaware	62,116	38,121	17,591	0.80	0.42	0.16
Maryland	176,050	111,828	67,003	0.37	0.19	0.10
East	8,905,963	3,891,039	2,482,644	1.22	0.42	0.22
United States	29,023,380	27,493,644	24,546,876	1.70	1.19	0.78

Source: Tryon, *Household Manufactures in the United States, 1640–1860*, 308–9, table 17.

cial production (see table 7.2). In the East, northern New England maintained the highest per capita values of household manufactures, consistent with the region's weaker commercial agriculture; by 1860 Vermont diverged as it shifted into dairy production. Industrial states in southern New England reached low per capita levels by 1840, and declines during the 1840s left them with trivial amounts. In the Middle Atlantic, New York was anomalous in 1840, probably because many farm families produced goods for sale through retail stores as part of rural outwork that transferred raw materials to farm households for production into goods such as brooms, or farm families still maintained wool manufacturing.[1] Other eastern states matched low per capita household manufactures in southern New England, revealing the breadth of the shift out of household manufactures in states of prosperous agriculture and growing urban populations; every Middle Atlantic state swiftly ended home manufacturing after 1840, and per capita rates approached zero.

The termination of home manufacturing supported industrial growth: textile manufacturing benefited first and in greatest volume, but rural dwellers also stopped making many durables (e.g., furniture or farm equipment) and nondurables (e.g., utensils or decorative items), and new urban residents

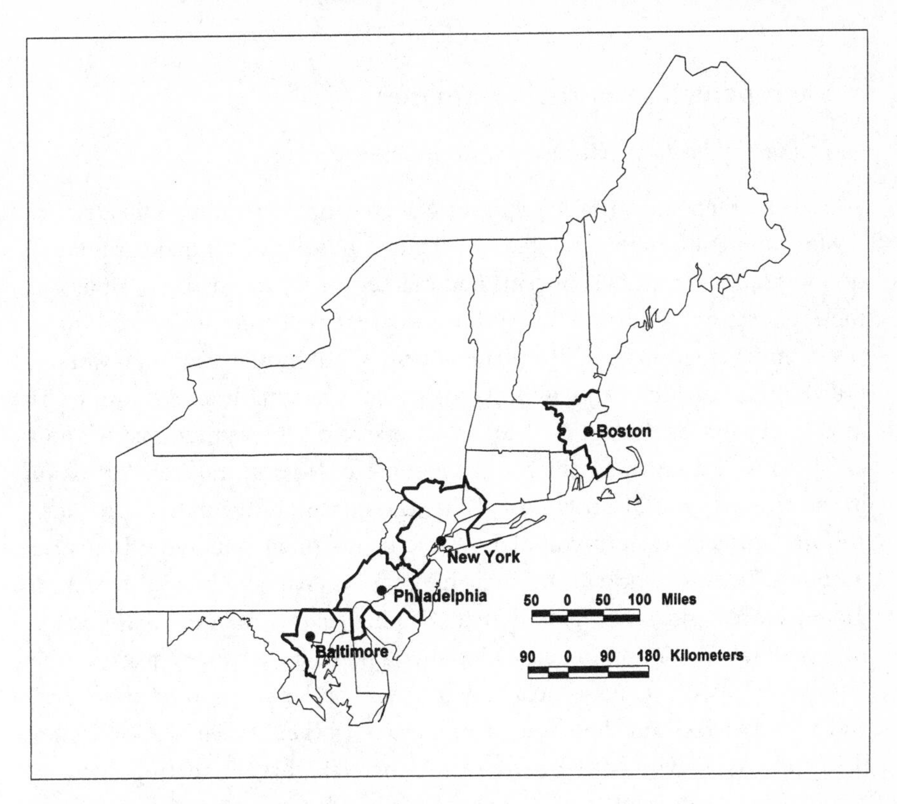

Map 7.1. Metropolitan Industrial Complexes

fueled industrialization because they produced little in the home. The metropolitan pivots of regional industrial systems—namely, Boston, New York, Philadelphia, and Baltimore—and their industrial satellites were the largest concentrated markets for manufacturers, and rich farmers nearby supplying vegetables, dairy products (e.g., fluid milk and butter), fruit, hay, and oats for the cities also were large sources of industrial demand. Each regional metropolis and its industrial satellites—collectively termed a "metropolitan industrial complex"—became the leading manufacturing agglomeration of their respective regional industrial system, as firms emerged to serve demand within the complex (map 7.1). Business intermediaries in a regional metropolis—including financiers, wholesalers, and commodity firms—provided industrial firms with market information and distribution services for accessing markets in their regional industrial system and in other regional systems in the East.

Metropolitan Industrial Complexes

Leaders of Sectoral Transformation

Metropolitan industrial complexes centered on New York, Philadelphia, Boston, and Baltimore accounted for a hefty 11 percent of the nation's population in 1840, a tempting, concentrated market for factory owners (table 7.3). Metropolises housed 22,738 agricultural workers, reflecting substantial intensive farming adjacent to built-up areas of cities, but most agricultural workers toiled in the satellites. Agricultural employment in complexes constituted almost 5 percent of the national total, underscoring the significance of those prosperous farmers; nevertheless, the complexes led sectoral transformation of the economy out of agriculture. With one-quarter of the nation's nonagricultural workers, eastern complexes were pivots of the emerging industrial economy, because workers and their households created a huge demand for manufactures and services, and they supplied hands to meet these needs. Metropolitan cores dominated national commercial exchange through their intermediaries of capital—including bankers, brokers, commodity dealers, and wholesalers—and they housed numerous workers to service local retail demands. As a share of the nation's commercial workers, metropolises accounted for 23 percent, and, together with nearby satellites, the complexes housed 29 percent of the total. Nonetheless, metropolises housed over 50 percent more workers in factories and workshops than in commercial establishments, whereas nearby counties had about seven times as many. With 10.5 percent of the nation's manufacturing workers, metropolises were industrial powerhouses, but their satellites surpassed them with 13.1 percent; collectively, complexes led the nation, with almost one-quarter of its manufacturing workers.[2]

Each complex presented a somewhat different economic face to the nation (see table 7.3). New York's complex contained 69,797 agricultural workers—greater than every other complex—underscoring the contribution of prosperous farmers to its economy. Yet with one-tenth of the nation's nonagricultural workers it towered as the leader of the sectoral transformation out of agriculture. New York's metropolis housed almost 60 percent more commercial employees than Philadelphia, confirming the stature of New York's decision makers who controlled commodity and financial capital. Its complex also contained the single greatest concentration of manufacturing workers, far above its share of population and employment, and both the metropolis, with its concentrated factories and workshops, and its satellites, with their sprawling array, housed similar numbers. Philadelphia's satellites, like New York's, contained numerous prosperous farmers supplying the metropolis, and manufacturing in the Philadelphia complex ranked close to New York's. Intermediaries of capital in Boston's metropolis amounted to less than one-fourth of those in Philadelphia and, even including commercial workers in satellites, left Boston about half the size of Philadelphia's complex. The satellites of Boston's metropolis held an industrial army, however, placing the complex on par with Philadelphia's and New York's complexes as centers of national manufacturing.[3] Yet even Boston's complex housed a huge agricultural labor force that provided food for industrial workers and served as a source of demand for some manufacturing output. Baltimore's complex was distinctive: an overwhelming share of industrial laborers worked in the metropolitan county, and agricultural workers nearby far outnumbered the few local manufacturing workers.

Share of the Nation's Manufactures

Eastern complexes dominated the nation's manufacturing, but industries concentrated differently, reflecting market orientation and the supplies of skilled entrepreneurs and workers (table 7.4). The superior transportation and distribution services of metropolises made them attractive for entrepôt manufacturing (processing raw materials), and nearby areas along transportation arteries—such as rivers, canals, and railroads—also offered suitable sites. Yet neither metropolises nor satellites housed disproportionate amounts of these manufactures. Raw material sites or interior transportation and distribution centers offered competitive production sites for liquor distilling, milling (e.g., flour, grist, lumber, and oil), leather tanneries, and tobacco goods, because high transport costs for low-value, bulky products encouraged processing near raw material sites to raise value of goods before shipping. Sugar, chocolate, and confectionery manufactures concentrated in metropolises, however, accounting for over half of national employment, and firms gained advantages from close proximity to markets—prosperous urban consumers and nearby rural dwellers. Commerce-serving manufactures tied to metropolitan intermediaries agglomerated in the complexes: metropolises employed one-third of the nation's printing/publishing workers, and the subsidiary manufacture of paper concentrated in satellites.

Table 7.3. Metropolitan Industrial Complexes of New York, Philadelphia, Boston, and Baltimore as a Share of National Population and Employment, 1840

	Total Number				% of Nation		
	Nation	Metropolis	Satellites	Complex	Metropolis	Satellites	Complex
				Four Metropolitan Industrial Complexes			
Population	17,120,000	889,801	957,703	1,847,504	5.2	5.6	10.8
Total Employment	4,799,499	173,674	273,050	446,724	3.6	5.7	9.3
Agriculture	3,735,901	22,738	154,158	176,896	0.6	4.1	4.7
Nonagriculture	1,063,598	150,936	118,892	269,828	14.2	11.2	25.4
Selected Employment							
Commerce	117,607	26,624	7,324	33,948	22.6	6.2	28.9
Manufacturing	387,303	40,583	50,594	91,177	10.5	13.1	23.5
				New York			
Population		401,612	323,395	725,007	2.3	1.9	4.2
Total Employment		88,111	88,362	176,473	1.8	1.8	3.7
Agriculture		13,084	56,713	69,797	0.4	1.5	1.9
Nonagriculture		75,027	31,649	106,676	7.1	3.0	10.0
Selected Employment							
Commerce		13,744	2,040	15,784	11.7	1.7	13.4
Manufacturing		15,718	13,076	28,794	4.1	3.4	7.4
				Philadelphia			
Population		258,037	285,545	543,582	1.5	1.7	3.2
Total Employment		45,577	74,194	119,771	0.9	1.5	2.5
Agriculture		3,704	48,170	51,874	0.1	1.3	1.4
Nonagriculture		41,873	26,024	67,897	3.9	2.4	6.4
Selected Employment							
Commerce		8,727	1,699	10,426	7.4	1.4	8.9
Manufacturing		14,979	11,162	26,141	3.9	2.9	6.7

			Boston			
Population	95,773	302,111	397,884	0.6	1.8	2.3
Total Employment	19,563	92,300	111,863	0.4	1.9	2.3
Agriculture	348	33,817	34,165	0.0	0.9	0.9
Nonagriculture	19,215	58,483	77,698	1.8	5.5	7.3
Selected Employment						
Commerce	2,088	3,299	5,387	1.8	2.8	4.6
Manufacturing	1,981	25,459	27,440	0.5	6.6	7.1
			Baltimore			
Population	134,379	46,652	181,031	0.8	0.3	1.1
Total Employment	20,423	18,194	38,617	0.4	0.4	0.8
Agriculture	5,602	15,458	21,060	0.1	0.4	0.6
Nonagriculture	14,821	2,736	17,557	1.4	0.3	1.7
Selected Employment						
Commerce	2,065	286	2,351	1.8	0.2	2.0
Manufacturing	7,905	897	8,802	2.0	0.2	2.3

Source: U.S. Bureau of the Census, *Compendium of the Sixth Census, 1840.*

Note: Metropolitan county is defined as the metropolis for each complex, and *outside* comprises nearby counties; those included in each complex were the following. New York: metropolis (New York, Kings, Queens, Richmond); outside (Connecticut: Fairfield; New York: Westchester, Putnam, Rockland, Orange; New Jersey: Bergen, Essex, Hudson, Middlesex, Morris, Passaic, Somerset). Philadelphia: metropolis (Philadelphia); outside (Pennsylvania: Bucks, Chester, Delaware, Montgomery; New Jersey: Burlington, Gloucester, Mercer; Delaware: New Castle). Boston: metropolis (Suffolk); outside (Essex, Middlesex, Norfolk, Plymouth). Baltimore: metropolis (Baltimore); outside (Anne Arundel, Harford).

Table 7.4. Metropolitan Industrial Complexes of New York, Philadelphia, Boston, and Baltimore as a Share of National Manufacturing Employment by Industry, 1840

	Number Employed	Four Metropolitan Industrial Complexes as Percent of Nation		
Manufacture	Nation	Metropolises	Satellites	Complexes
Entrepôt	108,768	3.9	6.6	10.5
Liquor	12,223	7.0	3.2	10.2
Mills (flour, grist, lumber, oil)	60,788	0.6	4.8	5.4
Leather tanneries	26,018	5.5	12.6	18.2
Sugar, chocolate, confectionary	1,355	54.2	2.4	56.5
Tobacco goods	8,384	10.5	6.7	17.1
Commerce serving	20,713	23.5	11.2	34.7
Printing, publishing	11,523	32.5	5.3	37.7
Paper	4,726	6.5	22.7	29.2
Cordage	4,464	18.2	14.4	32.6
Local, regional, national market	257,822	12.2	15.9	28.1
Bricks, lime	22,807	5.7	11.2	16.9
Granite, marble	3,734	31.0	7.8	38.8
Earthenware	1,612	7.3	15.7	23.0
Precious metals	1,556	55.3	9.6	65.0
Various metals	6,677	22.3	6.8	29.1
Soap, candles	5,641	9.1	4.8	13.9
Glass	3,236	13.1	22.6	35.7
Drugs, medicines, paints, dyes	1,848	51.0	8.7	59.6
Total textiles	111,761	11.4	18.7	30.2
Cotton textiles	72,119	7.3	21.8	29.1
Woolen textiles	21,342	8.0	19.7	27.7
Silk textiles	767	6.4	7.4	13.8
Flax textiles	1,628	6.1	12.7	18.8
Mixed textiles	15,905	35.7	4.3	40.0
Hats, caps, bonnets	20,176	12.1	20.7	32.8
Leather goods	17,136	4.1	11.6	15.7
Musical instruments	908	65.4	5.2	70.6
Furniture	18,003	15.4	13.7	29.1
Carriages, wagons	21,994	6.2	14.9	21.1
Gunpowder	496	9.5	39.7	49.2
Cannon, small arms	1,744	2.0	5.0	7.1
Hardware, cutlery	5,492	7.7	17.8	25.5
Machinery	13,001	26.5	16.3	42.8
Total	387,303	10.5	13.1	23.5

Source: U.S. Bureau of the Census, *Compendium of the Sixth Census, 1840.*

Nevertheless, the contributions of entrepôt and commerce-serving manufactures to total industrial effort in complexes paled next to manufactures selling to local, regional, and national markets and drawing on the special skills of entrepreneurs and labor. As leading concentrations of urban infrastructure, consumers, and rich farmers, metropolitan complexes presented unparalleled market opportunities. Building construction required immense supplies of materials: workers in metropolises dominated the production of granite and marble goods used for structural support and exterior and interior decoration, and satellites had 23 percent of the nation's glass workers. Some manufactures meeting broad demands from rural and urban consumers yet requiring minimal capital investment and skills, such as soap and candles, could be produced widely; complexes had no special attractions for them. Manufactures sold to prosperous consumers, however, or those requiring specialized knowledge and skills that individuals in metropolises could access better than others agglomerated in metropolises: they housed over half of the nation's workers in precious metals and in drug, medicine, paint, and dye manufactures, and they accounted for over two-thirds of musical instrument workers.

Complexes had substantial shares of the nation's employment in manufactures such as textiles and clothing (e.g., hats, caps, and bonnets) which shifted from home manufacturing into factories or workshops. Large markets also encouraged local producers of heavy consumer durables such as furniture to shift into factory production, and these firms leveraged that lead to supply furniture throughout the region. In contrast, bulky consumer durables heavily demanded by rural households (e.g., carriages and wagons) were produced throughout regional systems. Light, high-value metal manufactures serving intermediate industrial requirements and consumer demands such as hardware and cutlery faced large demands in complexes; nevertheless, they could be shipped over large market areas, thus allowing entrepreneurs elsewhere to start factories and workshops to compete with these in complexes. The machinery industry drew on specialized skills, much work still remained customized, and production required substantial capital investment. As the largest concentration of industry demanding machinery, complexes presented unique market opportunities, and they contained skilled workers and ample capital. Metropolises housed over one-fourth of the nation's machinery workers, and satellites had substantial employment; collectively, complexes dominated national production, with 43 percent of machinery employment.

Diversity and Specialization

Within metropolitan complexes the broad range of economic activity—including finance, wholesaling, transportation, manufacturing, and agriculture—and household consumers presented diverse markets to manufacturers, and large-scale demand for numerous goods allowed firms in many industries to meet market thresholds to operate profitably. New York's complex, with the greatest range and scale of economic activity, had the most diversified manufacturing (measured by the Herfindahl index), and Philadelphia's complex, the second largest economic center, was the next most diversified. Boston's complex was much less diverse (table 7.5).[4] The Baltimore complex's high diversity is misleading because manufactures of its metropolis heavily impacted the diversity index, and its satellites housed little industry. Each complex's metropolis had more diverse manufacturing—befitting its larger scale and broader range of economic activity—than its satellites. Satellites of New York's and Philadelphia's complexes also were diverse, because these rings were massive industrial centers in their own right; Boston's satellites, however, were less diverse.

Yet industrial subareas within complexes often specialized (as measured by location quotients).[5] Metropolises and satellites of each complex exhibited little specialization in entrepôt manufactures; sugar, chocolate, and confectionery manufactures—consumer products sold to large agglomerations of prosperous urban and nearby rural consumers—were exceptional, and each metropolis was highly specialized (see table 7.5). Complexes had high specialization in several commerce-serving manufactures; the New York, Philadelphia, and Boston metropolises, the leading headquarters of firms controlling exchanges of commodity and financial capital and pivots of information, specialized in printing and publishing. Although firms sold books outside the complexes, the major components of printing and publishing served local demands, and satellites of New York and Philadelphia specialized in paper manufacturing, the input to this industry. The cordage specialization of the New York, Boston, and Baltimore metropolises supported shipbuilding, repair, and supply for these shipping centers. As the nation's largest urban agglomerations, eastern complexes specialized in urban infrastructure manufactures. Satellites of New York, Philadelphia, and Baltimore and the latter's metropolitan county were moderately specialized in brick and lime manufacturing, suggesting this bulky, low-value production mostly operated from low-cost peripheral land in the complex. In contrast, every metropolis specialized in construction manufactures—cutting and shaping granite and marble products—which were difficult to transport and were made to order; either metropolises (in the case of New York and Boston) or their satellites (in the case of Philadelphia) specialized in glass manufacturing.

Manufactures sold to prosperous consumers or those that required specialized knowledge and skill to produce—such as precious metals; drugs, medicines, paints, and dyes; and musical instruments—were specialties of New York, Philadelphia, and Boston (see table 7.5). Firms probably supplied large shares of demand within their regional industrial systems because metropolises accounted for over half of national production (see table 7.4). No metropolis specialized in cotton or woolen textiles, which together accounted for 84 percent of textile employment (see table 7.5). Boston's satellites towered as specialists in cotton and woolen textiles; they supplied other industrial systems of the East, and surplus production also entered the Midwest. But the extensive cotton textile specialization of Baltimore's satellites represented a tiny share of the nation's employment (0.5 percent). New York's complex was a net importer of cotton and woolen textiles from other complexes, from the rest of its regional system, and from elsewhere in the East; as the nation's largest concentrated market, it offered lucrative opportunities for textile firms. Mixed-textile manufacturing in New York, Philadelphia, and Baltimore metropolises accounted for 36 percent of national employment in that sector. Philadelphia's metropolis and satellites, the second largest textile center among eastern complexes (with 10,744 workers), housed the widest array of textile specialties, whereas Boston and its satellites, the largest textile employers (with 13,289 workers, only 32 in the metropolis) focused on cotton and woolen textiles. The widespread demand for clothing—such as hats, caps, and bonnets—among prosperous urban and rural consumers lured firms in eastern complexes to meet the large local demand. Firms in the metropolis and satellites of every complex produced these goods, but the metropolises and satellites of New York and Boston had the greatest specialization; their firms reached throughout the East and elsewhere.[6]

In contrast, heavy, bulky consumer durables—such as furniture, carriages, and wagons—faced moderately high transportation costs, and the widespread demand for them opened up opportunities for firms in the metropolis and satellites of every complex to enter production (see table 7.5). New York's metropolis, with its huge local market, specialized in furniture, and its satel-

Table 7.5. Specialization (Location Quotients) of Metropolitan Industrial Complexes of New York, Philadelphia, Boston, and Baltimore, 1840

	New York			Philadelphia			Boston			Baltimore		
Manufacture	Metro	Sat	Comp	Metro	Sat	Comp	Metro	Sat	Comp	Metro	Sat	Comp
Entrepôt	0.38	0.62	0.49	0.26	0.72	0.46	0.58	0.35	0.37	0.54	0.57	0.54
Liquor	0.92	0.76	0.85	0.38	0.13	0.27	1.54	0.04	0.15	0.48	0.07	0.44
Mills (flour, grist, lumber, oil)	0.06	0.41	0.22	0.04	0.77	0.35	0.03	0.15	0.14	0.10	0.82	0.17
Leather tanneries	0.51	1.06	0.76	0.28	0.96	0.57	0.25	0.94	0.89	1.12	0.45	1.05
Sugar, chocolate, confectionery	5.95	0	3.25	2.56	0.23	1.56	26.55	0.26	2.16	3.22	0	2.89
Tobacco goods	0.64	0.63	0.63	1.20	0.56	0.93	0	0.44	0.41	1.58	0	1.42
Commerce serving	3.09	0.88	2.09	1.57	0.86	1.27	4.48	0.86	1.12	1.25	0.56	1.18
Printing, publishing	4.49	0.47	2.66	2.08	0.42	1.37	7.43	0.35	0.86	1.19	0.75	1.14
Paper	0.32	2.37	1.25	0.98	2.67	1.70	0	1.05	0.97	0.67	0.64	0.67
Cordage	2.42	0.38	1.49	0.87	0.08	0.53	1.62	1.96	1.94	2.02	0	1.81
Local, regional, national market	1.09	1.17	1.13	1.27	1.13	1.21	0.90	1.29	1.26	1.18	1.22	1.18
Bricks, lime	0.02	1.28	0.60	0.67	1.53	1.03	0.33	0.29	0.30	1.41	2.16	1.49
Granite, marble	3.77	0.21	2.15	2.80	0.47	1.81	2.41	0.64	0.76	1.76	6.94	2.29
Earthenware	0.18	2.15	1.08	0.63	2.02	1.22	0.24	0.38	0.37	1.98	0.54	1.83
Precious metals	9.60	2.23	6.25	3.57	0.25	2.15	2.64	0.22	0.39	0.60	0	0.54
Various metals	3.43	0.59	2.14	0.60	0.18	0.42	5.53	0.66	1.01	1.56	0	1.40
Soap, candles	1.09	0.15	0.66	0.70	0.14	0.46	0.76	0.60	0.61	0.76	0	0.68

Glass	1.59	0.39	1.05	0.85	7.37	3.63	4.35	0	0.31	0.56	0	0.50
Drugs, medicines, paints, dyes	6.85	1.09	4.24	5.15	0.58	3.20	1.69	0.49	0.58	1.17	0.23	1.07
Total textiles	0.57	0.96	0.74	1.64	1.14	1.42	0.06	1.80	1.68	1.36	1.46	1.37
Cotton textiles	0.18	1.01	0.56	1.04	1.31	1.16	0	2.16	2.00	1.23	1.95	1.30
Woolen textiles	0.42	0.91	0.64	1.40	1.32	1.37	0	1.92	1.78	0.42	1.07	0.49
Silk textiles	0.06	0.46	0.25	1.58	1.22	1.43	0	0.36	0.33	0	0	0
Flax textiles	0	3.02	1.37	1.59	0	0.91	0	0.37	0.35	0	0	0
Mixed textiles	2.57	0.59	1.67	4.68	0.20	2.77	0.39	0.26	0.27	3.43	0.03	3.08
Hats, caps, bonnets	1.83	1.78	1.81	0.75	0.32	0.56	2.17	2.08	2.09	0.34	0.19	0.33
Leather goods	0.39	0.63	0.50	0.41	1.15	0.73	0.34	0.86	0.82	0.38	2.34	0.58
Musical instruments	7.68	0.33	4.34	2.76	0	1.58	42.85	0.62	3.67	0.81	0	0.73
Furniture	1.94	0.71	1.38	1.06	0.66	0.89	1.25	1.42	1.41	1.36	0.12	1.23
Carriages, wagons	0.56	2.15	1.29	0.58	1.39	0.93	0.50	0.51	0.51	0.72	1.02	0.75
Gunpowder	0	0	0	0	11.68	4.99	0	0.92	0.85	4.64	0	4.17
Cannon, small arms	0.31	1.21	0.72	0.19	0.16	0.18	0	0.08	0.07	0	0	0
Hardware, cutlery	0.68	0.82	0.74	1.15	0.80	1.00	0.21	1.93	1.81	0.17	0.31	0.18
Machinery	3.03	2.78	2.92	2.16	1.39	1.83	1.76	0.43	0.52	2.45	0.23	2.22
Herfindahl index	0.07	0.09	0.06	0.10	0.11	0.09	0.10	0.20	0.17	0.10	0.19	0.10

Source: U.S. Bureau of the Census, *Compendium of the Sixth Census, 1840.*

Note: Location Quotient: $(M_iC_i/M_tC_i)/(M_iN/M_tN)$ where M_iC_i is employment in manufacture *i* in areal unit *i*; M_tC_i is total manufacturing employment in areal unit *i*; M_iN is employment in manufacture *i* in the nation; and M_tN is total manufacturing employment in the nation.

lites specialized in carriages and wagons. Light, high-value goods—such as hardware and cutlery—met broad industrial and consumer demands, and they were shipped long distances; each metropolis and its satellites produced hardware and cutlery, but only Boston's satellites achieved significant specialization. Because machinery confronted the widespread demand from large industrial markets in the complexes, each metropolis and its satellites housed producers; every metropolis was highly specialized, and New York's satellites also specialized.

Satellite Specialization

Satellites operated in a nest of industrial linkages and market relations within the complexes and outside them. New York's satellites met huge, broad-based demands for manufactures from prosperous farmers and urban populations in the complex, and interlinkages among subcomponents united diverse industrial sectors. Seven out of twelve—of the satellites (represented by counties) housed highly diversified industrial structures (with Herfindahl indexes below 0.15, in table 7.6), suggesting that many small factories and workshops in numerous industries met local demands; nevertheless, some satellites possessed moderately concentrated industrial structures, implying that firms met demands outside of their counties. Satellite specialization reached extreme levels (with a location quotient greater than 1.49) in twenty-five of thirty separate manufactures (see table 7.6), whereas only seven of the manufactures showed up as specialties when satellites were aggregated (see table 7.5). This extraordinary specialization reveals the finely tuned division of labor in manufacturing among New York's satellites, which also encompassed the metropolis as factories and workshops produced goods for intermediate or final demand in other satellites, the metropolis, and outside the complex. Besides numerous factory villages, satellites housed several sizable industrial towns and cities, including Elizabeth (4,184 people) and Newark (17,290) in Essex, Jersey City (3,072) in Hudson, and Paterson (7,596) in Passaic.[7]

New York's satellites specialized in entrepôt manufactures such as liquor processing, leather tanneries, and tobacco goods; nonetheless, collectively, satellites had only 17 percent of their employees in entrepôt industries, and the New York metropolis had even less (11 percent), whereas entrepôt constituted 28 percent of the nation's industry (see table 7.6). New York's industrial complex was a destination of processed goods, not a center of processing in its own right. Satellites possessed minimal specialization in commerce-serving manufactures, except paper, the main input to the metropolis' huge printing and publishing industry. Large urban infrastructure needs supported many satellites' specialization in bricks and lime, and granite and marble came from Putnam County, while Hudson (including Jersey City) specialized in glass. High-value, light manufactures selling in distant markets were specialties in satellites. Several specialized in drug, medicine, paint, and dye manufacturing, and Essex, albeit with only forty-eight employees, offered a glimmer of the drug and chemical manufacturing complex emerging near New York City. Numerous satellites specialized in cottons, woolens, silks, and mixed textiles, a continuation of pre-1820 manufacturing, but the flowering of satellite textile manufacturing came after 1840. Passaic textiles reached a large scale (with 1,690 workers), constituting 60 percent of its manufacturing employment. Virtually all workers (89 percent) produced cotton goods. Several satellites specialized in hats, caps, and bonnets, an indicator of the apparel industry, and Fairfield housed the famous Danbury hat industry.

New York's satellites specialized in consumer and producer durables (see table 7.6). Middlesex and Hudson specialized in furniture manufacturing, but they accounted for only 17 percent of satellite furniture employment; the appearance of furniture firms in most satellites suggests that they served local markets. Numerous counties specialized in carriage and wagon manufacturing; collectively, satellites employed 1,599 workers, whereas the metropolis employed 504. Satellites met prodigious demands for vehicles to transport passengers and commodities among specialized production and consumption centers of the complex. Putnam and Passaic housed two of the largest machinery concentrations in the nation; together, they employed 1,151 workers, just shy of the 1,597 employed in the great machinery works of the New York metropolis. Putnam (with 350 workers) contained the famous West Point Foundry at Cold Spring, on the Hudson River, and Passaic (with 801 workers) housed the equally famous Rogers Locomotive and Machine Works in Paterson as well as other machinery firms in that city and nearby towns.[8] Satellites of New York stood unsurpassed among eastern complexes in the sweep of their industrial specialties (twenty-five out of thirty), whereas the number of specialties in satellites of Philadelphia (sixteen) and Boston (twelve) fell far below New York's, even though their satellites housed comparable numbers of industrial workers; Baltimore (eight) had the fewest specialties.

Most Philadelphia satellites, like New York's, housed diversified industrial structures, and small factories and workshops competed effectively to meet

Table 7.6. Diversification (Herfindahl Indexes) and Specialization (Location Quotients) within the New York Metropolitan Industrial Complex, 1840

New York Metropolis		Fairfield, Conn.		Putnam, N.Y.		Bergen, N.J.		Hudson, N.J.		Middlesex, N.J.	
($E = 15{,}718$; $H = 0.07$)		($E = 1{,}642$; $H = 0.13$)		($E = 582$; $H = 0.39$)		($E = 308$; $H = 0.36$)		($E = 125$; $H = 0.24$)		($E = 324$; $H = 0.11$)	
5.95	Sugar, confec	1.90	Earthenware	1.69	Paper	5.06	Paper	2.69	Print, publish	2.15	Liquor
4.49	Print, publish	1.82	Woolen textiles	3.56	Granite, marble	3.10	Cotton textiles	12.49	Cordage	1.61	Leather tanner
2.42	Cordage	4.30	Hats, caps	17.92	Machinery	1.60	Carriage-wagn	71.12	Earthenware	1.57	Tobacco goods
3.77	Granite, marble	3.36	Carriage-wagn					38.30	Glass	3.79	Paper
9.60	Precious metals	1.80	Hardware-cutly	Rockland, N.Y.		Essex, N.J.		3.10	Furniture	2.10	Bricks, lime
3.43	Various metals			($E = 395$; $H = 0.24$)		($E = 3{,}238$; $H = 0.12$)				8.90	Earthenware
1.59	Glass	Westchester, N.Y.		4.30	Bricks, lime	2.19	Leather tanner	Passaic, N.J.		4.68	Silk textiles
6.85	Drugs-paints	($E = 901$; $H = 0.14$)		3.78	Precious metals	2.43	Paper	($E = 2{,}811$; $H = 0.37$)		3.59	Furniture
2.57	Mixed textiles	1.79	Tobacco goods	5.31	Drugs-paints	2.82	Earthenware	3.32	Paper	2.45	Carriage-wagn
1.83	Hats, caps	4.96	Bricks, lime	2.12	Cotton textiles	8.15	Precious metals	2.86	Cotton textiles	1.96	Hardware-cutly
7.68	Music instrum	1.87	Earthenware			1.74	Various metals	12.86	Flax textiles		
1.94	Furniture	1.86	Drugs-paints	Orange, N.Y.		3.11	Drugs-paints	5.53	Cannon, arms	Somerset, N.J.	
3.03	Machinery	1.55	Woolen textiles	($E = 1{,}575$; $H = 0.10$)		1.99	Mixed textiles	8.49	Machinery	($E = 399$; $H = 0.14$)	
		1.68	Silk textiles	2.58	Liquor	3.70	Hats, caps			1.67	Liquor
		3.79	Music instrum	2.32	Tobacco goods	3.61	Carriage-wagn	Morris, N.J.		1.52	Print, publish
		2.44	Carriage-wagn	2.34	Paper			($E = 776$; $H = 0.10$)		2.00	Bricks, lime
		4.15	Hardware-cutly	2.84	Bricks, lime			4.25	Liquor	5.96	Flax textiles
				2.63	Woolen textiles			4.33	Paper	2.07	Hats, caps
				1.79	Carriage-wagn			1.75	Bricks, lime	4.68	Carriage-wagn
								1.95	Silk textiles		
								1.93	Hats, caps		
								2.50	Carriage-wagn		

Source: U.S. Bureau of the Census, *Compendium of the Sixth Census, 1840.*

Note: Location Quotient greater than 1.49; number precedes name of manufacture.

E: number employed in manufacturing

H: Herfindahl index

Manufacture abbreviations: Cannon, arms: Cannon, small arms; Carriage-wagn: Carriages, wagons; Drugs-paints: Drugs, medicines, paints, dyes; Hardware-cutly: Hardware, cutlery; Hats, caps: Hats, caps, bonnets; Leather tanner: Leather tanneries; Music instrum: Musical instruments; Print, publish: Printing, publishing; Sugar, confec: Sugar, chocolate, confectionery.

Table 7.7. Diversification (Herfindahl Indexes) and Specialization (Location Quotients) within the Philadelphia Metropolitan Industrial Complex, 1840

iladelphia Metropolis		Bucks, Pa.		Chester, Pa.		Burlington, N.J.	
E = 14,979; H = 0.10)		(E = 827; H = 0.13)		(E = 1,811; H = 0.10)		(E = 851; H = 0.13)	
56	Sugar, confec	1.98	Leather tanner	1.54	Leather tanner	5.59	Paper
08	Print, publish	1.79	Tobacco goods	5.34	Paper	1.68	Bricks, lime
80	Granite, marble	2.98	Bricks, lime	2.14	Bricks, lime	2.26	Earthenware
57	Precious metals	10.17	Earthenware	2.12	Earthenware	4.78	Glass
15	Drugs-paints	12.21	Silk textiles	2.23	Leather goods	2.18	Leather goods
58	Silk textiles	2.08	Leather goods	1.69	Carriage-wagn	1.77	Furniture
59	Flax textiles	2.07	Carriage-wagn			4.04	Carriage-wagn
68	Mixed textiles			Delaware, Pa.			
76	Music instrum	Montgomery, Pa.		(E = 1,551; H = 0.35)		Gloucester, N.J.	
16	Machinery	(E = 2,079; H = 0.16)		2.85	Paper	(E = 1,291; H = 0.28)	
		3.71	Paper	1.86	Earthenware	60.54	Glass
		3.12	Bricks, lime	3.03	Cotton textiles	1.79	Drugs-paints
		1.60	Cotton textiles	2.53	Woolen textiles	1.96	Silk textiles
		1.68	Woolen textiles	1.50	Machinery	1.61	Carriage-wagn
		8.26	Gunpowder			1.97	Hardware-cutly
				Mercer, N.J.			
				(E = 1,001; H = 0.12)		New Castle, Del.	
				2.29	Leather tanner	(E = 1,751; H = 0.16)	
				1.87	Woolen textiles	1.74	Cotton textiles
				1.51	Carriage-wagn	64.66	Gunpowder
						5.09	Machinery

urce: U.S. Bureau of the Census, *Compendium of the Sixth Census, 1840.*
Note: See table 7.6 note for codes and abbreviations.

broad-based demands in prosperous local markets (table 7.7). Besides numerous factory villages, satellites included modest-sized industrial towns and cities: Camden (with 3,371 people) in Gloucester, Trenton (4,035) in Mercer, and Wilmington (8,367) in New Castle. Satellites specialized in entrepôt manufactures even less frequently than New York's satellites; leather tanneries, the only notable specialty, built on extensive beef cattle feeding, dairying, and slaughtering in the satellites, and some tannery specialists also produced leather goods. Counties around Philadelphia specialized in paper manufacturing for the metropolis' burgeoning printing and publishing industry, and its huge urban infrastructure requirements supported extensive satellite specialization in bricks and lime as well as glass. Across the Delaware River from Philadelphia, Gloucester's 653 glass workers accounted for 20 percent of national glass employment; its firms supplied markets along the East Coast.

High-value, light-industrial specialists included Gloucester, with its drug, medicine, paint, and dye production, and most satellites specialized in textiles. Cotton and woolen textile employment in Philadelphia's satellites (with 2,727 cotton textile workers and 811 woolen textile workers) numbered about the same as in New York's satellites (with 2,471 in cottons and 653 in woolens). Nevertheless, textile workers in Philadelphia's complex (10,744) significantly exceeded those in New York's complex (6,184), because Philadelphia's metropolis (with 7,086 workers) housed a much larger textile industry than New York's metropolis (2,566). Consumer and producer durables in Philadelphia's satellites mirrored those of New York's satellites. Every satellite housed furniture producers, revealing the widespread demand for this bulky good in prosperous local markets, but firms had difficulty capturing large market areas; only Burlington specialized. Most satellites specialized in carriage and wagon manufacturing to meet the demands for vehicles to transport people and commodities across this extensive agricultural and industrial complex.

With 521 machinery workers, Philadelphia's satellites had less than New York's satellites (with 1,221 workers), and this was comparable to the gap between the metropolises (1,084 in Philadelphia and 1,597 in New York). Philadelphia's satellites specialized in machinery, and New Castle housed the city of Wilmington; its 299 machinery workers augured future growth of this budding machinery center. Nearby, the great gunpowder works of DuPont gave New Castle its extreme specialization in that industry. On all accounts Philadelphia's satellites loomed as major industrial centers, but they lagged significantly behind New York's. That gap never closed, as New York's complex continued as the largest, most diversified industrial agglomeration in the nation, whereas Philadelphia's complex remained in second place in the East.[9]

Boston's satellites housed an enormous concentration of manufacturing; Middlesex (with 13,862 workers), the largest industrial satellite among eastern complexes, ranked among the least diversified (table 7.8). Nevertheless, even with its textile focus, Middlesex possessed significant employment in most industrial sectors, and neighboring satellites were moderately diversified; they met the widespread industrial demands of prosperous farmers and urban populations. Numerous industrial villages, towns, and cities ringed Boston, including Brockton (with 2,616 people) in Plymouth; Cambridge (8,409) and Lowell (20,796) in Middlesex, and Haverhill (4,336), Lynn (9,367), and Salem (15,082) in Essex.[10] Boston's satellites exhibited little specialization in entrepôt manufactures, yet satellites contained substantial commerce-serving industrial

Table 7.8. Diversification (Herfindahl Indexes) and Specialization (Location Quotients) within the Boston and Baltimore Metropolitan Industrial Complexes, 1840

Boston Complex					
Boston Metropolis (E = 1,981; H = 0.10)		Essex, Mass. (E = 4,133; H = 0.12)		Norfolk, Mass. (E = 5,222; H = 0.20)	
1.54	Liquor	2.31	Tobacco goods	2.06	Paper
26.55	Sugar, confec	3.95	Cordage	2.59	Cordage
7.43	Print, publish	3.98	Woolen textiles	1.74	Silk textiles
1.62	Cordage	2.30	Flax textiles	7.75	Hats, caps
2.41	Granite, marble	1.86	Music instrum	1.85	Hardware-cutly
2.64	Precious metals	2.30	Furniture		
5.53	Various metals			Plymouth, Mass.	
4.35	Glass	Middlesex, Mass.		(E = 2,242; H = 0.15)	
1.69	Drugs-paints	(E = 13,862; H = 0.39)		6.85	Cordage
2.17	Hats, caps	3.24	Cotton textiles	2.16	Hats, caps
42.85	Music instrum	1.95	Woolen textiles	4.56	Furniture
1.76	Machinery	1.69	Gunpowder	16.99	Hardware-cutly

Baltimore Complex					
Baltimore Metropolis (E = 7,905; H = 0.10)		Harford, Md. (E = 234; H = 0.16)		Anne Arundel, Md. (E = 663; H = 0.28)	
3.22	Sugar, confec	1.58	Mills	1.92	Bricks, lime
1.58	Tobacco goods	2.45	Paper	9.39	Granite, marble
2.02	Cordage	2.83	Bricks, lime	2.63	Cotton textiles
1.76	Granite, marble	2.05	Earthenware	2.05	Leather goods
1.98	Earthenware	3.33	Woolen textiles		
1.56	Various metals	3.19	Leather goods		
3.43	Mixed textiles				
4.64	Gunpowder				
2.45	Machinery				

Source: U.S. Bureau of the Census, *Compendium of the Sixth Census, 1840.*
Note: See table 7.6 note for codes and abbreviations.

specialization, including cordage, serving the large shipbuilding industry, and paper, serving Boston's vibrant printing and publishing industry.

Boston's satellites towered over the nation as specialists in high-value, light manufactures. They contained 14 percent of the nation's cotton textile workers and 13 percent of its woolen employees, making them dominant suppliers of textiles to the East and Midwest. Middlesex, the premier center (with 9,853 employees), alone accounted for 12 percent of the nation's cotton textile workers, and it also specialized in woolen textiles, as did nearby Essex; although Norfolk did not specialize in cotton or woolen textiles, it had sizable numbers of workers. These satellites also held commanding positions in hat, cap, and bonnet manufacturing: the complex as a whole accounted for 15 percent of the nation's employment, and Norfolk alone contributed 10 percent of the national total. Specialization in consumer and producer durables across Boston's complex differed from New York's and Philadelphia's. Compared to numerous furniture workers in the metropolises of New York (1,420) and Philadelphia (736), few worked in Boston (115); however, its satellites specialized in furniture (with 1,684 employed), and the total number of workers in Boston's complex (1,799) compared favorably to New York's (1,851) and Philadelphia's (1,080). Although no Boston satellites specialized in carriage and wagon manufacturing, they housed most of this sector (with 743 of 799 workers in the complex). Extraordinary specialization in hardware and cutlery production distinguished Boston's satellites from those of other complexes. Norfolk and Plymouth specialized, and together they housed 12 percent of the nation's employment, and Plymouth alone accounted for 10 percent, suggesting they supplied hardware and cutlery goods to other regional industrial systems. Boston's satellites did not specialize in machinery manufacture, and their total number employed (366) remained far below New York's (1,221) and somewhat below that of Philadelphia's (521) satellites.

Because Baltimore's satellites housed little industrial employment, they made insignificant contributions to specialization within the complex; most satellite manufacturers processed resources, and milling and paper were specialties (see table 7.8). Satellites housed typical urban infrastructure manufactures—bricks and lime; granite and marble—and, in keeping with this resource-related emphasis, they specialized in leather goods; specialization in cotton and woolen textiles represented outliers of the Baltimore County mills. This minuscule differentiation of Baltimore's satellites diverged from satellites of other metropolises because the complex mostly ended at the border of Baltimore County. In sum, metropolitan industrial complexes of the East Coast dominated the nation's manufacturing by 1840, prior to the onset of rapid industrial growth. Their highly differentiated manufacturing structures rested on a division of labor among firms and localities, and their prosperous farmers contributed to industrial demands, although large urban populations proved more important. Simultaneously, outside the complexes manufacturing expanded to serve local and subregional markets in areas of prosperous agriculture; this manufacturing reveals the impact of local and distant demands for goods on the East's industrial structure.

Manufacturing in Areas of Prosperous Agriculture

Broad-Based Industry

In prosperous agricultural areas broad industrial demands came from household consumers, urban infrastructure, natural resource sectors, and intra- and interregional trade, and manufactures meeting these demands required vibrant producer durables. Rich farmers, small wholesalers, large retailers, professionals, and manufacturers accumulated capital that workshops, nonmechanized factories, and a few large factories tapped to expand existing firms or start new ones. Firms generated significant productivity gains, and members of the citizenry, including manufacturing owners and mechanics, were highly inventive. In central and western Massachusetts, Connecticut, New York State along the Erie Canal, southeastern Pennsylvania, and Maryland, prosperous agricultural counties displayed significant industrial outcomes of this interaction of demand and supply by 1840 (table 7.9).

Agricultural employment still surpassed nonagricultural in every county, and the agriculture/nonagriculture ratio in Franklin (in Massachusetts) and in many counties along the Erie Canal reached 2.5 to 1 or higher, within range of the national ratio of 3.5 to 1. Nevertheless, the 50,766 manufacturing workers in these selected counties constituted 13 percent of the nation's industrial labor force, equivalent to 56 percent of the total in the metropolitan complexes. Virtually every county had more people employed in manufacturing than in commerce, and several housed over 3,000 industrial workers; the 23,959 residing in the Erie Canal corridor totaled almost as many as in each of the three greatest metropolitan complexes, New York, Philadelphia, and Boston (see table 7.3). Most prosperous agricultural counties contained diverse industrial structures serving broad demands for manufactures (see table 7.9). Consistent with large agricultural and other resource sectors, most counties' share of employment in entrepôt manufactures was at or above the 28 percent for the nation. Few, however, had over half of their manufactures in the entrepôt category; most had two-thirds or more of their employment in commerce-serving and manufacturing for local, regional, and national markets.

Massachusetts and Connecticut

Since the colonial period Massachusetts and Connecticut counties had been integrated into the wholesaling and trading webs of Boston or New York.

Table 7.9. Population, Employment, and Industrial Structure of Selected Prosperous Agricultural Counties of the East, 1840

	Number						Percentage of Total Manufacturing Employment			
		Employment			Selected Employment					
State/County	Population	Agricultural	Nonagricultural	Total	Commerce	Manufacturing	Entrepôt	Commerce	Local, Regional, National	Herfindahl Index[a]
Massachusetts										
Worcester	95,313	17,585	12,707	30,292	503	8,001	6.8	2.1	91.1	0.23
Franklin	28,812	6,017	1,738	7,755	182	1,185	20.4	1.3	78.3	0.13
Hampshire	30,897	6,547	2,808	9,355	236	1,587	16.3	8.4	75.3	0.11
Hampden	37,366	6,421	5,685	12,106	247	4,006	6.0	3.6	90.4	0.39
Berkshire	41,745	8,658	3,841	12,499	124	2,573	12.9	12.2	74.9	0.21
Connecticut										
Hartford	55,629	8,140	6,402	14,542	771	3,660	13.6	11.0	75.4	0.09
Litchfield	40,448	8,422	3,372	11,794	191	2,274	9.3	0.9	89.8	0.21
New York Erie Canal										
Albany, Schenectady, Rensselaer	146,239	19,455	13,417	32,872	1,737	6,672	13.5	5.4	81.1	0.12
Montgomery	35,818	6,226	4,847	11,073	411	305	63.6	0.0	36.4	0.18
Herkimer	37,477	12,596	3,672	16,268	353	1,276	35.9	3.0	61.1	0.10

Oneida	85,310	16,297	7,951	24,248	675	3,643	20.2	3.4	76.5	0.17
Madison	40,008	9,631	2,857	12,488	158	905	38.1	5.9	56.0	0.15
Oswego	43,619	8,605	2,318	10,923	87	555	61.8	6.1	32.1	0.25
Onondaga	67,911	11,741	6,488	18,229	681	1,437	26.7	3.5	69.9	0.10
Cayuga	50,338	11,020	3,915	14,935	478	1,838	18.7	3.1	78.2	0.10
Seneca	24,874	4,808	1,616	6,424	149	392	44.1	5.4	50.5	0.13
Ontario	43,501	10,137	3,406	13,543	438	990	46.0	4.0	50.0	0.15
Wayne	42,057	7,565	2,747	10,312	160	528	32.6	2.8	64.6	0.11
Monroe	64,902	10,045	6,729	16,774	745	2,437	37.5	4.7	57.8	0.09
Livingston	35,140	7,872	2,769	10,641	137	704	29.0	14.5	56.5	0.10
Genesee	59,587	11,412	2,893	14,305	252	762	35.6	2.9	61.6	0.13
Erie	62,465	11,022	5,317	16,339	893	1,515	36.4	4.8	58.8	0.13
Total Erie Canal	839,246	158,432	70,942	229,374	7,354	23,959	26.9	4.6	68.5	0.08
Pennsylvania										
Lancaster	84,203	10,393	4,766	15,159	250	1,617	46.7	3.0	50.3	0.09
York	47,010	5,558	3,909	9,467	175	1,254	61.6	4.4	34.0	0.13
Maryland										
Frederick	36,405	3,896	1,838	5,734	89	650	48.8	7.1	44.2	0.14
United States	17,120,000	3,735,901	1,063,598	4,799,499	117,607	387,303	28.1	5.4	66.6	0.09

Source: U.S. Bureau of the Census, *Compendium of the Sixth Census, 1840.*

Note:

[a] The Herfindahl index is based on the thirty selected manufactures. See table 7.4 for a list of manufactures.

Information about markets and technical information relevant to manufacturing flowed back and forth between the metropolises and their interior counties; thus, local manufacturing firms gained opportunities to acquire access to fixed capital, working capital as credit, and new technology. These counties became nationally prominent industrial centers; as a share of the nation, they housed 1.7 percent of agricultural workers, slightly below their 1.9 percent share of the population, but they contained 6 percent of manufacturing workers (see table 7.9). Worcester, Hampden, Berkshire, and Litchfield were among the least diversified of the prosperous agricultural counties, but this obscures their industrial structures; every county in Massachusetts and Connecticut contained 60 percent or more of the thirty selected manufactures (see table 7.4). Each county had firms in local and subregional (county or nearby county) market manufacturing—such as bricks and lime, milling, carriages and wagons, furniture, and machinery—and most had substantial employment, yet none specialized in them (table 7.10). Hartford, home of the subregional metropolis of the same name, possessed one of the most diversified industrial structures in the East, and it had related metropolitan specialties—tobacco goods, printing and publishing, and precious metals. These counties housed significant manufactures beyond their sizable local and subregional industries: firms leveraged access to capital, markets, and information to build powerful industrial sectors, and they specialized in manufactures selling in markets elsewhere in their regional industrial systems and, in some cases, in other systems.

Most counties in Massachusetts and Connecticut participated in early textile mill developments between 1790 and 1820; they contained 11 percent of the nation's textile employment by 1840, almost six times their population share. Many of their textile mills exported large shares of production to other parts of Boston's and New York's regional systems, to elsewhere in the East, and to the Midwest and South. Every county specialized in one or more types of textile manufacturing, and they contained famous firms on their roster (see table 7.10). In the town of Webster (Worcester County) Samuel Slater's descendants employed 175 to 200 textile workers, and in Chicopee (Hampden) consortiums of the same Boston Associates who had developed Lowell and other textile mill towns in eastern Massachusetts, southern New Hampshire, and Maine owned three firms with large mills that together employed 1,500 to 2,000 workers. Extreme specialization in silk textiles in Hampshire rested on firms in Northampton, including the Northampton Silk Company, capitalized at $100,000,

Table 7.10. Industrial Specialization (Location Quotients) in Prosperous Agricultural Counties in Massachusetts and Connecticut, 1840

		Massachusetts			
	Worcester		Hampshire		Berkshire
1.66	Cotton textiles	5.16	Paper	9.40	Paper
3.28	Woolen textiles	3.77	Woolen textiles	2.19	Cotton textiles
5.87	Hats, caps	18.14	Silk textiles	2.19	Woolen textiles
2.14	Cannon, arms	2.47	Hats, caps	1.78	Mixed textiles
		1.62	Leather goods	2.12	Gunpowder
	Franklin	1.69	Hardware-cutly	2.59	Cannon, arms
1.96	Various metals				
1.70	Silk textiles		Hampden		
3.29	Mixed textiles	2.54	Paper		
4.62	Hats, caps	3.27	Cotton textiles		
2.56	Hardware-cutly	2.10	Mixed textiles		
		5.07	Gunpowder		
		15.52	Cannon, arms		
		Connecticut			
	Hartford		Litchfield		
2.80	Tobacco goods	5.47	Precious metals		
1.77	Print, publish	1.56	Various metals		
4.52	Paper	1.90	Woolen textiles		
3.81	Precious metals	10.11	Mixed textiles		
5.13	Various metals	4.16	Hardware-cutly		
3.63	Woolen textiles				
5.24	Silk textiles				
5.12	Gunpowder				
2.61	Cannon, arms				
5.80	Hardware-cutly				

Source: U.S. Bureau of the Census, *Compendium of the Sixth Census, 1840.*
Note: See table 7.6 note for codes and abbreviations.

and it attracted as many as twenty-two subscribers from New York. These counties participated in a precursor of the apparel industry; Worcester, Franklin, and Hampshire specialized in hats, caps, and bonnets, and their total employment of 2,934, including Worcester's 2,445, constituted 15 percent of the nation's workers.

The production of light, high-value metal goods started before 1820 and achieved a solid foundation by 1840; southern New England would remain prominent for the rest of the century. Worcester, Hampden, Berkshire, and Hartford Counties specialized in firearms, accounting for one-quarter of the nation's employees, and Franklin, Hampshire, Hartford, and Litchfield specialized in hardware and cutlery. They employed 9 percent of the nation's hardware and cutlery workers; when Hampden and Worcester are included, this set housed 12 percent, and, when Norfolk and Plymouth counties, immediately south of Boston, are added, this larger set accounted for one-fourth of the nation's workers and supplied other regional industrial systems. Extraordinary specialization in paper manufacturing in prosperous agricultural counties—including Hampshire, Hampden, Berkshire, and Hartford—illustrates the capacity of entrepreneurs to exploit social networks and technological advances to produce high-quality paper, a product that withstood long-distance transportation. These counties' paper workers accounted for 15 percent of the national total, and they amounted to just over half (52 percent) of the total number of paper workers in the four complexes.[11]

Berkshire paper mills exemplify this production for distant markets: early mills dated from around 1800 and traced their lineage from the Crane family, but most production met local demands in the county or in nearby counties in New England and New York. Farmers and businesses in Berkshire had no more than a fifty-mile wagon trip to Hudson River ports and their wholesalers, especially in Albany. After 1819 paper producers looked more to New York City's burgeoning market, the nation's greatest consumption center for expensive, high-quality paper for business. The costs of the wagon transport of paper to Hudson River ports and steamboat (or sailing vessel) shipment to New York City and the reverse transport of rags to mills represented trivial shares of selling prices of high-priced paper. By the early 1830s most paper production headed to New York City, and its wholesalers sold paper, supplied rags, and provided credit for Berkshire mills. By 1840 Berkshire's nineteen mills forged a strong position in paper manufacturing; their 295 workers constituted 6 percent of national employment. Compared to national averages, typical Berkshire mills had more employees (15.5 workers in Berkshire vs. 11.1 workers in the nation), larger capitalization ($15,474 vs. $11,139), and a higher value of output ($19,737 vs. $14,444).

Rather than operating as isolated entities, ruthlessly competing with one another, Berkshire paper mill owners functioned within wide-ranging social networks. As individuals left one firm to start another, they maintained ties to former mentors, partners, and coworkers; many of these network ties rested on kinship, but they were not restricted to it. Individuals skilled in commercial accounting, marketing, and sales shared expertise with mill owners, who maintained contacts with paper mills in the Connecticut Valley, eastern Massachusetts, and New York State. By the late 1820s, and even earlier, local machine shops accumulated expertise in designing, constructing, and repairing paper equipment. Some leading shops also maintained network ties with other machine shop centers, such as in the Connecticut Valley, widening the access of the Berkshire mills to the latest technology. Therefore, the social networks of owners and skilled workers in the Berkshire paper industry avoided the traps of isolation and of heavily redundant ties among a close-knit group. Instead, information, support, and capital moved through network ties that kept mill owners abreast of markets and technology, making them formidable competitors in the New York City market and elsewhere.[12]

The Erie Canal Corridor

The canal corridor from the Albany-Schenectady-Rensselaer conurbation at the intersection of the Mohawk and Hudson Rivers to Buffalo on Lake Erie was one of the nation's richest agricultural areas outside specialized farming zones near East Coast metropolises. Much of this corridor exemplifies the powerful impetus that prosperous agriculture gave to local and subregional industrialization; it housed 4 percent of the nation's agricultural employment, whereas it accounted for 6 percent of all manufacturing workers (see table 7.9). The large farm sector generated vast outputs requiring processing; thus, the entrepôt manufacturing share of industrial employment was near or above the national share in every county except Albany-Schenectady-Rensselaer, Oneida, and Cayuga. Yet most counties also had 50 to 80 percent of their industrial employment in manufactures meeting local, regional, and national consumer and producer demands. With a population of 146,239, the conurbation counties of Albany, Schenectady, and Rensselaer ranked among the greatest concentrated industrial markets in the East, and their urban component (the cities of Albany, Schenectady, and Troy) totaled 59,836, making it the largest urban market in the East outside the four metropolises (see table 7.3). The conurbation was the biggest manufacturing district in the canal corridor and contained a diversified industrial sector that met urban and rural demands (see table 7.9). Albany's (with a Herfindahl index of 0.10) 2,958 industrial employees worked in diverse sectors, whereas Schenectady's (0.20) 285 workers and Rensselaer's (0.22) 3,429 workers were more specialized.

Conurbation counties housed distinctive industrial sectors (table 7.11). Albany's specialties, befitting its subregional metropolis, resembled large metropolises; its entrepôt manufacture—sugar, chocolate, and confectionery—

Table 7.11. Industrial Specialization (Location Quotients) in Prosperous Agricultural Counties of the Erie Canal Corridor in New York, 1840

Albany, Schenectady, Rensselaer		Albany		Schenectady		Rensselaer	
1.53	Cordage	2.61	Sugar, confec	1.88	Cotton textiles	2.09	Glass
2.47	Granite, marble	2.20	Cordage	2.78	Carriage-wagn	2.37	Cotton textiles
1.76	Earthenware	5.58	Granite, marble	5.23	Machinery	6.24	Flax textiles
2.34	Precious metals	3.49	Earthenware			5.69	Gunpowder
3.21	Flax textiles	2.36	Precious metals			3.29	Hardware-cutly
2.71	Hats, caps	3.65	Various metals				
4.54	Music instrum	1.63	Drugs-paints				
2.93	Gunpowder	5.07	Hats, caps				
1.50	Cannon, arms	8.80	Music instrum				
1.96	Hardware-cutly	1.62	Carriage-wagn				
		2.63	Cannon, arms				
Montgomery		**Herkimer**		**Oneida**		**Madison**	
4.36	Liquor	2.31	Leather tanner	2.30	Glass	1.92	Leather tanner
1.98	Mills	9.27	Granite, marble	1.92	Cotton textiles	1.79	Various metals
2.78	Leather tanner	2.00	Various metals	1.85	Woolen textiles	4.89	Woolen textiles
2.02	Woolen textiles	2.13	Leather goods	1.87	Music instrum		
1.66	Silk textiles	2.27	Hardware-cutly				
1.84	Mixed textiles						
1.55	Furniture						
Oswego		**Onondaga**		**Cayuga**		**Seneca**	
2.92	Mills	1.60	Bricks, lime	1.81	Woolen textiles	5.09	Liquor
2.39	Leather tanner	2.16	Woolen textiles	1.56	Furniture	2.09	Paper

3.25	Paper	1.86	Leather goods	2.61	Hardware-cutly	1.52	Bricks, lime
5.46	Silk textiles	2.41	Carriage-wagn	5.83	Machinery		
1.55	Furniture	1.60	Machinery				
2.29	Hardware-cutly						
Ontario		Wayne		Monroe		Livingston	
1.53	Mills	1.55	Leather tanner	2.00	Leather tanner	1.62	Sugar, confec
2.62	Leather tanner	1.51	Bricks, lime	1.88	Sugar, confec	9.66	Paper
1.70	Earthenware	4.55	Earthenware	1.93	Various metals	1.71	Earthenware
3.15	Woolen textiles	2.31	Various metals	1.81	Drugs-paints	2.08	Drugs-paints
1.53	Silk textiles	4.31	Glass	4.56	Silk textiles	5.02	Silk textiles
1.69	Leather goods	2.87	Silk textiles	1.61	Hats, caps	2.54	Leather goods
2.29	Carriage-wagn	2.14	Leather goods	2.72	Furniture	1.82	Music instrum
1.79	Cannon, arms	2.40	Carriage-wagn			2.33	Carriage-wagn
		3.33	Machinery			1.58	Cannon, arms
Genesee		Erie					
1.57	Mills	2.45	Sugar, confec				
1.76	Woolen textiles	1.62	Print, publish				
12.59	Silk textiles	3.03	Bricks, lime				
2.34	Leather goods	2.38	Earthenware				
3.03	Carriage-wagn	3.10	Music instrum				
		3.08	Furniture				

Source: U.S. Bureau of the Census, *Compendium of the Sixth Census, 1840.*

Note: See table 7.6 note for codes and abbreviations.

was sold to prosperous urban and nearby rural consumers, and Albany only narrowly missed qualifying as having specialized in tobacco goods (location quotient = 1.48) and liquor (1.39). Its eight breweries employed 130 workers and produced 9 percent of the nation's commercial beer, and their average output was 4.4 times greater than the nation's typical brewery. Printing and publishing (1.43), the quintessential commerce-serving manufacture, almost qualified as a specialty, and specialization in cordage reflected extensive shipping operations of the river and canal port. Manufactures selling to prosperous consumers or requiring specialized knowledge and skill to produce—such as precious metals; drugs, medicines, paints, and dyes; and musical instruments—also were Albany specialties. It had consumer and producer durables manufactures but only specialized in carriages and wagons. Hats and caps, accounting for 26 percent of the industrial labor force, employed the most workers; this apparel focus expanded in the conurbation and nearby counties after 1840. Albany did not participate extensively in textiles by 1840, but the famous Harmony Manufacturing Company, backed by New York City financiers, opened as a cotton mill in 1837 near the Cohoes Falls on the Mohawk River.

Schenectady remained a small industrial center in 1840, but it specialized in cotton textiles, carriages and wagons, and machinery; the fifty machinery employees, a significant number for this period, foreshadowed its future importance. Rensselaer, including the city of Troy, loomed far larger as an industrial center, positioning it among the East's leading counties; it specialized in glass for containers and urban infrastructure. Textile employment, however, dominated Rensselaer's industry, accounting for 54 percent of manufacturing workers, especially in cotton textiles (82 percent). Hardware and cutlery (with 160 workers) was an important specialty, and the booming iron and stove industry in Troy and Albany built stoves for large market areas of the East. Erastus Corning, a prominent Albany business leader and railroad magnate, possessed major stakes in these industries, but census data do not adequately identify them. The conurbation's manufacturing firms had large local markets to service, but employment totals and levels of specialization in hats and caps, cotton textiles, and stoves suggest that some firms sold in larger regional markets, and a few reached to other regional industrial systems by 1840.[13] Corridor counties farther west faced smaller local and subregional markets; nevertheless, they remained in the web of finance, trade, and information dominated by New York City and, secondarily, Albany. Entrepreneurs built strong industrial enterprises, and some looked beyond local markets to subregional or regional markets.

Corridor counties from Montgomery (adjacent to the conurbation) west to Oswego (bordering Lake Ontario) contained rich agricultural areas possessing access by the Mohawk River and by wagon to Albany prior to the Erie Canal, so long as goods had high value per unit weight. Processing industries were important, and many remained significant after the canal opened, because the reduction of low-value bulk remained cost-efficient (see table 7.9). Montgomery and Oswego, including the flour milling port with the same name, specialized in milling, and the industry employed 46 percent of Oswego's manufacturing workers. Although not specializing in milling, other counties—including Herkimer (with 232 employees), Oneida (367), and Madison (190)—had numerous milling employees (see table 7.11), and their extensive cattle raising and fattening and dairying generated large supplies of hides, thus supporting leather tanning specialization.

Manufacturers served the local markets of prosperous farmers and small town residents: every county contained woolen mills and consumer and producer durables such as carriages and wagons as well as furniture. Herkimer and Oneida housed some of the East's richest farmers (specialists in high-quality cheese) and had sizable industrial employment (see table 7.9), and some firms targeted subregional and larger market areas. Oneida County, where the city of Utica is located, specialized in cotton and woolen textiles, and these firms employed 46 percent of its manufacturing labor force; cotton textiles (with 1,302 employees) rooted in Rhode Island technology (see table 7.11). Herkimer and Oneida possessed sizable metal producer durables sectors: Herkimer specialized in hardware and cutlery, and, though Oneida did not specialize, its 66 workers surpassed Herkimer's 41 employees. Neither county specialized in machinery, but Herkimer's 45 workers and Oneida's 141 ranked them as significant machinery producers.

Herkimer's firearms manufacturing remained tiny, and the seven employees probably worked in Remington's foundry at Ilion; it started nearby around 1820, manufacturing gun barrels and some complete firearms. Production surged after 1844, when the Remington firm purchased the equipment of the Ames Manufacturing Company in Chicopee Falls, Massachusetts, and hired William Jenks, the inventor of a breech-loading carbine. Remington and Sons Armory subsequently dominated Ilion's industrial landscape. Prosperous farmers in a middle tier of counties from Onondaga (near the Mohawk Valley) to Seneca (near the Genesee Valley) also provided markets for diverse manufactures; some were sold in subregional and larger market areas. The major industrial counties of Onondaga (the home of Syracuse) and Cayuga (Auburn) specialized in woolen textiles; Onondaga firms also specialized in carriages and wagons and Cayuga firms in furniture (see tables 7.9 and 7.11). Onondaga (with 77 workers) and Cayuga (360) specialized in machinery—in fact, it ranked among the nation's greatest machinery centers. The city of Auburn achieved fame by producing agricultural machinery. The Cowing and Company Pump and Fire Engine Works started business in Seneca Falls, in Seneca County, in 1840, but this future nationally prominent firm probably did not have a census listing then.[14]

The Genesee region—including Monroe (with its subregional metropolis of Rochester), Ontario, Wayne, Livingston, and Genesee Counties—ranked among the nation's great wheat producers in 1840, yet the Genesee revealed that even prosperous wheat farming, a less-intensive agriculture than dairying and cheese production, spurred eastern manufacturing. The sizable total of 5,421 industrial workers in these rural counties amounted to one-fifth the number in the Boston, New York, or Philadelphia complexes (see tables 7.3 and 7.9). Processing, chiefly flour milling, accounted for 29 to 46 percent of manufacturing workers, and both Ontario and Genesee Counties specialized in milling (see table 7.11). Monroe dominated Genesee flour milling, and its thirty-six mills, averaging 11,358 barrels of flour annually—almost seven times greater than the nation's typical mill (1,697 barrels)—contributed 5.5 percent of national production, but it did not meet the criterion for specialization. Genesee region farms engaged in extensive livestock raising and cattle fattening—importing cattle from hill country to the south and from Ohio. Ontario, Wayne, and Monroe Counties specialized in tanning. Nevertheless, Genesee region manufacturing was highly diversified, and local and regional market manufactures employed over half of the workers. Employment levels and specialization suggest that most nonprocessing manufactures served local (county) and Genesee region markets (see tables 7.9 and 7.11).

Consumer manufactures—such as woolen textiles, hats and caps, and furniture—located across the region, but counties seldom specialized in these goods. Each county, except Monroe, specialized in carriage and wagon manufacturing, yet Monroe's firms (with 183 workers) employed the greatest number; they met prodigious demands for the transport of bulk agricultural goods, especially wheat, to processing centers and Erie Canal ports, and they provided passenger transport for dispersed rural dwellers. Every county, except Ontario, housed small machinery shops, but only Wayne (with 59 employees) specialized. Monroe's (Rochester) manufactures typified a subregional metropolis; besides extensive flour milling, the county specialized in leather tanning as well as sugar, chocolate, and confectionery (see table 7.11). Making drugs, medicines, paints, and dyes—requiring specialized knowledge and skill—was a specialty, and Rochester's large consumer market of 20,191 people supported sizable furniture manufacturing serving markets elsewhere in the subregion; although Monroe did not specialize in machinery, the 72 employees made it a large production center.[15] Livingston's large paper manufactures (with 83 workers) and their extreme specialization imply that those firms supplied both Rochester businesses and other commercial firms in the subregion.

Although Buffalo (pop. 18,213), the subregional metropolis in Erie County, housed just 10 percent fewer people than Rochester in 1840, Erie's manufacturing labor force was only 62 percent as large. In the midst of rapid change Buffalo's industrial surge and dramatic growth as a lake port and regional metropolis of western New York came in the next two decades. Milling made up almost one-quarter of manufacturing employment and dominated Erie's industrial structure, but it just missed the specialization criterion, and the nine flour mills averaging 11,639 barrels annually matched Rochester's mills. Erie's (Buffalo) specialization exemplified those of a subregional metropolis (see table 7.11): sugar, chocolate, and confectionery served prosperous urban dwellers and nearby farmers; printing and publishing served numerous businesses; bricks and lime supported urban infrastructure construction; musical instruments met the demands of the local elite; and furniture met the needs of urban and nearby rural dwellers. Erie employed only thirty-nine machinery workers, but the explosive growth of agricultural machinery firms and those in other metal fabricating industries was imminent.[16]

Manufacturing in the Erie Canal corridor remained small scale in 1840, but entrepreneurs in this land of prosperous agriculture built a formidable array of industrial workshops and small factories meeting local and subregional demands and occasionally demands in other regional industrial systems of the East. This economy generated substantial capital to fund manufacturing, and the accumulation of industrial skills and capital supported its participation in large-scale industrial expansion between 1840 and 1860. The corridor became home to some of the United States' greatest industrial corporations, but some would operate in small cities—such as Amsterdam, Canajoharie, Little Falls,

Table 7.12. Industrial Specialization (Location Quotients) in Prosperous Agricultural Counties in Pennsylvania and Maryland, 1840

Pennsylvania				Maryland	
Lancaster		York		Frederick	
3.76	Liquor	6.82	Liquor	3.60	Leather tanner
2.47	Sugar, confec	2.18	Leather tanner	2.52	Paper
2.31	Tobacco goods	2.10	Tobacco goods	2.12	Bricks, lime
1.93	Earthenware	2.16	Paper	2.59	Earthenware
2.15	Various metals	2.87	Earthenware	12.43	Silk textiles
3.44	Silk textiles	1.69	Carriage-wagn	1.63	Carriage-wagn
1.54	Carriage-wagn	2.81	Hardware-cutly		
4.67	Cannon, arms				
2.31	Hardware-cutly				

Source: U.S. Bureau of the Census, *Compendium of the Sixth Census, 1840.*
Note: See table 7.6 note for codes and abbreviations.

Herkimer, and Ilion in the Mohawk Valley—away from the publicity that firms in East Coast metropolises received.

Industry in Agricultural Pennsylvania and Maryland

Lancaster and York, both west of Philadelphia, and Frederick, west of Baltimore, underscore the impact of prosperous agriculture on industrial development. Manufacturing firms in Pennsylvania's counties benefited from the greater density of free-farm agriculture, whereas firms in Frederick faced the depressant on industrial demand of the local slave agriculture and of slavery in counties immediately southward; Frederick's population, however, contained a smaller share of slaves. Consistent with their prosperous agriculture, the counties of Lancaster, York, and Frederick possessed large shares of manufacturing in processing (see table 7.9); although milling narrowly missed the cutoff for specialization, it constituted the first or second largest sector in each county, and processing specialties included liquor, leather tanning, and tobacco goods (table 7.12). The small cities of Lancaster (with 8,417 people) and York (4,779) provided modest industrial markets, and sugar, chocolate, and confectionery manufacturing in Lancaster County probably met the demands of its small subregional center and nearby rich farms. Like grain areas in the Genesee region, each county specialized in carriage and wagon manufacturing, and both Lancaster and York specialized in hardware and cutlery production. These Pennsylvania and Maryland counties replicated industrial transformation elsewhere in the East, and their diversified manufacturing met the broad-based local and subregional demands for goods.[17]

Poised for the Late-Antebellum Industrial Surge

In much of the East household manufacturing fell to minor levels by 1840, and metropolitan complexes and areas of prosperous agriculture away from the complexes possessed finely honed industrial structures. Metropolitan complexes led industrialization because they provided huge markets, vast amounts of capital, marketing information, and distribution services (e.g., wholesaling and transportation), and some of their industries produced for other areas of the East. Industrial transformation also swept up prosperous agricultural areas in Massachusetts, Connecticut, the Erie Canal corridor, and southeastern Pennsylvania and Maryland; most of their firms met local and subregional demands, but some produced for regional markets. Although the 1840 benchmark measured an economic downturn, the subsequent two decades witnessed one of the most rapid periods of industrial growth in United States history, and metropolitan industrial complexes and budding manufacturing centers outside complexes in areas of agricultural prosperity participated. Many factory goods continued selling within regional industrial systems, but advances in production organization, technological change, innovations in wholesaling, and transportation improvements, as well as the emergence of new products selling in large market areas, combined to create a greater integration of eastern industrial markets and an increased penetration of midwestern markets by eastern firms. Firms in shoes, cotton textiles, and Connecticut manufactures led the move to large market areas before 1820, and they continued expanding afterward.

CHAPTER EIGHT

Building Competitive National Market Industries

> More than 400 additional shoemakers live in this town [Marblehead], but as they work for Lynn employers, are included in the report of that town.
>
> —BENJAMIN MUDGE, 1832

> The cost of goods manufactured is diminished 20 per cent since the establishment of the Chicopee Manufacturing Company, in consequence of a reduction of wages, cost of raw material, and improvements in machinery. The depreciation in their value in market, is more than 33½ per cent.
>
> —WILLIAM FOSTER, 1832

> The markets for three-fourths of all the manufactured articles [of Connecticut firms] are New York, Philadelphia, Boston, Providence, and Baltimore. A great part are sold at those depots for the interior; some are exported to foreign countries.
>
> —H. L. ELLSWORTH, 1832

In the early 1830s observers recognized that firms pursued various strategies to reach the national market. According to Benjamin Mudge, shoe manufacturers managed a division of labor across urban centers, such as Marblehead and Lynn, in eastern Massachusetts, whereas William Foster noted the Chicopee Manufacturing Company, a large cotton textile enterprise of the Boston elite, stressed cost reductions to achieve market dominance. H. L. Ellsworth claimed that "Connecticut" firms sold goods in the national market, as well as internationally, through metropolitan wholesaling centers.[1] These strategies built on skills and advantages acquired before 1820; nevertheless, firms needed to surmount new challenges because competitors elsewhere in the East maintained market shares in some local and subregional markets, and new markets emerged outside the East. When midwestern markets finally supported production at scales larger than craft shops in the 1820s and 1830s, eastern producers of shoes, cotton textiles, and Connecticut manufactures supplied quality goods at such low prices—because the transport costs for light, high-value goods were a small share of purchase price—that midwestern competitors failed to capture significant market shares.

Shoe Firms Dominate without Machinery

Capturing the Lead

An observer predicting the future dominance of shoe production based on shares of total national production in 1810 would fail: at that time the Middle Atlantic led with 45 percent of production, whereas New England had 39 percent, but by 1860 New England accounted for 60 percent of production and the Middle Atlantic only 25 percent (table 8.1). The South's share declined to a trivial level by 1840, whereas the West (mostly Midwest) rose from almost nothing in 1810 to 10 percent by 1860. Massachusetts, New York, and Pennsylvania captured similar shares of national production in 1810, but Massachusetts' share more than doubled, New York's fell by one-third, and Pennsylvania's dropped by half by 1860. Ohio contributed about 5 percent of national production from 1840 to 1860; its firms produced about two-thirds of the West's production in 1840, but Ohio's share declined to 40 percent by 1860 as the region's population almost tripled.

If local craft workers produced all shoes, then a territory's share of national production divided by its share of national population should be 1.0 for each territorial unit. This provides a simple, albeit imprecise, benchmark to identify areas of surplus and deficit production, and, based on this measure, competitiveness of regional and state manufacturers diverged (see table 8.1). New England firms rapidly boosted their lead, whereas Middle Atlantic firms fell steadily to a small deficit by 1860. The West never shifted from the large deficit when it housed few people, and the South, whose relative deficit widened during the period from 1810 to 1840, remained a huge market throughout the antebellum. Massachusetts firms speedily ratcheted up surplus production from triple the state's consumption in 1810 to almost thirteen times by 1860, and its share of interstate shipments rose from 80 percent in 1810 to over 95 percent during the decade from 1840 to 1850. New York and Pennsylvania reached a balance between production and consumption by 1840 and maintained it until 1860. Ohio always fared better in meeting state consumption needs than the aggregate West, but its firms never boosted production beyond

Table 8.1. Distribution of the Value of Shoe Production by Region and State, 1810–1860

	Percentage of Total Value of Production				Percentage of Production per Percentage of Population			
Region	1810	1840	1850	1860	1810	1840	1850	1860
New England	38.5	49.3	55.4	59.7	1.9	3.8	4.7	6.0
Middle Atlantic	45.0	36.8	31.1	25.0	1.5	1.2	1.1	0.9
West	1.5	8.1	8.8	10.3	0.4	0.4	0.4	0.3
South	15.1	5.6	4.6	5.0	0.3	0.1	0.1	0.2
Total	100.1	99.8	99.9	100.0				
Selected States								
Massachusetts	20.9	37.2	44.7	50.3	3.2	8.6	10.4	12.9
New York	17.5	18.4	14.4	11.9	1.3	1.0	1.1	1.0
Pennsylvania	18.1	10.1	10.4	9.2	1.6	1.0	1.0	1.0
Ohio	—	5.1	4.3	4.0	—	0.6	0.5	0.5

Sources: Thomson, *Path to Mechanized Shoe Production in the United States,* 67, table 6.1; U.S. Bureau of the Census, *Historical Statistics of the United States,* ser. A7, 172, 195.

Note: New England is Massachusetts, Connecticut, Rhode Island, Maine, New Hampshire, and Vermont; Middle Atlantic is New York, New Jersey, Pennsylvania, Delaware, Maryland, and District of Columbia; West is Ohio, Indiana, Michigan, Illinois, Wisconsin, Iowa, Missouri, Minnesota, Kansas, Nebraska, California, Oregon, and Washington; and South is the remainder of states.

meeting half of consumption demand. As the earliest midwestern state to build on prosperous agriculture and forge an industrial machine, the failure of Ohio firms to capture larger market shares underscores the power of New England firms—principally those in Massachusetts—to dominate interstate shoe markets outside the East.[2]

The Massachusetts Paradox

Shoe firms in Boston's regional system dominated interstate markets, yet according to various measures this industrial system differed little from others in the East. All parts of the East accessed large urban markets in the metropolitan complexes of their respective regional systems, and, if transport costs constituted important components of final selling prices, then firms nearer a complex possessed competitive advantages, yet that differentiation among firms applied to each industrial system. Prosperous agriculture existed in each regional system; thus, each system had its own extensive demand for shoes and generated capital for funding small-scale firms (shoe shops). Wholesalers in each metropolis provided marketing and distribution services, allowing firms to reach large market areas, and all firms accessed those services. Early in the nineteenth century, however, Boston's regional system acquired distinctive features. Although many farms successfully met growing urban food demands, the region also contained numerous farms occupying hilly, low-fertility land, and could not compete in post-1820 agricultural markets. Therefore, many women, children, and men either left New England or were available for alternative employment. Some women and children on productive farms looked for other work at various times of the year, because they contributed low labor productivity within a hay, dairying, and grain agricultural system. Areas with higher rural densities and greater shares of unproductive farms contained larger supplies of low-wage labor.

As the core of Boston's industrial system, Massachusetts contrasts with New York State, the core of New York City's industrial system. Massachusetts had rural densities of fifty people per square mile in 1820, whereas New York had twenty-five. And the greater share of unproductive farms in Massachusetts generated densities of low-wage rural labor over twice as high as those in New York, and Massachusetts manufacturers accessed numerous unproductive farms in nearby New Hampshire and Maine.[3] The necessary condition of high rural densities and unproductive farms appeared throughout Rhode Island, Connecticut, and Massachusetts, but the concentration of shoe manufacturing in eastern Massachusetts resulted from a sufficient condition with related components: a specialized wholesale, marketing, and distribution complex and a production complex.

The Wholesale, Marketing, and Distribution Complex

Because everyone needed shoes, consumer demand responded to price changes, and when producers made low-cost shoes, rather than expensive custom-made shoes by skilled craft workers, they sold more shoes per capita. Rising per capita income from 1820 to 1860 also increased the demand for shoes. Estimates of per capita shoe output based on sales suggest it rose from 2.0 in 1810 to 2.1 in 1840 to 3.5 in 1850, but this underestimates greater per capita output because low-priced shoes constituted a growing share of shoe production. Because shoes sold across wide market areas, marketing and distribution required wholesalers, but large producers might internalize those functions. After 1820 wholesalers in Boston and manufacturers in Lynn—some of whom wholesaled—expanded their wholesale and retail sales outlets in the South and Midwest, and Massachusetts' interior manufacturers and merchants joined them. Thus, the Batcheller brothers, manufacturers in Brookfield in the 1820s and 1830s, owned a jobbing house in Boston and acquired regular customers in the South and West.

This wholesaling, marketing, and distribution juggernaut centered in Boston moved beyond general wholesale services: in 1828 Boston's shoe jobbing trade totaled one million dollars, whereas New York City had four minor jobbers; by 1856 Boston's two hundred wholesale and jobbing houses generated fifty million dollars of business, yet in 1858 New York City had only fifty-six shoe wholesale houses selling fifteen million dollars' worth of goods. Tight bonds between merchants and manufacturers existed in the industry hub of Lynn; during the years 1832–60 fathers of about half of new shoe manufacturers were either shoe manufacturers or merchants. In 1859 Boston's wholesaling-marketing-distribution services blanketed the nation, shipping shoes to metropolises such as Cincinnati, Chicago, St. Louis, New Orleans, and San Francisco and directly shipping to small cities, including Dayton (Ohio), New Albany (Indiana), and Dubuque (Iowa).[4] This specialized service complex provided Massachusetts shoe manufacturers lower-cost services and access to larger markets than those available to other manufacturers, and they also built a formidable production machine, relentlessly driving down shoe prices and raising product quality.

The Shoe Production Complex

Shoe firms in eastern Massachusetts benefited from the shoe service complex. Because individual firms faced a perfectly elastic demand curve, they sold their entire production without impacting shoe prices. Until the sewing machine was adapted to shoe manufacturing in the early 1850s, however, they possessed no machines that significantly lowered production costs. They found other means to raise productivity and lower costs, and implementing a division of labor among workers and firms offered one approach. After 1820 central shops expanded rapidly: owners devised a modest division of labor within central shops and organized a division of labor among other producers, such as ten-footers, linked to central shops.

As the central shop system became more elaborate, unskilled workers increasingly were employed for individual tasks, reducing labor costs more. Lynn's central shops confronted inadequate supplies of females in the family system and ten-footers (which included families); females handled too many household tasks and could not expand production much. Therefore, central shops innovated reorganized labor tasks: by 1830 a shop in Lynn sent materials to unrelated females—thus increasing the female pool—and to unrelated males, and by the 1840s most firms copied this strategy. During the 1840s and 1850s the division of labor within ten-footers advanced as owners took advantage of new tools and methods of production to subdivide tasks, and specialization in shoe types—ladies, men's, or children's—further contributed to organizational efficiency; town-level specialization appeared as clusters of firms making similar products. Under the central shop system firms needed little fixed capital investment; most capital was tied up in materials, and funding came from owners or through credit arrangements with merchants. This system flexibly expanded production: low fixed capital requirements and access to ample working capital facilitated easy entrée into shoe manufacturing; and shop owners added outworkers, thus avoiding bottlenecks in one production unit. This labor organization encouraged the rise of large manufacturers, which gained discounts from wholesalers or formed their own selling units.[5]

The emergence of a producer goods industry tied to shoe production augmented productivity gains from the division of labor. Prior to 1840 most producer goods' innovations facilitated the growing use of unskilled workers and enhanced their capacity to make cheap, good-quality shoes. Methods of pegging uppers to soles arose around 1811, thus reducing costs because unskilled labor pegged shoes faster than skilled sewers bound uppers and soles. By the 1820s pegged shoes were common. The wooden last—the form around which shoes were made—hindered the division of labor among separate shops and homes. Although a shoemaker needed only one set of different-sized lasts and made adjustments in fitting as required, the inability to manufacture standard lasts meant different production units had difficulty making parts fit. When Thomas Blanchard of Millbury, Massachusetts, perfected the irregular turning lathe for making gunstocks in the early 1820s, he quickly adapted it for last making; this machine method spread in the 1820s, and most lasts were machine made by the 1830s. Innovations in producing patterns, such as the system of diagram patterns for cutting introduced in Boston in 1832, also aided the standardization of shoe manufacture with unskilled labor.

The greater range of styles which Massachusetts shoe producers innovated required specialized tools and machines, and toolmakers, who were gaining new skills, made new styles possible. Starting in the 1830s, machines for working leather and new tools for individual tasks proliferated. The number of patents for crimping and other upper-shaping devices jumped from three in the 1820s to twenty in the 1830s, and that number was patented in each of the

Table 8.2. Shoe Industry by Region and State, 1850

Region	Employment			Output ($) per Firm	Capital ($) per Firm	Value Added ($) per Worker[a]	Capital ($) per Worker
	Total	Percentage Female	Per Firm				
New England	62,217	40.7	19.0	9,144	1,858	299	123
Middle Atlantic	31,645	21.4	5.9	3,133	905	377	172
Midwest	7,256	5.9	4.4	2,848	696	433	164
South	4,036	7.7	4.0	2,476	806	409	209
Total	105,154	31.2	9.3	4,774	1,142	339	145
Selected states							
Massachusetts	51,562	43.3	37.0	17,302	3,370	295	116
New York	13,796	24.3	6.5	3,670	1,029	408	180
Pennsylvania	10,785	16.9	5.0	2,639	881	363	191
Ohio	3,826	7.5	4.8	2,889	679	400	148

Source: Secretary of the Interior, *Abstract of the Statistics of Manufactures, Seventh Census, 1850.*

Notes: New England is Massachusetts, Connecticut, Rhode Island, Maine, New Hampshire, and Vermont; Middle Atlantic is New York, New Jersey, Pennsy!vania, Delaware, Maryland, and District of Columbia; Midwest is Ohio, Indiana, Michigan, Illinois, Wisconsin, Iowa, and Missouri; and South is Alabama, Arkansas, Florida, Georgia, Kentucky, Louisiana, Mississippi, North Carolina, South Carolina, Tennessee, Texas, and Virginia.

[a] Value added is computed as value of output minus value of raw materials. For value added per worker and capital per worker, the number of workers is computed as total males plus 0.5 times total females.

next two decades. Specialized bottoming tools and pincers for lasting appeared, and tools to cut sole channels and trimmers for soles and heel edges were patented. By the mid-1830s Brockton, Massachusetts, was a leading toolmaking center. A Lynn manufacturer first used a sewing machine for stitching uppers in 1852, and within four years Lynn firms used as many as a thousand sewing machines. During the 1850s a widening range of machines became available for various parts of the production process, which offered efficiencies through combining machines in large factories to carry out a range of manufacturing steps, and during the 1860s large steam-powered factories appeared.[6]

Massachusetts Dominates without Machinery

Nevertheless, contributions of machinery to shoe manufacturing remained minor before 1850. Massachusetts firms dominated by integrating a wholesale-marketing-distribution service complex with a production complex—based on a division of labor using unskilled labor and on hand tools and simple equipment—which achieved substantial productivity gains. During the period 1820–50 national labor productivity in shoes rose at real annual rates of 1 to 1.3 percent, and total factor productivity—including gains from division of labor and other productivity enhancements—rose 0.8 to 1.6 percent annually. Because individual craft shoemakers contributed little to those productivity gains, they reflect the success of Massachusetts producers, in concert with Boston's efficient service complex, in dominating shoe manufacturing (table 8.2). In 1850 New England manufacturers accounted for about 60 percent of national employment in shoes, and Massachusetts firms alone housed half of them. New England firms built their industrial organization on a division of labor with females constituting 41 percent of employees (43 percent for Massachusetts firms), whereas in the Middle Atlantic they made up only 21 percent. The Midwest and South employed few females, confirming that their firms were uncompetitive in markets for cheap shoes. Typical New England firms employed about three times as many workers and produced about three times as much output by value as firms in other regions, and Massachusetts firms had five to six times as many employees and produced five to six times the output by value of firms elsewhere.

Value added per worker gives a misleading measure of firm dominance; on that account New England and Massachusetts firms had the least productive employees, but that measure does not consider product mixes (see table 8.2). Outside New England small craft shops served custom markets and employed skilled, high-wage workers producing small quantities of shoes, whereas Massachusetts firms made cheap shoes, employed numerous low-wage workers, and invested minimal capital per worker, because their productivity came from the division of labor, not from tools and machines. Massachusetts' firms transformed operations after 1850, deploying large cash flows from sales, far beyond those of their competitors in other regions, and accessing highly liquid credit markets in Massachusetts to purchase specialized machinery such as sewing machines. These changes generated soaring productivity in the shoe industry during the 1850s as labor productivity rose 4.4 to 6.0 percent annually and total factor productivity increased 2.0 to 3.3 percent annually.[7] Nevertheless, this transformation of Massachusetts' shoe industry came long after its firms had achieved dominance (see table 8.1).

Elaborate social networks boosted the competitive advantages of Massachusetts firms. The wholesaling-marketing-distribution complex proved adept at conveying information to manufacturers about new markets and changes in demand such as new styles. External information entered eastern Massachusetts via shoe wholesalers, and both external and internal information circulated within the production complex through Boston wholesalers' and jobbers' ties to retailers and shoe manufacturers. Wholesalers' and retailers' participation in shoe manufacturing augmented the transfer of information and enhanced their capacity to form cooperative ventures. In Lynn, the largest shoe manufacturing center, local Quakers remained major shoe industry participants up to 1850; their positions in banking, finance, manufacturing, commerce, and real estate supported the expansion of shoe manufacturing, and their collective effort and cooperation augmented that capacity. Social network ties to Quaker communities nationwide enhanced the competitive capacities of Lynn manufacturers, because those ties provided trustworthy information and customers and offered means to impose sanctions for malfeasant behavior. Yet Quakers did not operate in a closed, heavily redundant network; outside the Quaker community they participated in business networks supplying diverse information, enhancing their competitive capacity.[8]

Because Massachusetts shoe manufacturers faced a perfectly elastic demand curve, each firm expanded production without affecting its selling price, which shifted competitive battles to external markets. When individual manufacturers lowered costs or developed new styles, their advantages did not undercut other local firms; therefore, firms gained from sharing information about techniques, new tools, and the organization of a division of labor, because other manufacturers might reciprocate. Firms elsewhere could not replicate this piecemeal, because it took collective action against the existing competitive Massachusetts complex; however, such mobilization posed inordinate organizational and logistical problems. Therefore, most shoe manufacturers elsewhere continued as small craft shops, but the increasing capacity of Massachusetts firms to employ machinery to produce higher-quality shoes nibbled away at craft shop markets, leaving them with only wealthy consumers demanding custom-made shoes. Outside New England manufacturers could not enter factory production until systems of machines were developed after the 1850s. In contrast to shoe manufacturers, which combined a wholesaling-marketing-distribution complex and a division of labor using simple technology, New England cotton textile firms combined their wholesaling-marketing-distribution complex with massive capital investments and sophisticated machinery technology.

New England Cotton Textile Firms Dominate

As of 1820, the cotton textile industry consisted of three cores based on different models: a Providence core with outliers in southern New Hampshire and upper New York State followed Slater's lead; a Philadelphia core of skilled immigrant mule spinners and skilled handloom weavers produced fine-quality yarns and fabrics; and an incipient Boston core combined substantial capital with machinery technology. Based on industrial measures, astute observers could not predict which core (or cores) would dominate after 1820. Rhode Island, Massachusetts, and New York each had two thousand to twenty-four hundred employees and capital investment of $1.5-$2 million in 1820. Maryland had the most capital-intensive firms (averaging $121,418), whereas New York's typical firm had about 40 percent as much capital, and Massachusetts and Rhode Island firms had about 20 percent as much.[9] This indeterminacy ended swiftly: the Boston core separated from the rest during the 1820s and by the early 1830s took a commanding lead, which it maintained for the rest of the century. Boston's investors shrewdly targeted the mass market for everyday cloth.

Supply Prices Drive Demand

From 1805 to 1820 cotton textile production surged, but the small initial base left production modest at the end of that period (fig. 8.1). Francis Lowell's brilliant ploy in the cotton textile tariff of 1816 priced cheaper British coarse fabrics out of the U.S. market and allowed domestic firms to ramp up produc-

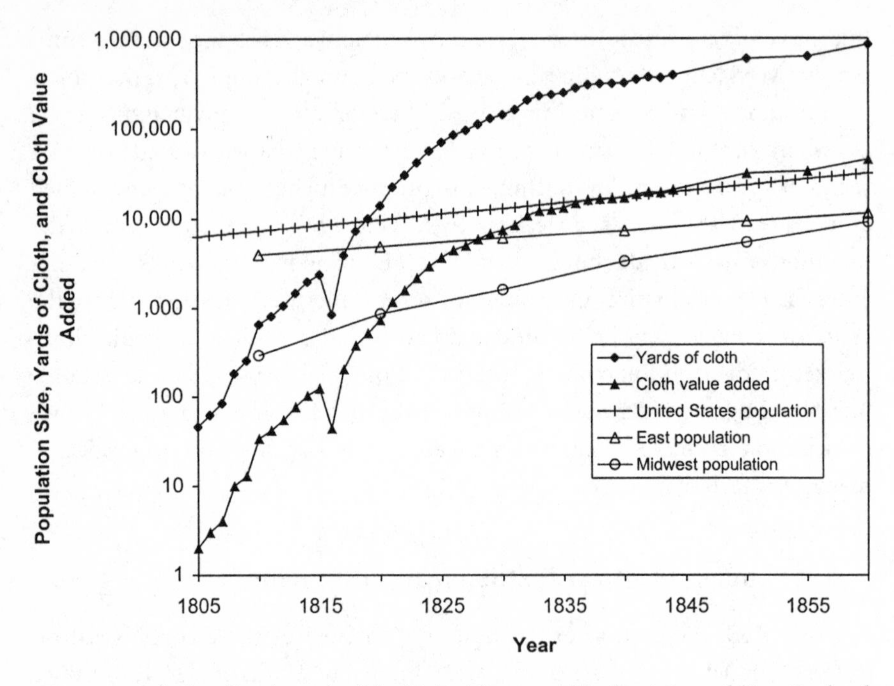

Fig. 8.1. Population Size (1,000) of the United States and Its Regions and New England Yards of Cotton Cloth (1,000) and Cloth Value Added ($1,000), 1805–1860. *Sources:* U.S. Bureau of the Census, *Historical Statistics of the United States,* ser. A7, 172, 195; Zevin, "Growth of Cotton Textile Production after 1815," 123–24, table 1.

tion of those fabrics, precisely the goods most demanded, and the tariff on coarse fabric remained sufficiently high for the next forty years to continue protecting American producers. Production of cotton cloth, measured by volume and value, was sixty-two times larger in 1860 than in 1820, but growth differed markedly over time; during the period from 1820 to 1835 production jumped eighteen times, whereas over the next twenty-five years it rose just over three times. Economic difficulties from 1837 to 1843 barely register in the growth of cloth output and of cloth value added.

An annual population growth of about 3 percent throughout the antebellum spurred the demand for cotton textiles, yet during the period from 1820 to 1835 the growth of output from cotton mills far surpassed national population growth, and it substantially exceeded population growth in the East, the largest market, and in the Midwest, the most rapidly growing market (see fig. 8.1). Population growth before 1835 accounted for less than one-third of the growth of cotton cloth production, and urban growth had even less impact. In 1820 only 7 percent of the nation's population lived in urban areas, and this figure rose to just 11 percent by 1840, a minuscule contribution to cloth demand, even accounting for urban residents' higher propensity to consume manufactures. Rising per capita income also stimulated demand, but gross domestic product per capita increased less than 1 percent annually before 1840 (see table 2.1), thus contributing trivially to the growth of cotton cloth production. Because cotton cloth has a high value relative to its weight, tiny reductions in cloth prices resulting from transportation improvements spurred little demand.[10]

Plunging prices of cotton cloth powered the vertical jump in cloth production after 1820, and much of this decline of about 50 percent—from twenty-one cents per yard to ten cents—ran its course around 1830; then the growth of cloth production slowed markedly (see figs. 8.1 and 8.2). After 1820 organizational and technical conditions were transformed—most new cotton mills were vertically integrated production systems of machine spinning and machine weaving—and all other important processes were mechanized by 1830. This technical progress accounted for about 80 percent of the decline in the price of cloth from 1815 to 1833; declining prices of raw cotton, the largest material expense, accounted for less than 20 percent of falling cloth prices. Reductions in textile workers' wages did not lower cloth prices; instead, their real wages rose during the period 1825–33, while labor costs declined (see fig. 8.2). Vertical integration of production and the mechanization of processes generated dramatic labor productivity gains—about 6.5 percent annually from 1820 to 1832, among the highest rates achieved in any industry during the antebellum. Plunging cloth prices drove hand loom weavers of coarse cloth out of business and encouraged households to cease manufacturing cotton cloth. These changes permeated production and consumption sectors by the early 1830s; subsequently, increased demand from population growth, urbanization, and rising per capita income supported most output growth, which accounts for the slower growth of cotton cloth production after the early 1830s.[11] The decision of New England entrepreneurs and investors to shift aggressively into cotton textiles during the 1820s set future contours for the industry.

New England Leads the Textile Industry

New England firms contributed slightly greater national shares of employment and capital investment in cotton textiles than Middle Atlantic firms in

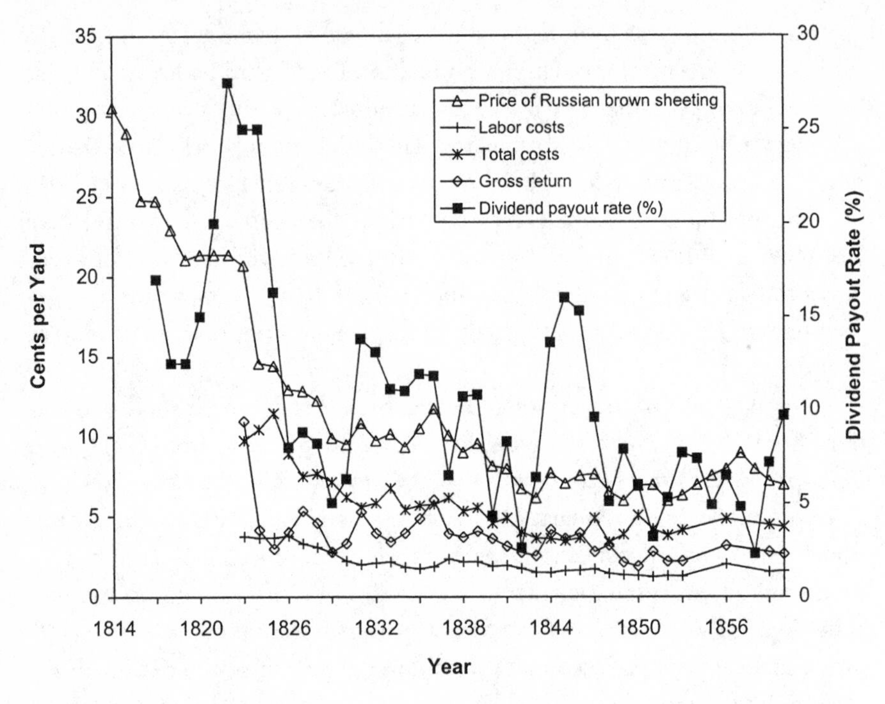

Fig. 8.2. Prices, Costs, and Returns for Cotton Cloth (Cents per Yard) and Dividend Payout Rate (%), 1814–1860. *Source:* Zevin, "Growth of Cotton Textile Production after 1815," 134, 144, tables 3, 5.

1820, and each region housed one of the leading textile states; Rhode Island and New York each accounted for just under 20 percent of employment and capital investment (table 8.3). When cotton textile output soared and prices plunged during the 1820s, textile shares diverged, and by 1830 New England dominated. The concentration of cotton textile capital remained stable from 1830 to 1860, with New England accounting for 65–70 percent and the Middle Atlantic for 20–25 percent. Massachusetts held about one-third of the nation's capital, and with its New Hampshire and Maine satellites the Boston core held about half.

The scale of New England mills in 1820 did not foreshadow their dominance; typical Middle Atlantic firms had 50 percent more employees and twice as much capital investment, but by 1831 the average New England mill more than doubled its capital investment and had 17 percent more capital than a Middle Atlantic firm (table 8.4). Based on employment and capital investment, Maine, New Hampshire, and Maryland contained the largest firms; however,

Table 8.3. Employment and Capital Invested in Cotton Textiles by Region and State as Share of Nation, 1820–1860

	Percentage of Total Employment					Percentage of Total Capital Investment				
Region/state	1820	1831	1840	1850	1860	1820	1831	1840	1850	1860
New England	54.1	51.4	64.9	67.1	66.7	49.6	69.8	68.4	72.3	70.3
Maine	0.8	0.5	2.0	4.1	5.5	1.6	1.9	2.7	4.5	6.1
New Hampshire	4.9	8.1	9.7	13.1	10.4	5.6	13.1	10.8	14.7	12.8
Vermont	1.5	0.8	0.4	0.3	0.3	1.0	0.7	0.2	0.3	0.3
Massachusetts	16.4	21.4	29.0	31.1	31.5	14.3	31.7	34.1	38.2	34.2
Connecticut	10.7	6.9	7.1	6.7	7.4	11.6	7.0	6.2	5.7	6.7
Rhode Island	19.9	13.7	16.8	11.8	11.5	15.4	15.4	14.3	9.0	10.2
Middle Atlantic	40.5	48.0	25.2	21.2	23.8	46.2	29.5	22.7	17.3	19.0
New York	18.6	8.9	10.3	6.8	6.3	18.8	9.0	9.6	5.6	5.5
New Jersey	6.6	8.7	3.3	1.9	2.1	4.7	5.0	3.4	2.0	1.3
Pennsylvania	6.1	23.9[a]	7.7	8.3	12.3	6.3	9.3	6.5	6.1	9.3
Delaware	3.3	2.2	0.8	0.9	0.9	4.0	0.9	0.6	0.6	0.6
Maryland	5.8	4.2	3.2	3.3	2.2	12.4	5.3	2.6	3.0	2.3
Rest of nation	5.4	0.7	9.8	11.7	9.5	4.3	0.7	9.0	10.4	10.7
Nation	100	100	100	100	100	100	100	100	100	100
Total employed	12,108	62,208	72,119	92,286	122,028					
Total capital (in thousands)						10,783	40,613	51,102	74,501	98,585

Sources: Jeremy, *Transatlantic Industrial Revolution*, 276, 279, app. D, tables D.1, D.5; U.S. Bureau of the Census, *Compendium of the Sixth Census, 1840; Report on the Manufactures of the United States at the Tenth Census, 1880.*

Note: [a] Census employment figures for Pennsylvania in 1831 boost Middle Atlantic's figures, but employment data for Pennsylvania are not consistent with capital investment data. The 9.3 percent Pennsylvania share of capital investment is consistent with the state's 9.7 percent share of cotton spindles.

Table 8.4. Employment and Capital Investment per Establishment in Cotton Textiles by Region and State, 1820–1860

	Employment per Establishment					Capital Investment per Establishment				
Region/state	1820	1831	1840	1850	1860	1820	1831	1840	1850	1860
New England	30	60	65	110	143	$24,851	$53,369	$48,115	$95,448	$121,509
Maine	19	36	157	312	356	34,000	95,625	155,333	277,475	316,754
New Hampshire	19	126	113	276	289	19,368	132,500	89,084	248,875	286,065
Vermont	16	28	37	27	47	10,064	17,382	16,871	22,500	33,900
Massachusetts	34	52	70	135	177	26,599	50,355	58,047	133,595	155,321
Connecticut	32	46	42	48	70	30,570	30,053	25,836	32,962	51,372
Rhode Island	35	73	53	69	92	24,142	53,986	32,416	42,247	65,701
Middle Atlantic	48	113	50	56	85	48,330	45,567	31,650	36,711	55,292
New York	55	49	57	73	97	49,487	32,763	37,990	48,569	68,145
New Jersey	58	107	43	82	58	36,214	39,758	30,764	70,643	30,013
Pennsylvania	26	222 [a]	38	37	81	24,157	56,097	22,777	21,774	49,746
Delaware	45	137	51	70	101	47,778	38,450	30,045	38,342	52,955
Maryland	64	114	95	126	134	121,418	93,217	54,350	93,167	112,725
Rest of nation	17	60	26	61	64	11,848	41,429	16,560	43,481	58,137
Nation	34	78	53	84	112	$30,205	$50,703	$37,328	$68,100	$90,362

Sources: Jeremy, *Transatlantic Industrial Revolution*, 276, 279, app. D, tables D.1, D.5; U.S. Bureau of the Census, *Compendium of the Sixth Census, 1840; Report on the Manufactures of the United States at the Tenth Census, 1880.*

Note: [a] See table 8.3 note for a discussion of high census employment figures for Pennsylvania in 1831, which boosts the Middle Atlantic's employment figures and, in this case, employment per establishment.

their few firms and the formation of new, large firms substantially raised their averages. Massachusetts housed many older, small mills; its average firm size in 1831 obscured the surging investment in large mills during the 1820s, and among states with numerous cotton mills Massachusetts reached the apex based on average employment and capital investment by 1840. Many small Massachusetts mills collapsed during the 1840s, reducing its number from 300 to 213 and raising firms' average employment and capital investment. By the 1850s Massachusetts and its Maine and New Hampshire satellites not only accounted for half of the cotton textile industry, but also their firms were much larger than firms elsewhere.

Firms Battle for Market Areas

During the 1820s unprecedented changes swept through the cotton textile industry: cotton goods prices plunged, productivity surged, firms vertically integrated and mechanized all their important processes, and new firms started at a large scale. Fierce market area battles raged, and the outcomes set future market contours; few firms focused on their local markets and survived, because competitors usurped those markets. By the early 1830s most cotton spinning mills collapsed or vertically integrated to weaving, and surviving spinning mills served either specialized weaving firms or local markets of rural dwellers (table 8.5). Spinning mills in eastern Pennsylvania sold to Philadelphia weaving firms, and those in western Pennsylvania sold to local rural dwellers or met small-scale midwestern farm demand through Pittsburgh wholesalers. The Connecticut Valley firm of Monson and Brimfield claimed to sell about one-third of its cotton goods in Massachusetts, and the rest went to wholesalers in East Coast metropolises. Oneida Manufacturing Society and Utica Cotton Factory in central New York, however, specifically identified the local market of Oneida and Herkimer Counties as important outlets, although both looked to New York City wholesalers to dispose of the rest of their merchandise. Their local sales suggest that some mills in rich agricultural areas successfully controlled substantial shares of local and subregional markets.

Nevertheless, most cotton textile firms sent the majority of their output to wholesalers in Boston, New York City, Philadelphia, and Baltimore, which in turn distributed cloth to the East, Midwest, and South (see table 8.5). Firms following this distribution pattern ranged from the huge Amoskeag Manufacturing Company in Manchester, New Hampshire, to the tiny Walley and Company mill in the Berkshire Mountains at Williamstown, Massachusetts, and the

Table 8.5. Selected Cotton Textile Firms by Location, Capital, and Market, 1832

Location	Name of Firm	Capital (dollars)	Market
New England			
Manchester, N.H.	Amoskeag Manufacturing Co.	200,000	Boston, New York City, Philadelphia, and Baltimore
Peterborough, N.H.	Union Manufacturing Co.	100,000	Boston
Ludlow, Mass.	Monson & Brimfield Cotton Manufacturing Co.	180,000	Mass. (1/3), New York City, Philadelphia, and Baltimore
Williamstown, Mass.	Walley & Co.	9,000	New York City
Fall River, Mass.	Pocasset Manufacturing Co.	120,000	New York City and Philadelphia
Sutton, Mass.	Maunchang Cotton Manufacturing Co.	52,000	Providence
Scituate, R.I.	Hope Factory	90,000	New York City, Philadelphia, and Baltimore
Eastern New York			
Hudson	James Wild Cotton Manufacturing Co.	40,000	New York City and Philadelphia
Stuyvesant	Columbia Manufacturing Society	160,000	New York City and Philadelphia
Central New York			
Utica	Oneida Manufacturing Society	100,000	Local (1/3) and New York City
Utica	Utica Cotton Factory	70,000	Local (2/3) and New York City
Otsego	Hope Manufacturing Co.	80,000	New York City and Albany
Otsego	Oaksville Manufacturing Co.	50,000	New York City
New Berlin	Farmers & Mechanics' Manufacturing Co.	70,000	New York City
Eastern Pennsylvania			
Manayunk	Wagner & Brothers	70,000	Philadelphia
Bucks County	W. P. Jenks & Co.	55,000	Philadelphia
Western Pennsylvania			
Allegheny County	Hope Cotton Factory	77,000	Pittsburgh
Beaver County	Pugh, Wilson & Co.	16,000	Locally

Source: McLane, *Documents Relative to the Manufactures in the United States.*

Note: Capital refers to the amount invested in fixed capital: real estate, buildings and fixtures, power, tools, and machinery.

Oaksville Manufacturing Company at Otsego in central New York. Because eastern wholesalers distributed cotton cloth to midwestern wholesalers at minimal transport cost markup, potential textile entrepreneurs in the Midwest faced abundant supplies of low-cost cloth, a strong disincentive to start cotton mills after 1830. Some small mills remained competitive in New England, central New York state, and eastern and western Pennsylvania; they survived through low-cost owner management, personal attention to small workforces to maximize productivity, and strict controls on the fixed costs of land, buildings, and machinery. The year 1840, however, defined a watershed in the number of cotton textile firms: their numbers peaked at about 1,400 firms, following a rise from 357 in 1820; after 1840 the count dropped to about 1,100 in 1850 and 1860.[12]

By 1840 the concentration of the cotton textile industry in the East reached extraordinary levels; the thirty-five counties, each with over $200,000 of capital investment, accounted for 85 percent of the nation's textile capital and 94 percent of the East's investment (table 8.6).[13] In this group average capital per firm was 12 percent greater than that of New England firms, 70 percent greater than that of Middle Atlantic firms, and 44 percent greater than the national average. In Maine and New Hampshire, satellites of the Boston capital, several counties within fifty miles of the Massachusetts border contained most of the states' capital. Massachusetts' four leading counties—Bristol, Hampden, Middlesex, and Worcester—housed 87 percent of the state's capital and 29 percent of the nation's, and Middlesex, the focus of Boston capitalists, held about half of the state's capital and 18 percent of the nation's. Similarly, Windham, Connecticut, held over half of that state's capital, and its southern neighbor, New London, chipped in 18 percent; both counties bordered Rhode Island and were satellites of the Providence core. Providence County contained two-thirds of Rhode Island's textile capital, and, with 10 percent of the nation's capital, Providence ranked second after Middlesex. They each housed more textile capital than every one of the Middle Atlantic states, and each county's capital exceeded the entire investment in cotton textiles outside the East. New York's six leading counties accounted for 75 percent of the state's textile capital; they formed two clusters, which emerged before 1820: the Hudson Valley (Albany, Columbia, Dutchess, and Rensselaer Counties) and central New York (Oneida and Otsego Counties). Similarly, New Jersey's textile cluster (in Passaic and Essex Counties) in the New York City region was formed before 1820. Over half of Pennsylvania's textile capital was located in Philadelphia County, or, with its satellites (Delaware and Montgomery Counties), three-fourths. New Castle,

Table 8.6. Counties with Over $200,000 of Capital Investment in Cotton Textile Manufacturing in the East, 1840

State and County	No. of Firms	Capital ($)	Capital per Firm ($)	County Capital as Percentage of State	Capital of State/County as Percentage of Nation
Maine	6	1,398,000	233,000		2.7
York	5	1,287,000	257,400	92.1	2.5
New Hampshire	58	5,523,200	95,228		10.8
Hillsborough	25	2,205,700	88,228	39.9	4.3
Merrimack	3	398,000	132,667	7.2	0.8
Rockingham	6	816,300	136,050	14.8	1.6
Strafford	17	2,015,700	118,571	36.5	3.9
Selected counties	51	5,435,700	106,582	98.4	10.6
Massachusetts	278	17,414,099	62,641		34.1
Berkshire	25	582,000	23,280	3.3	1.1
Bristol	55	2,526,999	45,945	14.5	4.9
Essex	9	773,000	85,889	4.4	1.5
Hampden	17	2,097,000	123,353	12.0	4.1
Middlesex	35	8,952,500	255,786	51.4	17.5
Norfolk	38	514,700	13,545	3.0	1.0
Worcester	71	1,507,400	21,231	8.7	2.9
Selected counties	250	16,953,599	67,814	97.4	33.2
Connecticut	116	3,152,000	27,172		6.2
Hartford	12	351,000	29,250	11.1	0.7
New London	16	553,100	34,569	17.5	1.1
Tolland	11	203,000	18,455	6.4	0.4
Windham	54	1,711,500	31,694	54.3	3.3
Selected counties	93	2,818,600	30,308	89.4	5.5
Rhode Island	209	7,326,000	35,053		14.3
Kent	45	1,504,000	33,422	20.5	2.9
Newport	10	496,000	49,600	6.8	1.0
Providence	130	4,977,000	38,285	67.9	9.7
Washington	23	269,000	11,696	3.7	0.5
Selected counties	208	7,246,000	34,837	98.9	14.2
New York	117	4,900,772	41,887		9.6
Albany	2	225,000	112,500	4.6	0.4
Columbia	11	893,300	81,209	18.2	1.7
Dutchess	11	865,000	78,636	17.7	1.7
Oneida	13	805,750	61,981	16.4	1.6
Otsego	8	245,600	30,700	5.0	0.5
Rensselaer	21	651,150	31,007	13.3	1.3
Selected counties	66	3,685,800	55,845	75.2	7.2
New Jersey	43	1,722,810	40,065		3.4
Essex	2	384,000	192,000	22.3	0.8

Table 8.6. Continued

State and County	No. of Firms	Capital ($)	Capital per Firm ($)	County Capital as Percentage of State	Capital of State/County as Percentage of Nation
Mercer	7	206,000	29,429	12.0	0.4
Passaic	20	933,000	46,650	54.2	1.8
Selected counties	29	1,523,000	52,517	88.4	3.0
Pennsylvania	106	3,325,400	31,372		6.5
Allegheny	5	580,000	116,000	17.4	1.1
Delaware	21	306,600	14,600	9.2	0.6
Philadelphia	45	1,923,600	42,747	57.8	3.8
Montgomery	11	324,700	29,518	9.8	0.6
Selected counties	82	3,134,900	38,230	94.3	6.1
Delaware	11	330,500	30,045		0.6
New Castle	11	330,500	30,045	100.0	0.6
Maryland	21	1,304,400	62,114		2.6
Baltimore	13	1,099,200	84,554	84.3	2.2
Total selected counties	808	43,514,299	53,854		85.2
Nation	1,369	51,102,359	37,328		100.0

Source: U.S. Bureau of the Census, *Compendium of the Sixth Census, 1840.*

Delaware (including the industrial city of Wilmington), a Philadelphia satellite, contained the state's entire cotton textile capital, and, similarly, Baltimore housed virtually all of Maryland's.

Clusters of huge firms existed, especially in New England. The five firms in York, Maine, averaged $257,400 of capital, and three of four New Hampshire counties had capital investment per firm greater than $100,000; Middlesex, Massachusetts, the national leader, held thirty-five firms averaging $255,786 of capital, an extraordinary scale at this point in eastern industrialization (see table 8.6). Several Middle Atlantic counties—Albany (New York), Essex (New Jersey), and Allegheny (Pennsylvania)—housed small numbers of firms averaging over $100,000 of capital investment, whereas others—Berkshire (25 firms), Norfolk (38), and Worcester (71) in Massachusetts—contained numerous small firms with averages below $25,000 of capital. Those clusters gave Massachusetts a low state average, even though it housed the nation's greatest concentration of huge firms. Providence's 130 cotton firms averaged only $38,285 of capital, revealing the sharp distinction between the Providence and Boston cores. Philadelphia, the third textile core, also stood out, with 45 firms that averaged only $42,747 of capital. Thus, distinctions among the three cores of the cotton textile industry—Boston, Providence, and Philadelphia—sharpened noticeably after 1820, and they led the growth and structural transformation of the industry.

The Boston Core

As of 1820, the Boston Manufacturing Company at Waltham remained the chief effort of the Boston merchant elite, and its average annual dividend payout rates of 14 percent during the period 1817–20 and of 24 percent from 1821 to 1824 were two to four times the typical interest rates in the East (see fig. 8.2). Information about those extraordinary profits swiftly reached New York's, Philadelphia's, and Baltimore's merchant elite through well-oiled trade and financial ties; certainly, the new form of cotton textile manufacturing—large-scale, vertically integrated mills—must have enticed them.[14] This leaves a conundrum: why did they fail to follow the path of Boston's merchants, who threw large amounts of capital into cotton textiles? Their rapid achievement of dominance during the 1820s rested on three ingredients: effective social networks of capital, access to managers and mechanics technically proficient in new industrial processes, and a business plan facilitating the quick formation of numerous firms and their efficient management.

Social Networks of Capital

Members of Boston's merchant elite achieved legendary status for their capacity to extend their cooperative approach from trade to diverse industrial, transportation (canals and railroads), and financial (banks and insurance companies) enterprises. Their social networks of capital—that is, their ongoing social, political, and economic relations—built trust, allowing them to enter ventures together, and their networks provided mechanisms to punish malfeasance. No individual dominated network ties, which maximized the diverse information flowing through networks, and variegated links within, and bridges across, networks minimized redundant information. Effective social networks were a necessary condition for Boston's elite to make large investments in cotton textiles, but they were not sufficient. Merchant elite in New York, Philadelphia, and Baltimore also participated in equally powerful networks and formed large cooperative enterprises, yet they did not invest huge sums in cotton textiles.[15]

Networks of Managers and Mechanics

Boston's investors had a competitive advantage, however, over investors in other metropolises; not only did some merchants, such as Patrick Jackson and Kirk Boott, possess technical skills, but they also accessed managers and mechanics technically proficient in new industrial processes rooted in Arkwright cotton spinning technology. Those processes embodied a package of labor organization (unskilled workers tending machines), machines, and integrated production system, and the package extended to vertically integrated mills combining spinning and weaving. Because most technically proficient managers and mechanics had roots in the Providence core, they concentrated in New England. When Boston investors decided to expand rapidly after 1820, demand for those managers and mechanics exceeded supply; thus, they had little incentive to migrate to other eastern regions, and Boston investors paid high wages or salaries to retain them. Because specialized machinery firms providing a full package of machines and service assistance did not exist in the 1820s, investors could not buy machinery from outside suppliers and run cotton mills on their own. They needed to hire skilled mill managers and mechanics to construct, install, operate, and maintain machinery, and most firms not following that procedure failed. Saco Manufacturing Company, organized in 1826 by Boston merchants and local Saco (Maine) investors, failed by 1829, in part because they did not hire a skilled manager and mechanics. After investors reorganized the firm as the York Manufacturing Company and, in 1831, hired Samuel Batchelder, the reorganized York firm succeeded. Batchelder was superintendent of the Hamilton Manufacturing Company in Lowell, and before 1825 he had owned and was the technically skilled manager of a cotton textile firm in New Ipswich, New Hampshire.[16]

Rapid developments in textile machinery and power transmission during the 1820s—when real costs of new integrated mills fell by over one-third—required investors, mill managers, and mechanics to participate in New England social networks to stay abreast of changes; failure to keep up gave new and old firms uncompetitive cost structures, and plunging coarse cloth prices meant firms sold at losses (see fig. 8.2). Dividend payout rates of New England cotton mills fell to 7 percent annually from 1826 to 1830, which was a disincentive for investors outside New England to start large-scale, vertically integrated mills. They could not anticipate that dividend payout rates would jump to 12 percent annually during the period 1831–36, and, when they recognized that profitability had recovered, they faced another disincentive to starting mills: steady declines in cloth prices during the 1830s, albeit at a slower pace than the 1820s. By the 1830s specialized firms drew on New England's supply of managerial and technical talent to construct fixed capital (such as buildings and dams), commence factory operations, and train workers, thus smoothing processes for investors to start new mills. Lack of access to that talent, coupled with cloth price declines and uncertain profitability, had dissuaded merchants and other wealthy investors in New York, Philadelphia, and Baltimore from large commitments of capital to cotton manufacturing.[17] Access of Boston's merchant elite to networks of technically proficient managers and mechanics constituted a necessary condition for successful investments during the 1820s, but, as with access to social networks of capital, it was not sufficient to achieve dominance in cotton textiles quickly.

A Business Plan

The Boston merchant elite's rapid dominance rested on a third necessary condition—a business plan facilitating quick formation of numerous firms and efficient management of them. By 1816 they codified the plan's eight components: corporate management structure, extensive liquid capital, accounting systems, vertically integrated production, a focused effort on technology and products, labor organization, a marketing strategy, and a protective tariff. Although individual components rooted elsewhere, Boston elite made an innovative leap: they combined components in the organization, management, and operation of the Boston Manufacturing Company, and the plan enabled them to rapidly form new companies and manage them efficiently during the 1820s and early 1830s, when cloth production grew swiftly. However, the plan's sophistication and skillful management required to implement it hindered efforts of business elite in other metropolises to copy it. Its organizational and management components foreshadowed modern business organization and management, but few firms replicated Boston investors' strategy until the 1850s.[18]

The marketing strategy deserves emphasis, because surging production required assured markets. Initially, mills consigned cloth production to general commission agents handling other types of goods; Boston Manufacturing Company used Ward and Company, but soaring output from the Boston Associates' mills created a dilemma during the 1820s. Unspecialized merchant wholesalers neither found sufficient markets rapidly enough to absorb the supply nor effectively handled fluctuations in demand and supply around the trend line of this larger production and consumption. By 1828 investors—led by Nathan Appleton—in Waltham and Lowell firms started James W. Paige and Company, a specialized agency (selling house) to handle all cloth production of multiple firms.

The shift to selling houses solved the Associates' dilemma: by 1829 A. and A. Lawrence and Company started, and shortly thereafter James K. Mills and Company and Francis Skinner and Company formed. By 1836 Boston housed twenty-one domestic selling houses, and each handled output of four-to-twenty cotton textile firms; selling houses also forwarded large quantities to New York dealers for wider distribution. Boston houses received commissions based on sales, took financial charge of marketing goods, and provided advances on sales to manufacturers, thus serving as creditors. They owned stock in textile firms, and owners influenced textile management through personal relations with treasurers of textile firms. Many principals in selling houses founded textile firms and owned stock in them; textile merchants and their firms contributed 17 percent of initial equity in eleven leading cotton textile firms of Boston's elite. Average equity holdings varied with the industry's growth: when selling houses first started in the late 1820s, they held 6 percent of textile equity, but, consonant with their growing importance, equity holdings grew to 19 percent during the 1830s, and, as textile growth continued decelerating after 1840, their holdings declined to 7 percent by the late 1850s. This overlapping ownership, combined with the profitability of selling houses, suggests that some Boston elite garnered greater profits from selling houses than from their textile firms. Attention to marketing and the formation of selling houses closely tied to textile firms assured Boston's elite that they could sell huge cloth output from swift expansion during the 1820s and early 1830s.[19]

The Associates astutely approached founding cotton textile firms: they relied on multiple social networks of investors; each network had one or more hub individuals leading efforts, and they bridged to other investor networks. Consequently, no individual monopolized investment decisions, and information flowed within and among networks; this maximized information flow, minimized redundancies, and permitted alternative checks on decisions. They maintained social network ties to hinterland investors because some merchants, such as Cabot, Jackson, Lee, and Appleton, had served as wholesalers for Boston's hinterland, and several, such as Dwight and Appleton, grew up in the hinterland or had worked there during part of their career. These ties provided the Associates with local allies to invest in firms, information about suitable mill sites, and assistance in purchasing land and rights to water and power.

Nathan Appleton, the undisputed "dean" of cotton textile investors, participated in many firms but did not use his hub position to control them; instead, Appleton facilitated or initiated projects (table 8.7). Other Associates—such as Perkins, Dwight, Jackson, Lyman, Thorndike, and Lawrence—were hubs of investor networks, and this multiple network structure and many bridges across networks provided flexibility to mobilize investors for projects. The Associates may have numbered as many as seventy-seven over the antebellum, and the investor pool available for initial stock offerings probably numbered over five hundred during the period from 1820 to 1840. Most firms started with five to fifteen shareholders; thus, if initial calls for capital amounted to $100,000 to $300,000, investors spread risks so none of them contributed more than a small share of personal wealth. Average stockholder contributions to the initial offerings of eleven major cotton mills amounted to $11,600 from merchants and mercantile firms, $20,000 from textile merchants and their firms, and $11,000 from financiers; collectively, they accounted for 62 percent of first-year equity holdings in those mills, underscoring their significance in founding large firms. From about 1820 to the mid-1840s the number of shareholders in individual firms rose to one hundred to four hundred as the Associates passed on stock as gifts or through estates and as other investors purchased stock; nevertheless, until the mid-1840s control remained with a few investors and their families.[20]

Most attention to the Boston Associates focuses on Lowell, arguably the greatest urban industrial project of the antebellum and a site for the repeated implementation of their business plan. Their comprehensive approach included a planned city with streets, bridges, operatives' housing, civic buildings, a canal system, water power, and building sites for multiple firms. Beginning in 1821, the Associates purchased stock of Locks and Canals Company, giving them a canal and water power rights, and they also bought adjacent farmland. The Lowell project unfolded rapidly: incorporation of Merrimack Manufacturing Company in 1822, transfer of most machine shop assets of the Boston Manufacturing Company to the Merrimack company in 1823, production of Merrimack's first cloth in 1824, incorporation of Hamilton Manufacturing

Company in 1825, and that same year the launch of the revitalized Locks and Canals Company. Locks and Canals housed a machine shop and controlled land and water power rights; this separated those activities from cloth manufacture and foreshadowed that division at other textile projects. Between 1828 and 1835 at least seven more firms incorporated (see table 8.7), and, as of 1835, the Lowell machine shop employed 300; Lowell's mills contained over 100,000 spindles and annually produced 32.8 million yards of cotton cloth, accounting for 13 percent of New England production (see fig. 8.1).

Their simultaneous, multilocational replication of their business plan at other New England sites was as astounding as their Lowell project, suggesting a conscious strategy to dominate production of coarse cloth during the 1820s and into the 1830s. Around 1823 Boston investors commenced efforts in the Cocheco and Nashua companies, both in New Hampshire, and in the Taunton company in Massachusetts. Under the leadership of Springfield investors led by the Dwight family, the Boston and Springfield Manufacturing Company incorporated in 1823 and rechartered as the Chicopee Manufacturing Company in 1828. Edmund Dwight soon played a prominent role; in 1809 he married a leading Boston merchant's daughter and moved to Boston by 1816, when he joined with James K. Mills in a merchant partnership. By the late 1820s Chicopee operated three mills totaling about 14,000 spindles, similar to the Hamilton company at Lowell, and by 1831 the Associates controlled it. During the 1830s the Associates expanded at earlier sites and took control of other firms. In 1831, led by Samuel Batchelder, they acquired control of York Manufacturing Company at Saco, Maine, and, led by Oliver Dean and Willard Sayles, who were involved in the predecessor firm, they incorporated Amoskeag at Manchester, New Hampshire. Between 1831 and 1836 the Associates poured investments into ten new firms, and this flurry of investment sweeping through Manchester, Saco, Lowell, Chicopee, and Taunton corresponded with the last rapid expansion of cloth production and a jump in the dividend payout rate (see figs. 8.1 and 8.2). During the 1837–43 economic contraction the Associates did not shy from taking greater control of existing firms or starting them: between 1836 and 1838 at least five Associates became directors of Amoskeag, a group of them founded the Stark Mills (1838) and Manchester Mills (1839) at Manchester, and some started Massachusetts Cotton Mills (1839) at Lowell and the Dwight Manufacturing Company (1841) at Chicopee.[21] Thus, the Boston elite brilliantly manipulated all three necessary condi-

Table 8.7. Cotton Textile Firms That the Boston Associates Founded or Later Gained Control Of

Date Founded	Firm, Location	Level of Capitalization (dollars)	Selected Stockholders
1813	Boston Manufacturing Co., Waltham, Mass.	400,000	N. Appleton, B. Gorham, C. Jackson, P. T. Jackson, F. C. Lowell, J. A. Lowell, I. Thorndike Sr.
1822	Merrimack Manufacturing Co., Lowell, Mass.	600,000	N. Appleton, W. Appleton, J. W. Boott, K. Boott, W. Dutton, B. Gorham, P. T. Jackson, P. Moody
1823	Chicopee Manufacturing Co., Chicopee, Mass.	500,000	W. Appleton, E. Dwight, J. Dwight, S. Henshaw, G. W. Lyman, H. G. Otis
1823	Cocheco Manufacturing Co., Dover, N.H.	unknown	E. T. Andrews, W. Appleton, E. Francis
1823	Nashua Manufacturing Co., Nashua, N.H.	600,000	W. Amory, D. Sears
1823	Taunton Manufacturing Co., Taunton, Mass.	600,000	S. Crocker, E. Dwight, H. G. Otis, W. H. Prescott, I. Thorndike Sr.
1825	Hamilton Manufacturing Co., Lowell, Mass.	600,000	E. Appleton, N. Appleton, W. Appleton, S. Batchelder, J. W. Boott, B. Gorham, P. T. Jackson, G. W. Lyman, W. Pratt, I. Thorndike Sr.
1828	Appleton Co., Lowell, Mass.	500,000	N. Appleton, S. Appleton, W. Appleton, E. Francis, P. T. Jackson, J. Lowell, J. Lowell Jr., G. W. Lyman, T. H. Perkins, W. Sturgis, I. Thorndike Sr., I. Thorndike Jr.
1828	Lowell Manufacturing Co., Lowell, Mass.	500,000	P. T. Jackson, G. W. Lyman, T. H. Perkins, I. Thorndike Sr., I. Thorndike Jr.
1831	Amoskeag Manufacturing Co., Manchester, N.H.	1,000,000	W. Amory, N. Appleton, W. Appleton, O. Dean, I. Gay, P. T. Jackson, F. C. Lowell Jr., J. A. Lowell, J. K. Mills, L. Pitcher, W. Sayles, Lyman Tiffany, J. Tilden
1831	Lawrence Manufacturing Co., Lowell, Mass.	1,200,000	N. Appleton, W. Appleton, B. R. Nichols

1831	Suffolk Manufacturing Co., Lowell, Mass.	450,000	S. Appleton, W. Appleton, H. Cabot, G. W. Lyman, J. Tilden
1831	Tremont Mills, Lowell, Mass.	500,000	W. Appleton, G. Hallet, Abbott Lawrence, Amos Lawrence, B. R. Nichols, W. Pratt, D. Sears
1831	York Manufacturing Co., Saco, Maine	500,000	S. Batchelder, E. Dwight, Abbott Lawrence, Amos Lawrence
1832	Cabot Manufacturing Co., Chicopee, Mass.	1,000,000	N. Appleton, P. C. Brooks, H. Cabot, S. Cabot, E. Dwight, G. Kuhn, G. W. Lyman, J. K. Mills, H. G. Otis, T. H. Perkins, I. Sargent, I. Thorndike Sr.
1833	Hopewell Co., Taunton, Mass.	unknown	S. Crocker, E. Dwight, S. A. Eliot, J. K. Mills, H. G. Otis, C. Richmond
1835	Boott Mills, Lowell, Mass.	1,150,000	N. Appleton, K. Boott, H. Cabot, W. Dutton, Abbott Lawrence, J. A. Lowell, G. W. Lyman, G. W. Pratt, P. T. Jackson, J. H. Wolcott
1835	Whittenden Mills, Taunton, Mass.	unknown	E. Dwight, S. A. Eliot, C. A. Mills
1836	Perkins Mills, Chicopee, Mass.	500,000	W. Appleton, H. Cabot, A. Thorndike
1838	Stark Mills, Manchester, N.H.	500,000	W. Amory, N. Appleton, W. Appleton, S. Henshaw, F. C. Lowell Jr., G. W. Lyman, W. Sayles
1839	Manchester Mills, Manchester, N.H.	1,000,000	W. Amory, N. Appleton, O. Dean, S. Frothingham, J. C. Howe, I. Livermore, F. C. Lowell Jr., W. Sayles, D. Sears
1839	Massachusetts Cotton Mills, Lowell, Mass.	1,500,000	O. Goodwin, Abbott Lawrence, J. A. Lowell
1841	Dwight Manufacturing Co., Chicopee, Mass.	500,000	W. Appleton, S. Cabot, E. Dwight, W. H. Gardiner, G. Kuhn, T. H. Perkins, I. Sargent, W. Sturgis

Sources: Dalzell, *Enterprising Elite*, 233–38, app.; Anonymous, *Manchester*, 269–302; Shlakman, "Economic History of a Factory Town," 39–42, table 1; Ware, *Early New England Cotton Manufacture*, 301–2, 320–21, app. A, K; White, *Memoir of Samuel Slater*, 255–56.

Note: Levels of capitalization may be greater than actual capital invested, and the list of stockholders indicates those who were involved at some point. Information is only an approximation and does not necessarily correspond to the founding date of firm.

tions for dominance of the cotton textile industry, and these conditions were sufficient; no other industrial group in the East challenged them during the rest of the antebellum.

The Providence Core

Surprisingly, cotton textile investors in the Providence core failed to challenge the Boston Associates after 1820. Led by the Brown family, Providence's merchant elite possessed ample capital to fund existing mills, and investors accessed many technically competent mill managers and mechanics. Textile firms followed a business plan including an on-site manager and ties to Providence wholesalers who supplied inputs (primarily cotton) and distributed outputs (e.g., yarn and cloth) within New England or to wholesalers in other East Coast metropolises. The Providence core boasted two of the largest cotton textile firms in the nation—Almy, Brown, and Slater in Slatersville, Rhode Island, housing 5,170 spindles before 1815; and Blackstone Manufacturing Company in Blackstone, Massachusetts, housing 5,000 spindles by 1809—whereas the Boston Manufacturing Company in Waltham, the signature effort of the Boston Associates, housed 5,800 spindles in 1820, a minor size advantage.

From 1820 to 1831 Rhode Island investors rapidly boosted capital in cotton textiles at 12 percent annually, but Massachusetts investors outpaced them, raising their capital 20 percent annually (see table 8.3). Consequently, Massachusetts' share of the nation's textile capital more than doubled to 32 percent from 1820 to 1831, whereas Rhode Island's share stabilized at 15 percent. Nonetheless, measured by firm size, Rhode Island's investors seemed to keep pace with Massachusetts' investors—both doubled their capitalization per firm (see table 8.4). The Providence core housed some of the nation's largest cotton textile firms in 1832; its ten biggest firms averaged capital investment of $203,245, four times larger than the national average for 1831 (tables 8.4 and 8.8). The top ten firms in the Boston core averaged $663,330 of capital, however, and this excludes the mammoth Lawrence Manufacturing Company, incorporated in 1831, and its $1.2 million invested in four mills in Lowell by 1835.[22]

Social Networks of Capital

Because Providence investors possessed less capital, this might explain their failure to keep pace with the Boston Associates. Wholesalers and financiers in Providence controlled trade and finance in a subregion of New England including Rhode Island and adjacent parts of Massachusetts and Connecticut, but Boston's wholesalers and financiers accumulated far more capital because they operated over much of New England. Therefore, Providence core investors could not fund the same number of large, vertically integrated firms as the Associates; most Providence core firms lacked sufficient capital to start at a large scale and funded growth out of retained earnings. Owners of cotton mills often sold shares and could have reallocated capital to large firms, but they did not follow that course. Rhode Island investors sometimes clustered firms: by 1832 eight Warwick firms had aggregate capital of $1,026,300, and four Smithfield firms had total capital of $612,000. Owners of sizable mills within each cluster certainly knew one another and could have reorganized to form huge firms; nevertheless, none appeared during the 1820s.

Compared to the Boston core, the Providence core's smaller merchant and financial capital prevented investors from funding the same number of huge textile firms, but that does not explain their failure to fund any equivalent to the Associates' top half-dozen firms by 1832 (see table 8.8). In the 1820s some wealthy Providence core merchants—including Seth Wheaton, Edward Carrington, and members of the Brown family—could have funded several firms equivalent to the Associates' leading ones. Instead, their largest—Almy, Brown, and Slater—possessed only $240,000 in capital by 1832, and the Blackstone Manufacturing Company, with a value of $216,000 around 1813, had just $150,000 of capital in 1832. Wheaton and Carrington participated in several firms, including Providence Manufacturing Company and Blackstone Manufacturing Company, but they never dramatically enlarged any of them. The Blackstone River in Rhode Island and Quinebaug River in eastern Connecticut possessed several large water power sites; therefore, power availability did not constrain them. A decision to spread risk may have motivated their refusal to commit more capital for individual firms or to participate in founding new, huge firms, but their failure to fund even one, which they could readily accomplish, suggests that other factors figured in their decision making.[23]

Social networks of capital in the Providence core hindered the formation of huge firms. Although Providence's merchants and financiers participated in local social networks leading to cooperation in funding firms, their network bridges to the Boston core did not generate cooperative ventures in large textile firms. Yet merchants and financiers maintained network bridges between Providence and Boston to exchange commodity and financial capital. Failure of those bridges to extend to cotton textile investment stemmed from differences between commodity-financial and industrial-investment networks.

Table 8.8. Ten Most Highly Capitalized Cotton Textile Firms in the Boston and Providence Cores and in the States of New York and New Jersey, 1832

Firm	Capital ($)
Boston core	
Boston Manufacturing Co., Waltham, Mass.	333,493
Merrimack Cotton Manufacturing Co., Lowell, Mass.	1,329,846
Hamilton Cotton Manufacturing Co., Lowell, Mass.	705,404
Appleton Cotton Manufacturing Co., Lowell, Mass.	353,555
Lowell Manufacturing Co., Lowell, Mass.	175,205
Chicopee Manufacturing Co., Chicopee, Mass.	486,000
Amoskeag Manufacturing Co., Manchester, N.H.	200,000
Nashua Manufacturing Co., Nashua, N.H.	549,800
Cocheco Manufacturing Co., Dover, N.H.	1,500,000
Great Falls Manufacturing Co., Somersworth, N.H.	1,000,000
Total for Top Ten Firms	6,633,303
Providence core	
Almy, Brown & Slater, Slatersville, R.I.	240,000
Albion Mills, Smithfield, R.I.	142,000
Natick Mills, Warwick, R.I.	250,000
Crompton Mills, Warwick, R.I.	274,300
Scituate Manufacturing Co., Scituate, R.I.	150,000
Jenkins and Man, Cumberland, R.I.	180,000
Pawtucket Cotton Manufacturing Co., Pawtucket, R.I.	196,150
Blackstone Manufacturing Co., Blackstone, Mass.	150,000
Thames Manufacturing Co., Norwich, Conn.	300,000
Windham Manufacturing Co., Willimantic, Conn.	150,000
Total for Top Ten Firms	2,032,450
New York state (county)	
J. & G. Pierson & Brothers, Ramapo (Rockland)	100,000
Mattewan Factory, Fishkill (Dutchess)	150,000
Joseph & Benjamin Marshall Factory, Hudson (Columbia)	300,000
Columbia Manufacturing Society, Stuyvesant (Columbia)	160,000
N. Wild, Kinderhook (Columbia)	84,000
Farmers' Cotton Manufacturing Co., Schaghticoke (Rensselaer)	116,000
Troy Woolen & Cotton Factory, Troy (Rensselaer)	100,000
Oneida Manufacturing Society, Utica (Oneida)	100,000
New York Mills, Utica (Oneida)	250,000
Jefferson Cotton Factory, Watertown (Jefferson)	100,000
Total for Top Ten Firms	1,460,000
New Jersey state (county)	
John Colt & Co., Paterson (Essex)	118,700
Daniel Holsman & Co., Paterson (Essex)	142,000
Collett and Smith & Co., Paterson (Essex)	125,000
John W. Berry & Co., Paterson (Essex)	80,000
A. & R. Carrick & Co., Paterson (Essex)	100,000
Godwin, Clark & Co., Paterson (Essex)	150,000

Table 8.8. Continued

Firm	Capital ($)
Munn & Whitehead & Co. (Bergen)	54,160
James Hoy & Co., Trenton (Burlington)	100,000
Lewis Waln & Co., Trenton (Burlington)	169,000
Lippincott & Richards & Co. (Gloucester)	120,000
Total for Top Ten Firms	1,158,860

Source: McLane, *Documents Relative to the Manufactures in the United States.*

Note: Capital refers to amount invested in fixed capital: real estate, buildings and fixtures, power, tools, and machinery.

Commodity-financial networks have local components within each metropolis (e.g., Boston or Providence) and social network bridges between metropolises, and repeated interfirm transactions such as cooperative ventures and exchanges of capital build trust. Because intermediaries of capital must be recognized as trustworthy to participate in these networks, members possess potent mechanisms to enforce sanctions for malfeasance within and across metropolises; excluding actors from these networks terminates their business.

Merchants and financiers can build local networks of industrial investors—including some individuals participating in commodity-financial networks, but skilled wholesalers or bankers may not understand how to establish and manage an industrial enterprise. Industrial networks convey distinctive information related to product markets, technology, and firm organization and management; thus, investment networks have strong local bases, because members require personal contact to evaluate the capacities of others for forming industrial firms and successfully managing them and to hire technically proficient managers and mechanics. Sanctions for malfeasance can be imposed in a local industrial network, because a range of formal and informal mechanisms exist such as shunning in social groups and exclusion from business and social organizations, but sanctions cannot readily be imposed across industrial social networks, especially in other metropolises, because sanctions are unavailable or ineffective. If industrial investors need capital, they might acquire it from another metropolis, but that increases risk because investors have less understanding of trustworthiness and capabilities of investors and because sanctions for malfeasance cannot be imposed. Boston's and Providence's industrial investor networks were asymmetric: Providence's elite needed capital from Boston if they wanted to finance many huge firms, whereas the Boston Associates possessed ample capital to fund numerous large firms within their local industrial network, eliminating risks on joint ventures with Providence investors.

Although the Providence core housed few wealthy people, leading investors could have sought many individuals with small capitals—professionals (e.g., lawyers and physicians), mechanics, retailers, and farmers—to join investment groups to start huge firms, but little evidence exists that wealthy investors tried this approach. To minimize risk a merchant firm such as Brown and Ives had to expend considerable effort to reach numerous investors operating in different social networks of capital, and it had few bridges to them. Not only did small investors have less experience with large-scale investments, but also Brown and Ives lacked information about their competence and trustworthiness, and, if malfeasance occurred, little recourse existed to enforce sanctions. Those with small capitals had difficulty funding large firms, because they lacked financial experience with major investment projects and they needed many investors. Moreover, networks of professionals, mechanics, retailers, and farmers typically had local bases in towns and villages and few bridges to other networks, limiting the network size for funding a huge firm. Therefore, Providence core investors financed numerous small firms, and their mills grew large through reinvestment of earnings; most survivors did not reach a huge size until after 1860.[24]

An Incomprehensible Business Plan

The small supply of wealthy elite and fragmented capital pool hindered the formation of huge firms in the Providence core, but this cannot explain the complete absence of firms equivalent to the Boston Associates' largest firms during the 1820s and 1830s. The actions of Providence's elite suggest that they comprehended neither the significance of the Associates' business plan for competing in the national market for coarse fabrics nor the salience of its interrelated components for successfully organizing and managing large, vertically integrated firms.

Almy, Brown, and Slater and Blackstone Manufacturing Company—among the greatest cotton textile firms in 1820—could have restructured to the huge, vertically integrated form with around $300,000 of additional capital for each, an amount within the financial means of Providence merchants. Investors in mills descending from Natick Manufacturing Company of Warwick, Rhode Island—first established in 1808—divided mills among different owners and expanded them separately, whereas they could have reorganized them as one huge firm. Similarly, Lyman Cotton Manufacturing Company in North Providence, which led the switch to power loom weaving in 1817, sedately added mills and then divided them up under a reorganization in 1828. Yet family members, under patriarch Daniel Lyman, were the economic, social, and political elite of Rhode Island with access to a large pool of capital. In the mid-1820s Providence core investors—Samuel Slater, an industry leader, and Larned Pitcher, a talented builder of cotton textile machinery—took stakes in the predecessor of Amoskeag Manufacturing Company in Manchester, New Hampshire. At that time Boston Associates Oliver Dean and Willard Sayles became involved, and under the 1831 incorporation they capitalized the firm at one million dollars. Slater never became seriously involved, however, and no other Providence merchants joined Amoskeag, even though Slater and Pitcher were prominent in the Providence core, and by the late 1830s the Associates poured vast sums of capital into the Manchester mills without the participation of the Providence core investors.[25]

Providence investors missed the significance of a focused effort on technology. As experienced practitioners of Arkwright cotton spinning technology, owners of Providence core firms understood the concept of vertically integrated production. Yet firms slowly adopted power loom weaving during the 1820s, and only one-third of Rhode Island's mills had mechanized weaving by 1826. Installation of power looms added steps utilizing new technologies, and they created disjunctions in output: power looms consumed large quantities of yarn, which required greater numbers of spindles, and, if spindle speed increased, then power loom speed had to increase or more looms had to be added. Therefore, adding power looms required reconstituting the entire production organization, and that system needed regular adjustment to realize the full capabilities of new technologies. To implement this system owners required a sophisticated, multitiered administrative structure: a board (made up of the directors of the Associates' firms) to focus on strategic decisions about the organizational structure, finance, and markets; an executive officer (the treasurer of the Associates' firms) to manage the firm; and a factory manager to supervise the production systems. Providence core owners, however, did not fully grasp the importance of this structure. Slater, the preeminent mill manager and mechanic of the period 1790–1820, handled overall strategic decisions as owner and inserted himself into the operational details of factories at different locations. Under the corporate form of Samuel Slater and Sons, his sons did not separate these administrative functions until after Samuel died in 1832, and they waited until around 1839 to institute cost accounting (allocating costs to functions) as a financial control system. Thus, Slater corporation lagged the sophisticated management approach of the Associates' firms by fifteen to twenty years, but that family, often tarred as too conservative, was not exceptional; Providence firms frequently retained a partnership form, and owners intervened in managing the mill. Mill managers were not viewed as critical officers, and their modest compensation levels reflected this status, whereas the Associates' factory managers were richly remunerated.[26]

Providence core investors failed to comprehend changes in corporate strategy which a regime of large-scale production required. The Boston Associates quickly shifted to a marketing strategy of consigning all production to one wholesaler before 1820, and by the late 1820s they formed specialized selling agencies, whereas Providence firms failed to develop coherent marketing strategies until at least the 1850s. The Slater family firm tried to set prices and credit conditions and guaranteed sales, but it was uncompetitive in many markets and took on greater risk rather than transferring it to a specialized selling agency. Even the astute merchant firm of Brown and Ives did not coordinate sales of its separate cotton textile firms, and the firm consigned cloth to different selling agencies in the same city, thus competing against itself. Providence investors did not implement a labor strategy permitting a rapid, large-scale expansion and quick adjustment to business fluctuations. Their family labor system entailed longer-term obligations by owners and families and substantial fixed costs for infrastructure, such as village facilities and parcels for farming to supply food to workers. Parents negotiated with owners over work assignments for themselves and their children, reducing managerial flexibility to maximize productivity. This family labor system remained mostly intact through the 1840s and began to unwind with the influx of immigrants to mills during the 1850s, whereas the Associates formalized their boardinghouse system for young, single women before 1820 and replicated it—including boardinghouses for married workers—at most firms through the 1840s, until the wave of immigrants required a different approach. The Associates mobilized large workforces quickly, retained substantial flexibility to organize production, and gained high productivity from a workforce heavily weighted toward educated, young women.[27]

Although wealthy elite in the Providence core did not grasp the Boston Associates' business plan, social networks of textile mechanics bridged cores as mechanics moved among firms and as cotton textile machinery and other machinery firms exchanged information and products. Ira Gay and Larned Pitcher, leading Providence core mechanics, participated in the Amoskeag Company when the Associates started enlarging efforts in Manchester; this arrangement was not unprecedented.[28] The exchange of technological information and skilled workers between Providence and Boston cores, yet the absence of exchange of a sophisticated understanding of the Associates' business plan, points to significant network differences. Skilled cotton textile mechanics and firms within both cores shared a technical language, machine problems needing solutions, and values about improving machinery, whereas elite social networks in each core operated within their own industrial networks. Because the Associates' business plan was sophisticated, hearing or reading descriptions of specific components did not convey their full meaning; the plan was more than the sum of its parts—it incorporated an unprecedented vision of how large manufacturing enterprises could penetrate the mass market. The inability of Providence core investors to participate in local industrial networks of the Boston core, not their conservatism or lack of astuteness, explains their failure to start a single huge, vertically integrated firm comparable to the greatest of the Associates' firms before 1840. This explanation seems plausible because New York and New Jersey investors exhibited the same failure to establish those firms, whereas they, along with Providence core investors, built leading firms in other industries.

New York and New Jersey

Cotton textile firms in New York State and New Jersey, across the Hudson River from Manhattan, were derivative of Providence core firms. Mechanics from Providence migrated to New York and New Jersey mills before 1820, and, if they did not have direct links to Providence, few firms innovated enough to form a separate branch of the textile industry. Investors in New Jersey and the Hudson Valley (in Rockland, Dutchess, Columbia, and Rensselaer Counties) accessed social networks of capital in New York City, because its wholesalers and financiers had tight bonds with those areas; their vast capital could have funded numerous huge, vertically integrated firms such as those of the Boston Associates. Nevertheless, as of 1832, capital of the top ten firms in New York and New Jersey averaged one-fifth or less of the Boston core's behemoths; even Providence core firms were 30 percent or more larger than New York and New Jersey firms (see table 8.8). Because New England housed most of the technically skilled mill managers and mechanics, New York's business elite faced difficulty expanding rapidly, and the Associates matched them in paying high wages and salaries. Although cotton textile wholesalers in New York City distributed the output of the Associates' firms, New York's elite, like that of Providence, could not participate in the Associates' industrial networks and thus did not comprehend their business plan, which explains New York elite's failure to form huge firms.

Consider the investment activities of Benjamin Marshall, who, along with his older brother Joseph, owned a cotton mill in Hudson, New York, capitalized at $300,000 (see table 8.8). Benjamin, a wealthy New York City cotton merchant, possessed extensive bonds with New York's financial networks, providing access to immense capital; he participated in starting an early packet line between New York and Liverpool, England. Benjamin and other network members could have funded ten cotton textile firms as large as the Associates' greatest, and he accessed talented mill managers and sophisticated mechanics with roots in the Providence core. In 1825 he formed a partnership with Benjamin Walcott Jr., a skilled mill manager and mechanic from the Providence area who had arrived in Utica, New York, in 1809. The first factory of Benjamin Marshall and Walcott transformed into the New York Mills by 1832, but its capital stayed far smaller than most of the Associates' mills (see table 8.8); the firm finally reached premier national ranking toward the end of the century.[29] Although Walcott continued as a distinguished mill manager and investor and Marshall could access vast capital, their inability to participate in the Associates' industrial networks meant that their business plan remained incomprehensible. In contrast to New York and Providence core investors, who continued running cotton textile firms competing with the Associates' firms, most Philadelphia core investors followed a different path that took root during the period from 1790 to 1820.

The Philadelphia Core

The Philadelphia core—Philadelphia County and nearby counties in Pennsylvania and Delaware—accounted for most of the cotton textile industry of the two states; their collective share of the nation's total capital investment stayed around 7–10 percent during the years 1820 to 1860, confirming that Philadelphia core investors operated successfully in an industry that the Boston Associates dominated (see tables 8.3 and 8.6). As of 1820, cotton textile firms in the Philadelphia core specialized in fine-quality yarns and fabrics, and during the next thirty years most firms specialized in a few parts of the production process and responded flexibly to customer preferences and shifts in demand. Except for the Globe Mill's $200,000 of fixed capital, every Philadelphia core firm in 1832 possessed less than $81,000 and most had less than half that; in contrast, nine of the top ten Boston core firms had more capital than Globe (tables 8.8 and 8.9). Globe Mill did not operate a vertically integrated spinning and weaving mill with mechanical-powered machines, like those of the Associates; instead, it chiefly made yarn, and housed hand loom weavers. Most Philadelphia core firms focused on one production stage: dyeing cotton or cloth, spinning yarn, weaving cloth, or printing cloth (see table 8.9). In this integrated production complex, firms manufacturing before the final stage typically sold output to other local firms, and numerous spinning mills supplied skilled hand loom weavers, who remained pivotal to the production complex through the 1840s. The strategy of small-scale, flexible specialized production of cotton textiles succeeded for the industry as a whole, but individual firms frequently entered and exited, typical of small firms with limited capital.[30]

News of Waltham's fabulous success tempted some Philadelphia core investors to compete with Boston core firms. After the Schuylkill Navigation Company completed a canal and dam through Manayunk, six miles from Philadelphia's merchant heart, in 1819 investors commenced efforts to build cotton mills; however, they moved slowly. By 1824 Manayunk's largest factory, an integrated cotton mill, housed only 1,500 spindles and 60 power looms, but the pace quickened in the late 1820s, probably because investors recognized that coarse cloth pouring from the Associates' mills met insatiable demand. By 1828 Manayunk had five cotton mills containing a total of 14,154 spindles (2,831 spindles per mill) and 525 workers, and their hundreds of power looms demonstrated to investors integrated power spinning and weaving; by 1840 Manayunk had eight mills averaging 2,247 spindles. At that time four Lowell firms of the Associates each housed more spindles than the total for all eight Manayunk firms, making them pale replicas. The reluctance of Philadelphia's merchants to invest in small-scale, flexible manufacturing firms was strategic: they required close managerial supervision, detracting from merchant exchanges with greater profit margins; and capital invested per firm was too small to be deployed efficiently. Yet their failure seriously to commit capital to cotton textiles, even though Philadelphia housed a thriving industry, seems paradoxical. They possessed detailed information about the investment scale and pro-

Table 8.9. Cotton Textile Firms in the Philadelphia Core, 1832

Place and Firm Name	Process/product	Capital	Market
Philadelphia County, Pa.			
William Whitaker	spin yarn	$45,000	New York City and Philadelphia
Steel & Co.	weave cloth	50,000	—
Holmesburg Works	weave cloth	80,000	—
Knight & Co.	dye cotton	10,000	—
Comly & Co.	print cloth	75,000	—
Brown's & Co.	dye, print, and finish cloth	5,000	—
Shuttleworth & Co.	dye yarn	4,000	—
Ripka & Co.	weave cloth	25,000	—
Pennipack & Co.	print cloth	70,000	—
Washington & Co.	dye cloth	5,000	—
Wilson's & Co.	dye cloth	6,000	—
La Grange & Co.	print cloth	50,000	—
Richardson & Co.	spin yarn	17,000	—
Shallcross & Co.	dye cotton	12,000	—
Blacks & Co.	dye cotton	20,000	—
Pilling & Bolton & Co.	bleach cloth	30,000	—
Large & Co.	dye cotton	—	—
Horrocks & Co.	dye cotton, dye and finish cloth	15,000	—
Wagner & Brothers	spin yarn and weave cloth	70,000	Philadelphia, export
Globe Mill	spin yarn and weave cloth	200,000	—
Bucks County, Pa.			
William P. Jenks & Co.	spin yarn	55,000	Philadelphia
Delaware County, Pa.			
Thirteen spinning mills	spin yarn	—	—
Three weaving mills	weave cloth	—	—
Crozier & Co.	spin yarn	15,000	local and Philadelphia
J. & J. Riddle & Co.	spin yarn	33,000	local and Philadelphia
Ronaldson & Co.	spin yarn	39,000	local and Philadelphia
New Castle County, Del.			
Robert Hilton & Sons	spin yarn and weave cloth	65,892	Philadelphia, Baltimore, and New York City
Jacob Pusey & Co.	spin yarn	30,000	Philadelphia
Garret & Pusey & Co.	spin yarn	38,000	Philadelphia
J. B. Hutchinson & Co.	spin yarn	72,000	Philadelphia
Joseph Bancroft & Co.	spin yarn and weave cloth	42,000	Philadelphia, Baltimore, and New York City
Breck & Swift & Co.	spin yarn	18,000	Philadelphia
John Connelly & Co.	spin yarn	16,000	—

Table 8.9. Continued

Place and Firm Name	Process/product	Capital	Market
Trump & Co.	spin yarn	10,000	—
Mitchell & Guin & Co.	spin yarn	13,000	—

Source: McLane, *Documents Relative to the Manufactures in the United States.*
Note: Capital refers to amount invested in fixed capital: real estate, buildings and fixtures, power, tools, and machinery.

duction of the Associates' firms at Waltham, Lowell, and elsewhere, because Philadelphia wholesalers exported cotton cloth of New England mills to the Midwest (see table 8.5). Their diversion of capital to transportation projects cannot explain their failure to invest in huge mills during the 1820s and early 1830s. Large investments in canals did not commence until after 1825 and in railroads not until after 1835, and wealthy investors often sold shares after initial stock offerings, reallocating capital elsewhere.[31]

Many of the Philadelphia core's mills used hand-powered mule spinning and spinning jennies; however, competitive coarse cloth production required more sophisticated approaches than simply adding water-powered spinning and weaving. Limited supplies of skilled mill managers and mechanics familiar with Arkwright cotton spinning technology hindered the Philadelphia elite's capacity to make large cotton textile investments. Therefore, investors could not rapidly build many large, vertically integrated cotton mills; they slowly adopted power looms; and owners had difficulty operating mills efficiently. They also faced the same problem as Providence and New York investors—an inability to participate in the Associates' industrial networks. Thus, their business plan remained obscure. Philadelphia investors typically combined the functions of mill manager, purchasing, and marketing; they did not implement focused efforts at reducing costs, other than trying to keep labor costs low, and they maintained partnership or sole proprietor forms of organization. Wealthy investors among merchant, financial, and professional elite could underwrite large firms, but textile investors did not collect substantial capital from them; therefore, investors funded expansion through retained earnings, which was a slow process.[32] Consequently, Philadelphia investors never seriously challenged the Associates for dominance of the mass market for coarse cloth. Small, flexible specialized firms in the Philadelphia core successfully pursued niche markets, but these remained small during the antebellum.

The mammoth cotton textile mills of the Boston Associates and cities such as Lowell captured the imagination of the East Coast elite, seemingly epitomizing urban industrial growth during the period 1820–60, and this overshadowed Connecticut's textile firms and industrial cities. Nevertheless, building on an earlier base, many cotton textile firms reached a large size in eastern Connecticut after 1820, but that date marked the apogee of the state's share of national cotton textile employment (11 percent) and of capital investment (12 percent); its share on both measures fell to 7 percent by 1860 (see table 8.3). In other Connecticut industries, however, firms achieved national prominence through diverse strategies: some operated small workshops and factories, and others built large factories and employed machinery to undercut competitors.

Connecticut's Firms Capture National Markets

Before 1820 Connecticut's manufacturers possessed competitive advantages: a highly educated population; strong social networks of farmers, professionals, and businesspeople communicating information about markets and technological change; ample capital; market intelligence from the merchant elite in subregional metropolises (e.g., Hartford, New Haven, and New London) directly trading with merchants in other East Coast metropolises; direct links to New York City's merchant and financial elite providing market intelligence, sales outlets, technology, and capital; and peddlers marketing and selling directly to consumers. These advantages remained salient, but they did not guarantee industrial leadership; manufacturers continually instituted organizational and technological changes within existing industries and entered new industries. They demonstrated technological prowess by the first decade of the nineteenth century, when half of the state's counties ranked among the nation's most inventive counties, and by the 1830s two-thirds of the counties were in that top group.[33]

Industrial Prominence

After 1820 Connecticut's firms achieved notable production levels in a wide variety of light, high-value manufactures selling in large market areas outside the state (table 8.10). By 1832 capital investment in axes, brass foundries and brassware, and carpets each exceeded $250,000, and production values in axes and carpets each reached almost $350,000; firms also manufactured sizable amounts of Britannia and plated ware as well as clocks, but the absence of surveys of leading producers of hats and caps and also tinware understates their significance. In most industries employment levels exceeded one hundred workers, underscoring the considerable presence of manufactures selling in large market areas. Between 1832 and 1845 various industries—such as brass foundries and brassware, Britannia and plated ware, carpets, and clocks—grew strongly, and hats and caps, tinware, and hardware must have expanded, given their sizable capital investment, value of production, and employment in 1845. The ax industry stagnated, whereas cutlery and edge tools, pins, and rubber goods entered early growth stages, and that growth from 1832 to 1845—during which the nation experienced a six-year economic contraction, beginning in 1837—testified to firms' increasing market power. From 1845 to 1850 the Britannia and plated ware, button, and tinware industries stabilized, but most others experienced large jumps in capital investment, value of production, and employment.

From 1832 to 1850 firms in most industries increased capital investment and hired more employees; however, hat and cap manufactures and tinware firms remained small (table 8.11). By 1832 firms divided into two industrial groups—those that were capital intensive and those that were labor intensive. In the capital-intensive industries—such as axes, brass foundries and brassware, carpets, and firearms—firms had high ratios of capital per firm and capital per employee. Ax and carpet firms employed over one hundred employees each, whereas brass foundries and brassware and firearms firms employed ten to twenty employees each. Firms in labor-intensive industries—Britannia and plated ware, clocks, hats and caps, and tinware—had lower ratios of capital per firm and capital per employee. Firms divided into three groups based on employees per firm: firms in clocks and hats and caps operated moderate-sized factories of thirty to forty employees; firms in Britannia and plated ware ran small factories of nineteen employees each; and tinware firms were small workshops of eight employees.

During the half-decade from 1845 to 1850 the scale of firms—measured by capital per firm and employees per firm—surged in most industries, implying that firms were aggressively expanding their reach to the national market. By 1850 Connecticut's firms captured disproportionate shares of the nation's manufacturing in numerous, diverse industries: primary metals (e.g., copper and brass); producer durables such as brass foundries, cars for railroads, hardware, and cutlery and edge tools; and consumer goods such as Britannia and plated ware, carpets, clocks, combs, hosiery, India rubber goods, pins, and suspenders (table 8.12). Entrepreneurs concentrated production in fewer fac-

Table 8.10. Capital Investment, Value of Production, and Employment in Selected Industries in Connecticut, 1832, 1845, and 1850

	Capital Investment (dollars)			Value of Production (dollars)			Employment		
Selected Industries	1832	1845	1850	1832	1845	1850	1832	1845	1850
Axes	256,750	366,935	—	345,500	268,656	—	351	302	—
Brass foundries and brassware	323,650	780,900	1,174,300	137,000	1,401,494	2,333,481	147	648	673
Britannia and plated ware	19,750	133,500	127,550	257,550	400,367	441,050	190	320	496
Buttons	—	192,250	201,500	—	428,762	562,274	—	637	548
Carpets	288,000	584,000	910,000	342,000	587,407	1,079,292	324	934	1,178
Clocks	40,000	369,600	478,800	350,500	771,115	1,103,200	194	656	742
Combs	—	176,363	290,750	—	243,638	636,600	—	252	451
Cutlery and edge tools	—	67,525	698,300	—	91,837	841,301	—	176	916
Firearms	105,000	91,600	170,200	60,000	155,825	245,750	90	164	305
Hardware	—	427,000	1,293,250	—	644,269	2,360,190	—	869	2,285
Hats and caps	28,500	409,330	401,350	177,151	920,806	1,509,479	288	1,461	1,626
Pins	—	68,000	164,800	—	170,000	297,550	—	158	265
Rubber goods	—	87,000	537,500	—	190,000	1,218,500	—	116	1,347
Tinware	6,650	258,100	242,100	62,000	487,810	487,085	42	414	443

Sources: McLane, *Documents Relative to the Manufactures in the United States;* Tyler, *Statistics of the Condition and Products of Certain Branches of Industry in Connecticut, for the Year Ending October 1, 1845;* Secretary of the Interior, *Abstract of the Statistics of Manufactures, Seventh Census, 1850.*

Note: Data for 1845 only include firms with more than five employees.

Table 8.11. Capital Investment per Firm, Capital Investment per Employee, and Employees per Firm in Selected Industries in Connecticut, 1832, 1845, and 1850

Selected Industries	Capital Investment per Firm ($)			Capital Investment per Employee ($)			Employees per Firm		
	1832	1845	1850	1832	1845	1850	1832	1845	1850
Axes	85,583	15,289	—	731	1,215	—	117.0	12.6	—
Brass foundries and brassware	26,971	15,937	35,585	2,202	1,205	1,745	12.3	13.2	20.4
Britannia and plated ware	1,975	3,814	5,102	104	417	257	19.0	9.1	19.8
Buttons	—	4,577	6,948	—	302	368	—	15.2	18.9
Carpets	96,000	97,333	303,333	889	625	772	108.0	155.7	392.7
Clocks	6,667	11,550	28,165	206	563	645	32.3	20.5	43.6
Combs	—	4,767	10,769	—	700	645	—	6.8	16.7
Cutlery and edge tools	—	7,503	25,863	—	384	762	—	19.6	33.9
Firearms	35,000	13,086	21,275	1,167	559	558	30.0	23.4	38.1
Hardware	—	10,675	13,471	—	491	566	—	21.7	23.8
Hats and caps	4,071	2,057	3,064	99	280	247	41.1	7.3	12.4
Pins	—	11,333	41,200	—	430	622	—	26.3	66.3
Rubber goods	—	29,000	67,188	—	750	399	—	38.7	168.4
Tinware	1,330	2,555	2,989	158	623	547	8.4	4.1	5.5

Sources: McLane, *Documents Relative to the Manufactures in the United States;* Tyler, *Statistics of the Condition and Products of Certain Branches of Industry in Connecticut, for the Year Ending October 1, 1845;* Secretary of the Interior, *Abstract of the Statistics of Manufactures, Seventh Census, 1850.*

Note: Data for 1845 only include firms with more than five employees.

Table 8.12. Connecticut's Industries as a Percentage of the Nation's Industries and the Surplus Production Ratio of Connecticut's Industries, 1850

Selected Industries	Factories	Employment	Capital	Value of Production	Value Added[a]	Surplus[b]
Brass foundries	14.9	25.7	40.8	34.5	30.3	21.6
Britannia and plated ware	27.5	38.9	21.5	28.7	30.9	18.0
Buttons	49.2	50.4	51.3	58.3	53.9	36.6
Cars, railroad	9.8	16.4	25.4	16.8	12.7	10.6
Carpets	2.6	19.0	23.6	20.0	26.4	12.5
Clocks	73.9	92.8	95.8	93.4	93.3	58.5
Clock cases	100	100	100	100	100	62.7
Clock springs	100	100	100	100	100	62.7
Combs	17.9	25.2	45.9	39.4	25.9	24.7
Copper and brass	6.3	10.1	18.5	21.9	11.5	13.7
Cutlery and edge tools	6.7	21.4	30.1	22.1	21.9	13.8
Firearms	2.5	19.7	29.5	21.0	22.7	13.1
Hardware	28.2	32.5	36.5	33.9	36.3	21.3
Hats and caps	12.5	10.7	9.1	10.5	10.3	6.6
Hosiery	1.2	17.3	36.7	21.6	22.0	13.5
India rubber goods	23.5	52.5	36.9	40.3	28.2	25.3
Percussion caps	100	100	100	100	100	62.7
Pins	100	100	100	100	100	62.7
Suspenders	100	100	100	100	100	62.7
Tin and sheet iron works	3.6	6.0	5.9	5.5	5.2	3.4
Selected total	9.6	21.5	25.2	22.8	22.1	14.3
Total	3.0	5.3	4.9	4.6	5.1	2.9

Source: Secretary of the Interior, *Abstract of the Statistics of Manufactures, Seventh Census, 1850.*

Notes:

[a] Value added equals the value of production minus the cost of materials.

[b] Surplus production is computed as the following ratio: Connecticut's percentage of national value of production divided by Connecticut's percentage of national population (1.5949 percent).

tories than expected based on their share of manufacturing; except for Britannia and plated ware and hats and caps, Connecticut's share of factories in every industry fell below the state's share of employment, capital, value of production, and value added. The state achieved total or near-total dominance of the manufacture of clocks, percussion caps, pins, and suspenders, and it accounted for 30 to 50 percent of the nation's manufacture of brass foundry goods, Britannia and plated ware, buttons, combs, hardware, and India rubber goods. Firms in most industries produced enormous surpluses ranging from ten to sixty times Connecticut's share of national demand (see table 8.12).

In virtually every leading Connecticut industry in 1850 the state's firms had more capital, employed more workers, and produced a larger value of output than the typical national firm in that industry, and, relative to total national manufacturing, Connecticut's typical firm had 60 percent more capital, 74 percent more employees, and a 52 percent greater value of production (table 8.13). Firms in most of Connecticut's top industries invested greater amounts of capital per employee than firms elsewhere, and this investment translated into greater values of production per employee. The state's firms achieved their industrial prominence in older sectors—Britannia and plated ware, buttons, and clocks—by the early 1830s, according to Ellsworth's observations about market areas.[34] In newer industries—such as brass foundries, cutlery and edge tools, hardware, India rubber goods, pins, and suspenders—firms must have achieved dominance quickly between the late 1830s and mid-1840s, given their large shares of the nation's employment, capital, value of production, and value added in various industries in 1850 (see table 8.12). The brass industry's growth exemplifies how firms translated competitive advantages into industrial dominance.

The Brass Industry
Growth and Constraints

Connecticut's brass industry (brass foundries and brassware) grew rapidly after 1832 (see table 8.10), and its firms achieved national dominance no later than the 1840s; the state's brass firms accounted for about one-third of the nation's capital investment, value of production, and value added by 1850 (see table 8.12). Compared to the nation's typical brass foundries in 1850, Connecticut's had 175 percent greater capitalization, 73 percent more employees, and a 132 percent larger value of production, and for each employee Connecticut's foundries invested 58 percent more capital and achieved a 34 percent greater value of production (see table 8.13). Relative to firms in other industries in Connecticut, brass firms invested far more capital per employee, because they often ran foundries and rolling mills alongside fabricating brass into diverse products; as of 1850, however, individual brass firms employed small labor forces.

The status of Connecticut's brass firms in 1820 did not foreshadow their future dominance. At that time only one Waterbury firm—Leavenworth, Hayden and Scovill—manufactured brass buttons, and it did not roll its own brass sheets; an iron rolling mill in nearby Bradleyville did that. Since 1817

Table 8.13. Industry Comparisons between the United States and Connecticut, 1850

Selected Industries	Capital per Firm ($)		Employees per Firm		Value of Product per Firm ($)		Capital per Employee ($)		Value of Product per Employee ($)	
	U.S.	Conn.	U.S.	Conn.	U.S.	Conn.	U.S.	Conn.	U.S.	Conn.
Brass foundries	10,710	29,405	11.3	19.6	24,497	56,840	945	1,497	2,161	2,895
Britannia and plated ware	6,507	5,102	14.0	19.8	16,877	17,642	464	257	1,204	889
Buttons	6,661	6,948	18.4	18.9	16,345	19,389	361	368	886	1,026
Cars, railroad	21,854	56,875	37.9	63.8	60,818	105,000	577	892	1,605	1,647
Carpets	33,215	303,333	53.3	392.7	46,574	359,764	623	772	873	916
Clocks	21,730	28,165	34.8	43.6	51,370	64,894	625	645	1,477	1,487
Clock cases	2,520	2,520	10.6	10.6	7,600	7,600	238	238	717	717
Clock springs	3,717	3,717	12.7	12.7	12,015	12,015	293	293	949	949
Combs	4,196	10,769	11.8	16.7	10,701	23,578	354	645	904	1,412
Copper and brass	16,291	47,945	13.7	21.9	28,245	98,455	1,193	2,188	2,068	4,494
Cutlery and edge tools	5,790	25,863	10.7	33.9	9,509	31,159	543	762	892	918
Firearms	1,822	21,275	4.9	38.1	3,700	30,719	373	558	758	806
Hardware	10,409	13,471	20.7	23.8	20,464	24,585	503	566	990	1,033
Hats and caps	4,225	3,064	14.5	12.4	13,664	11,523	291	247	942	928
Hosiery	6,409	200,000	27.4	402.0	12,095	222,000	234	498	442	552
India rubber goods	42,815	67,188	75.5	168.4	88,951	152,313	567	399	1,178	905
Percussion caps	2,500	2,500	6.0	6.0	15,000	15,000	417	417	2,500	2,500
Pins	41,200	41,200	66.3	66.3	74,388	74,388	622	622	1,123	1,123
Suspenders	4,160	4,160	72.4	72.4	34,200	34,200	57	57	472	472
Tin and sheet iron works	1,811	2,989	3.2	5.5	3,918	6,013	559	547	1,208	1,100
Selected total	5,351	14,021	10.9	24.3	11,559	27,478	493	577	1,065	1,131
Total	4,334	6,924	7.8	13.6	8,284	12,608	557	510	1,065	929

Source: Secretary of the Interior, *Abstract of the Statistics of Manufactures, Seventh Census, 1850.*

Leavenworth, Hayden and Scovill rented a room of their factory to Daniel Hayden, the brother of David Hayden, one of the partners, for manufacturing lamps and other brass articles. This decision by a predecessor to Scovill and Company—a future industry giant—to rent space rather than to commence production epitomized the brass industry's limited prospects. Aaron Benedict owned a small Waterbury shop making bone and ivory buttons; this insignificant firm was the predecessor of the other future brass giant, Benedict and Burnham. British brass imports posed formidable competition for domestic firms, because British firms utilized more sophisticated machinery and employed workers with greater skills. Even the tariff of 1816 setting a 20 percent ad valorem rate (of the invoice value) did not protect Connecticut brass firms enough to allow them to displace British imports significantly. The tariff of 1818 raised the rate to 25 percent on articles manufactured wholly or principally from copper, which included brass sheet and wire and articles made from them, and this rate rose to 30 percent in the tariff bills of 1842 and 1846.[35] Because the tariff rate after 1820 stayed in a small range (25–30 percent) for the next forty years, an explanation for the changes in the brass industry's growth rate and its concentration in Connecticut must be sought outside the tariff wall.

From 1820 to the early 1830s several brass firms began operating in Waterbury and vicinity, and existing ones increased production, but their expansion paled next to the contemporaneous, explosive growth of the Boston Associates' cotton textile firms. Most of the top ten textile firms of the Boston core had greater capitalization than the entire brass industry in Connecticut in 1832 (see table 8.8). At that time Waterbury, the center of the brass industry, contained three brass button firms whose capital (fixed and working) totaled about $125,000 and three brass rolling mills whose capital totaled about $75,000. Although these firms may have earned generous profits behind the tariff wall, profits were not so lucrative that they enticed many firms into the industry or encouraged existing firms to raise production sharply. A shortage of capital cannot explain this reluctance to expand rapidly because Connecticut's prosperous citizens possessed ample capital, as exemplified by numerous banks and insurance companies, and by ninety-four cotton mills totaling $4.2 million of capital and two carpet factories totaling $268,000 of capital in 1832.

Inadequate machinery and a shortage of skilled workers caused the slow expansion of brass manufacturing from 1820 to the early 1830s. Around 1820 Waterbury's Leavenworth, Hayden and Scovill began addressing these problems; they hired James Croft, an Englishman trained in the brass industry, and they added other skilled English workers during the 1820s. A. Benedict, formed in 1823 as a partnership and successor to Aaron Benedict's bone and ivory button firm, also obtained skilled artisans from England. These skilled workers and accumulated experience of brass firms encouraged Benedict and Company and J. M. L. and W. H. Scovill to build brass rolling mills in Waterbury around 1830, and in that year Holmes and Hotchkiss started one in Waterbury, and Wolcottville Brass Company, founded in 1834, opened one in nearby Torrington. Israel Holmes, active in both firms, traveled to England several times during the 1830s and brought back machinery and skilled workers. Even as those firms gained sophistication producing sheet brass and brass wire by the mid-1830s, adequate production-quality machine tools—such as gear-cutting machines, grinding machines, milling machines, and lathes—increasingly became available to manufacture many brass products cheaply enough to compete with iron or wood products. Henceforth, other industries' proliferating demands for brass ingots, sheet, and wire and the internal development of brass products to meet growing demands powered the growth of Connecticut's brass firms.[36]

Demand for Brass

Chauncy Jerome's invention of the one-day brass clock in 1837 provided one substantial market; within several years production costs plunged to such low levels that completed clocks sold for two dollars or less—a mass market product. Because Connecticut firms already dominated markets for cheap wooden clocks and eight-day brass clocks, its firms possessed competitive advantages to add one-day brass clocks or shift production to them; by 1850 they manufactured virtually every clock in the nation, reaching a total value of just over one million dollars (see table 8.12). Between the mid-1820s and mid-1830s several inventors designed commercially successful pin machines, thus generating a demand for brass. Wholesaling and other network ties between New York City and Connecticut facilitated the concentration of pin manufacturing in the state. John Howe, a Connecticut native and pin machine inventor, moved to New York City, and in 1835 Howe Manufacturing Company was organized there. Through the influence of Anson Phelps—a Connecticut native and the leading partner in Phelps, Dodge and Company—the Howe company relocated to Derby, Connecticut, in 1838. Phelps, Dodge, one of New York City's largest commodity firms, imported tin, tin plate, iron, copper, and other metals and supplied them to Connecticut manufacturers. Phelps had been involved in Wolcottville Brass Company in Torrington (1834), and his other Derby firms included a copper rolling mill (1836), Ansonia Brass and Battery Company (1844), and Ansonia Manufacturing Company (1845), a producer of rolled brass and copper as well as brass and copper wire. In 1846 the brass firms of Benedict and Burnham and of Brown and Elton founded American Pin Company in Waterbury, capitalized at $50,000; this firm, along with the Howe firm, controlled key patents to manufacture pins and to stick them into crimped paper. By 1850 Connecticut housed all four of the nation's pin firms, which produced almost $300,000 worth of pins annually (see tables 8.10 and 8.12).[37]

A lengthening list of products from other industries added to brass demand, and brass firms decided to manufacture many products themselves; several became large businesses and were spun off as separate companies: Benedict and Burnham started American Pin Company in 1846 ($50,000 capital), Waterbury Button Company in 1849 ($30,000), and Waterbury Clock Company in 1857 ($60,000). Increased capitalization of Benedict and Burnham from 1823 to 1856 reflects the expansion of the Waterbury brass industry; capital rose at an almost constant 12 percent compound annual rate, from $6,500 to $400,000, excluding spin-offs. Some large firms, such as Holmes, Hotchkiss, Brown and Elton and Waterbury Brass Company, started before 1850, whereas others—including Brown and Brothers ($200,000 capital) and Holmes, Booth and Haydens ($110,000)—formed after 1850 (both were founded in 1853). Even with this burst of activity, the brass industry was at an early growth stage; patent rates in Waterbury, although not restricted to brass, first jumped around 1850. During a thirty-year period (1820–50) Waterbury people received only 21 patents, but from 1851 to 1855 they took out 26 patents; this figure surged to 62 patents during the second half of that decade. Their annual patent rates accelerated from 1861 to 1890, and they received a total of 1,136 patents (38 per year). Waterbury's (and nearby towns') large and small brass firms raised their industry share to commanding dominance: by 1880 Connecticut's brass and copper rolling mills accounted for 74 percent of the nation's $9,057,600 capital investment in those mills, and the state's brassware firms constituted 81 percent of the nation's $594,582 capital investment in that sector.[38] Yet the question remains: how did Connecticut's brass firms achieve such dominance?

A Market-Driven Strategy

Connecticut's brass firms self-consciously pursued market-driven strategies to sell goods to the national market and dominate the industry. Initially, their marketing juggernaut rested on the backs of peddlers distributing goods directly to consumers; they continued using peddlers and devised new selling strategies. Although brass industry leaders were not large-scale merchant wholesalers, as were some Boston Associates, most early leaders—such as Frederick Leavenworth, the Scovill brothers, and Charles D. Kingsbury—owned retail stores, and other prominent investors and managers of brass firms initially served as clerks in those stores; this retail experience sensitized them to consumer demands and to the importance of marketing.

Nevertheless, many small industrialists acquired experience as retailers, but only a small share of their firms penetrated markets outside their region. Numerous brass industry leaders acquired retail experiences beyond Waterbury, however, providing insights into markets throughout the East and South. Israel Holmes ran a sales store in Augusta, Georgia, for a Connecticut hat manufacturer from 1816 to 1818; Charles Kingsbury sold clocks in southern states and books in Virginia for a Philadelphia publishing house and served as an agent in Philadelphia for a Naugatuck button manufacturer from 1814 to 1821; Green Kendrick engaged in mercantile pursuits in Charlotte, North Carolina, during the years 1818 to 1822; Benjamin DeForest sold buttons for Aaron Benedict in New York in the 1820s; Israel Coe sold buttons and sheet metal in eastern cities in the late 1820s; William Brown worked in South Carolina for several years around 1830; and from 1821 to 1834 Gordon Burnham sold goods in large eastern cities and worked as a tinware peddler for the Yale firm in Meriden. They recognized that general wholesalers could not effectively distribute their products to the national market; therefore, when brass manufacturers started expanding in the mid-1830s, they followed the Associates' model and created selling houses. In 1835 Gordon Burnham, a partner in Benedict and Burnham Manufacturing Company, moved to New York City and founded the commission house of Baldwin, Burnham and Company to represent his brass firm; this house became a prominent distributor of brass products for Waterbury manufacturers. Charles Benedict moved to New York City in 1834 to work as a bookkeeper in a dry goods jobbing house and switched the next year to work for Baldwin, Burnham and Company.

In the mid-1840s, when brass manufacturing shifted to greater production, firms enlarged distribution networks (see table 8.10). In 1844 Burnham opened a commission house under the name of Burnham and Baldwin in Boston, and by 1846 it became Burnham and Welton. Arad Welton, who was active in various Waterbury firms, and Charles Scott became partners. Scott stayed in Boston for thirty years, and the firm became Burnham, Scott and Company. Arad Welton founded a commission house in New York City by 1854, and that year Waterbury native Thomas Porter started work in Welton's firm. Porter took over the firm and founded Porter Brothers, which distributed for many Waterbury firms. In 1852 brass manufacturing giants—Benedict and Burnham and Scovill—cofounded Benedict and Scovill with a capitalization of $50,000; this commission house served both firms from its New York headquarters. The next year Holmes, Booth and Haydens Manufacturing Company was established, and its partner Henry Hayden immediately took charge of a selling agency in New York City. Because these commission houses specialized in brass products and the principals had deep knowledge of the brass industry, Connecticut's manufacturers possessed sophisticated means to move their soaring output into the national market, thus contributing to their dominance by 1850.[39] Although this market-driven approach was a necessary condition for achieving dominance, it was not the only one.

A Network Strategy

The network strategy employed by Connecticut brass firms constituted the necessary condition for dominance, because their networks provided the capacities to build technologically sophisticated, well-capitalized firms that successfully competed against firms elsewhere. Although many brass firm principals were born in Waterbury and vicinity, they did not create closed, heavily redundant networks. Instead, brass firm networks possessed multiple hub individuals or families—the Leavenworths, Aaron Benedict, Charles Kingsbury, Gordon Burnham, the Scovill brothers, Israel Coe, Israel Holmes, John Elton, Arad Welton, the Brown brothers, and the Hayden family. They connected individuals in their group, and they bridged to other hub people, facilitating information flow and allowing them to mobilize substantial capital to expand firms or start new ones.

Shifting coalitions formed different enterprises, and some members left one firm and joined others to establish a new firm. Israel Coe joined Aaron Benedict and others to form the partnership of Benedict and Company in 1829, but, when Gordon Burnham entered the partnership in 1834, Coe left to start

Wolcottville Brass Company with Israel Holmes; Anson Phelps—the New York City commodity merchant—participated in that firm. Israel Holmes had been employed by the J. M. L. and W. H. Scovill Manufacturing Company, and after leaving their employ he joined others to start Holmes and Hotchkiss Manufacturing Company in 1830. When Waterbury Brass Company was organized in 1845, John Elton took the lead, and Philo Brown and Israel Holmes served as directors. In 1853, when Holmes, Booth, and Haydens Manufacturing Company organized, Gordon Burnham invested in it and joined the board of directors. Social networks in Waterbury and vicinity remained open; members reached externally to incorporate new individuals and to bridge to other networks outside the Waterbury core, enhancing information flow and multiplying the number of talented individuals involved in brass firms. The Hayden family of eastern Massachusetts relocated to Williamsburgh, Massachusetts, in the Connecticut Valley, and several Haydens moved to Waterbury just prior to 1820; they became hub individuals in the Waterbury brass networks. Hub individuals—including Israel Coe, Charles Kingsbury, and William Brown—began as clerks in retail stores or worked in sales for firms outside Connecticut and later were incorporated as partners in firms. Others, such as Green Kendrick and John Buckingham, were born outside Waterbury and married daughters of prominent Waterbury industrialists; then they entered the brass industry.

Several leaders were members of Yale alumni networks, but most attended common schools in the Waterbury vicinity, and many attended academies in the state, providing access to those alumni networks. Besides service as local government officials, many brass leaders were elected to the state legislature, which brought them into the capital networks of wealthy individuals, merchants, banks, and insurance companies and into the networks of industrial firms in which important technical advances in machine tools were being developed and implemented.[40] The market-oriented, network strategy of Connecticut's brass firms made them formidable competitors in the national market for brass products, and they leveraged the powerful position they had achieved by the 1840s into virtually complete dominance after 1860.

Mass Market Dominance

New England firms penetrated the mass market for shoes, cotton textiles, and Connecticut manufactures by the 1840s, yet the organizational and technological bases of those industries differed. In shoes an efficient, integrated production complex organized around small firms employing a division of labor reduced production costs, and innovations in tools and small equipment helped, but the machinery remained simple. In contrast, large, vertically integrated cotton textile firms employed a sophisticated business plan and technologically advanced machinery to drive down production costs. In the brass industry, the most fully documented Connecticut industry, technical talent arrived from England during the 1820s and early 1830s, but breakthroughs in low-cost production awaited the accumulation of machine tools by the mid-1830s, and large production awaited markets for diverse goods using expensive brass.

Nevertheless, these industries shared features: in each entrepreneurs realized they needed an efficient wholesaling, marketing, and distribution complex. The early emergence of this complex meant their goods reached markets efficiently as declining production costs opened ever-larger shares of the mass market. Because products had high value relative to weight, transportation costs added little to final selling prices; therefore, potential competitors in other regions needed to attain efficiencies equivalent to those of New England production complexes. The competitiveness of those industrial juggernauts also rested on their social networks. In shoes networks permitted firms to cooperate in a finely tuned division of labor, and networks linking the wholesaling-marketing-distribution complex to the production complex transmitted information about changes in demand and new styles. In cotton textiles not only did the Boston Associates' social networks provide avenues to assemble large sums for investment, but they also created industrial social networks conveying information about their sophisticated business plan. Merchant elite elsewhere could not participate, placing them at a severe competitive disadvantage. The Associates also benefited from separate social networks among skilled mill managers and mechanics which reached throughout New England and even outside. This network transmitted technological information that leading firms incorporated into mills, whereas firms elsewhere had difficulty hiring these managers and mechanics, because they were initially concentrated in New England, demand for their services exceeded supply, and they were well compensated there. Leaders in the brass industry in Waterbury and vicinity developed a network strategy keeping firms linked to information about national markets and technology, and within the Waterbury core they maintained multiple hubs of individuals and families who bridged within the core; they avoided the dangers of closed networks stifling information flow and hindering the formation of coalitions of firms for mobilizing capital. Thus, in shoes, cotton textiles, and brass, firms drove to national market dominance through strategies designed to target markets beyond New England.

CHAPTER NINE

The East Anchors the Manufacturing Belt

> The North American Manufacturing Belt originated at its eastern end along the Atlantic, and . . . has grown towards the west from Boston, from New York, Philadelphia and Baltimore. —STEN DE GEER, 1927

Following the economic disruption of the Revolutionary War and downturn during the 1780s, the United States economy entered an extended growth path, punctuated by episodes of contraction such as the Embargo (1807), the War of 1812, and that of 1837–43. By 1860 real (constant prices) gross domestic product was about fifteen times larger, and real GDP per capita had doubled, generating substantial wealth effects. The East led the transformation of agriculture and manufacturing, and significant changes during the period from 1790 to 1820, when there was slow commercial, agricultural, and manufacturing growth, set the stage for faster growth from 1820 to 1860. The East's farmers increased productivity, thus reducing food costs for rural and urban consumers, and rising real farm income translated into greater demand for manufactures. Farmers accumulated capital and funneled it into farm improvements, and their capital helped underwrite industrial development. Prosperous farmers generated potent demands for commercial services, stimulating the growth of villages and small towns and of subregional centers housing small-scale wholesalers and financiers. Urban centers contributed to demand for manufactures, and, as points of capital accumulation (with retailers, wholesalers, and professionals clustered there), their firms and individuals helped finance agriculture, commerce, and industry. Greater agricultural productivity freed farm labor to work in commerce and to expand workshops and factories. Eastern farmers continually altered agricultural production, first in response to new, better-quality farmland entering production in the East, and later in response to growing volumes of midwestern farm products entering the East after 1835. Farmers near the metropolises of Boston, New York, Philadelphia, and Baltimore increased their output of vegetables, fruit, dairy products, and hay for urban markets, and, as other cities, towns, and villages grew, nearby farmers began producing for urban markets.

Already by 1820 rising agricultural prosperity swept up large parts of the East, including areas within fifty to one hundred miles of East Coast metropolises as well as the Connecticut Valley, Hudson-Mohawk Valleys, and southeastern Pennsylvania, and wagon transportation provided efficient means to bring farm goods to markets. Prosperous agricultural areas offered growing markets for diverse manufactures produced in workshops or in small factories; demand spurred invention and innovation in manufacturing to supply new or better products, and fiercer competition in local and subregional markets galvanized firms to raise productivity, driving prices down and broadening markets for goods. This local and subregional industrial change was under way during the period from 1790 to 1820 and continued afterwards. Coincidental with widespread industrial expansion across the East after 1820, manufacturers of light, high-value goods (e.g., shoes, cotton textiles, and Connecticut manufactures such as tinware, clocks, buttons, and hardware) leveraged access to capital, market information, and distribution capacity and participation in industrial social networks to shift into greater production for large market areas. These firms used division of labor and better methods to organize and carry out tasks as means to improve productivity, but machinery was important only in cotton textiles. Their efforts generated big declines in prices of manufactured goods, and cheaper prices, rather than rising incomes, explain their sensational growth; those firms developed potent marketing organizations to funnel surging output into the national market.

The metropolitan complexes of Boston, New York, Philadelphia, and, secondarily, Baltimore were the nation's industrial powerhouses by 1840, and each formed the core of a regional industrial system including manufacturers in their hinterland. As of 1860, the East housed 36 percent of the nation's population, but its factories produced 74 percent of value added in manufacturing (table 9.1).[1] In every manufacture its share of national value added exceeded its population share, and in most sectors the East's industrial share was over twice its population share. The prominent exceptions—food, tobacco, and lumber—were processing industries in which early-stage raw material manufacturing prepared goods for long-distance shipment; the East's share of them ranged from 40 to 54 percent, but share of production understates the East's dominance. Factories in New England and the Middle Atlantic states accounted for virtually all interstate sales of manufactures in most industrial

Table 9.1. Percentage of National Value Added in Manufacturing by Region, 1860 and 1900

	1860				1900			
	Manufacturing Belt				Manufacturing Belt			
Industry	E	MW	Total	SO	E	MW	Total	SO
Processing								
Food	54	36	91	9	42	46	88	8
Tobacco	40	14	54	46	46	31	77	22
Lumber and wood products	46	33	78	20	30	42	71	29
Paper	90	9	99	1	79	21	100	0
Chemicals	74	12	85	15	60	27	87	12
Petroleum and coal products	79	14	93	7	65	30	95	3
Leather	86	9	94	6	76	20	96	3
Local and regional market								
Apparel	87	12	99	1	73	25	99	0
Furniture	67	29	96	4	51	46	96	3
Printing and publishing	83	13	96	4	60	33	93	5
Stone, clay, and glass	74	18	92	7	63	32	95	4
Primary metals	79	16	95	5	65	27	92	6
Fabricated metals	76	17	93	7	52	45	98	1
Machinery	67	24	91	9	62	35	97	3
Transportation equipment	72	17	89	11	47	41	89	8
National market								
Textiles	94	3	97	3	87	4	90	10
Rubber	100	0	100	0	87	14	101	0
Leather products	82	13	95	5	70	25	95	4
Instruments	99	2	101	0	77	22	99	0
Miscellaneous	91	8	99	1	76	24	100	0
Total	74	18	92	8	60	32	91	7

Source: Niemi, *State and Regional Patterns in American Manufacturing, 1860–1900*, 14–15, 45–54, tables 4–5, 12–13.

Notes:

E (East): New England (Maine, New Hampshire, Vermont, Massachusetts, Connecticut, Rhode Island), Middle Atlantic (New York, New Jersey, Pennsylvania, Delaware, Maryland)

MW (Midwest): Ohio, Indiana, Illinois, Michigan, Wisconsin, Minnesota, Iowa, Missouri

SO (South): Virginia, West Virginia, North Carolina, South Carolina, Georgia, Florida, Kentucky, Tennessee, Alabama, Mississippi, Arkansas, Louisiana, Oklahoma, Texas

Totals may not sum due to rounding.

sectors (table 9.2); elsewhere most factories served local or subregional (intrastate) markets. The East anchored the American manufacturing belt in 1860.

The Midwest's Window of Opportunity

Eastern manufacturers of products such as shoes, cotton textiles, tinware, and brass products (clocks, buttons, and pins) captured national markets by the 1830s and 1840s, but that accomplishment did not portend that eastern firms in many other industrial sectors would quickly follow suit. In most manufactures firms exhausted economies of scale at small factory sizes of about fifteen to twenty workers, and the primitive conditions of much industrial machinery precluded firms from reducing production costs significantly through mechanization. Instead, firms gained most efficiencies in production through the division of labor, the organization of tasks, and the use of simple machines. Heavy or bulky products—furniture, structural iron, pipes, planed lumber for construction, carriages and wagons, and agricultural machinery—remained costly to transport long distances; therefore, firms had difficulty selling them outside subregional or regional markets. Moreover, products such as industrial machinery required experts for selling and demonstrating their use; these goods likewise remained subregional or regional market goods. Although the railroad network dramatically enlarged after 1850, an integrated, highly efficient railroad system throughout the East and Midwest did not exist for several more decades. Eastern factories aiming to reach midwestern farm markets not only faced long-distance transportation from the East, but they also confronted low rural densities, which raised distribution costs. In 1860 the Midwest's rural densities were barely one-third those of the Middle Atlantic. Limitations of production efficiencies, distribution, and transportation restricted market area expansion in many industries until the 1870s, providing a window of opportunity for midwestern factories to build production and marketing skills with which to capture local, subregional, and regional markets in preparation for competition from eastern factories.[2]

The population of the Midwest followed the model of the East and built a powerful economic development engine on prosperous agriculture. Yet the Midwest's vast landscape of flat, fertile farmland, unparalleled in the world, distinguished it from the much smaller prime agricultural land in the East. Agriculture and other resource-processing sectors produced predominantly for intraregional markets through the 1830s, although some flour and meat

Table 9.2. Percentage of National Interstate Trade in Manufacturing Value Added by Region, 1860 and 1900

	1860					1900				
	Manufacturing Belt					Manufacturing Belt				
Industry	NE	MA	MW	Total	SO	NE	MA	MW	Total	SO
Processing										
Food	3	66	31	100	0	2	34	61	97	0
Tobacco	2	10	9	21	79	0	43	21	64	36
Lumber and wood products	20	29	38	88	5	14	0	69	82	18
Paper	70	30	0	100	0	53	43	5	100	0
Chemicals	16	69	0	84	16	8	71	10	89	11
Petroleum and coal products	24	76	0	99	1	7	79	14	100	0
Leather	33	67	0	100	0	21	67	12	100	0
Local and regional market										
Apparel	28	71	1	100	0	8	82	10	100	0
Furniture	31	49	20	100	0	17	36	47	100	0
Printing and publishing	13	87	0	100	0	17	67	16	100	0
Stone, clay, and glass	18	81	0	99	0	14	67	19	100	2
Primary metals	21	77	2	100	0	5	76	18	98	0
Fabricated metals	63	36	1	100	0	29	25	46	100	0
Machinery	44	46	11	100	0	26	49	25	100	0
Transportation equipment	40	60	0	100	0	6	49	45	100	0
National market										
Textiles	91	9	0	100	0	70	28	0	97	3
Rubber	43	57	0	100	0	82	17	1	100	0
Leather products	90	10	0	100	0	83	14	3	100	0
Instruments	49	51	0	100	0	30	59	12	100	0
Miscellaneous	49	51	0	100	0	44	49	7	100	0

Source: Niemi, *State and Regional Patterns in American Manufacturing, 1860–1900,* 60–66, tables 16–17.

Notes:

NE (New England): Maine, New Hampshire, Vermont, Massachusetts, Connecticut, Rhode Island

MA (Middle Atlantic): New York, New Jersey, Pennsylvania, Delaware, Maryland

MW (Midwest): Ohio, Indiana, Illinois, Michigan, Wisconsin, Minnesota, Iowa, Missouri

SO (South): Virginia, West Virginia, North Carolina, South Carolina, Georgia, Florida, Kentucky, Tennessee, Alabama, Mississippi, Arkansas, Louisiana, Oklahoma, Texas

Totals may not sum due to rounding.

exited earlier down the Ohio-Mississippi River system. By the 1840s many midwestern farmers cleared and improved land for commercial farming, and intraregional canals lowered transport costs for farm products. Soon soaring amounts of these goods were headed for eastern markets, especially over the Erie Canal. Prosperous farmers offered lucrative markets for manufacturers, and those firms, following eastern peers, improved the productivity of workshops and small factories, thus reducing costs and boosting market sales. Resource-processing manufactures served large intraregional markets for food, lumber, and other construction materials, and some processing (e.g., flour milling and meatpacking) prepared goods for interregional markets. Thus, midwestern firms developed thriving manufactures, especially during the years of rising food exports to the East from 1840 to 1860. By the end of that period the Midwest's factories accounted for 18 percent of the nation's value added in manufacturing, a remarkable achievement considering few firms competed in eastern markets. The Midwest had small shares of national market manufactures (see table 9.1).

As a rapidly growing agricultural region, the Midwest had significant shares of the nation's processing industries, and it housed robust local and regional market manufactures. Few factories distributed products to regional market areas exceeding the size of a state; except for the processing industries of food, lumber, and tobacco and the regional market industries of furniture and machinery, the Midwest had little or no share of interstate trade in manufactures (see table 9.2). Low rural population densities prevailing across the Midwest as of 1860 made it too costly to distribute in large market areas. Midwestern urban-industrial production concentrated heavily (about half) in its regional metropolises—Chicago, Cincinnati, Louisville, St. Louis, Cleveland, Detroit—consistent with their large markets and prosperous urban market farmers on nearby land; those cities were hubs of regional rail networks.[3]

The Manufacturing Belt

After 1860 eastern factories fared well even as settlement intensified in the Midwest and pushed into the Great Plains, Rockies, Southwest, and Pacific Coast and as those regions experienced substantial economic growth and development. By 1900 the East's share of national manufacturing value added had declined modestly from 74 percent in 1860 to 60 percent in 1900, testimony to the advantages of its early industrialization (see table 9.1). Even with a huge expansion in exploitation of natural resources outside the East, it continued to house substantial shares of most processing manufactures in 1900; this reflected the advantages of having available resources in the East, which reduced transport costs. The East retained significant shares of most local and regional market manufactures, corresponding to its share of the nation's industry. Its shares of national market manufactures eroded slightly, but those declines were insignificant because broad categories disguise the individual manufactures that predominated in the East. Across the entire range of industries, factories in New England and the Middle Atlantic continued accounting for significant shares of interstate trade in 1900, reflecting their capacity to reach large regional markets or the national market (see table 9.2). In processing, New England factories kept large shares of paper markets, and Middle Atlantic factories matched them; that region retained major shares of interstate trade for most processing manufactures. Declines in shares of interstate trade across most local and regional manufactures for New England and the Middle Atlantic comport with declines in their share of manufacturing value added between 1860 and 1900. Over that period New England's share fell from 28 to 16 percent, and the Middle Atlantic's share decreased from 45 to 43 percent. Both regions maintained overwhelming shares of interstate trade in national market manufactures, testimony to the difficulties competitors elsewhere encountered in trying to dethrone dominant clusters of firms once they achieved production efficiencies and marketing power.

After 1860 in many industrial sectors eastern factories confronted vigorous competitors in the Midwest. Its factories made substantial gains in their share of the nation's manufacturing during the 1860s, and large regional industrial systems surrounded most midwestern metropolises by 1880. These financial and wholesaling centers maintained their 50 percent share of total manufactures in midwestern urban places from 1860 to 1880. By the early twentieth century the sprawl of the manufacturing belt from the Atlantic seaboard to the edge of the Great Plains awed foreign observers. The Midwest not only significantly increased its share of value added in manufacturing from 18 to 32 percent between 1860 and 1900, but it also gained greater shares across the industrial spectrum (see table 9.1). Befitting the resource-rich Midwest, its factories achieved advances in every processing sector, and by 1900 it dominated the nation's food and lumber industries. The growth of its prosperous agriculture and of its surging urban populations—urban dwellers increased almost eightfold from 1.3 million to 9.5 million between 1860 and 1900 and raised the Midwest's share of the nation's urban population from 20 to 32 percent—stimulated robust increases in shares of every local and regional market manufacture. Greater shares of several national market manufactures—leather products (shoes), instruments, and miscellaneous goods—point toward the emergence of factories penetrating large market areas by 1900.[4]

Dominant shares of national interstate trade which midwestern factories held in food and lumber suggest that by 1900 its firms had built solid national positions even in processing manufactures (see table 9.2). From low or nominal levels of shares of interstate trade in many local and regional manufactures in 1860, midwestern factories boosted their shares substantially by 1900, indicating that many of them had extended their market reach in the Midwest. Participation of its factories in interstate trade in national market manufactures went from zero to modest shares in several sectors, notably instruments and miscellaneous, between 1860 and 1900, confirming that its firms had by then acquired the capacity to compete with eastern factories in national markets. Based on the timing of the emergence of regional industrial systems in the Midwest, the manufacturing belt was probably consolidated by the 1870s, and interlinkages among regional systems through supplier networks and licensing had become widespread—although much earlier in the case of the East's regional systems.

The share of the nation's manufacturing located in the East and Midwest stayed virtually constant, at just over 90 percent, between 1860 and 1900 (see table 9.1). That dominance continued until the 1950s, with only a slight erosion brought about by the shift of textiles to the South, the rise of regional industrial markets on the Pacific Coast protected by vast distances, and the growth of processing manufactures in the West and South. Industrial technology skills, capital networks supporting manufacturing, labor skills, and marketing expertise in the manufacturing belt were too great for competitors outside the area to overcome until the second half of the twentieth century, when new industrial bases such as aerospace and high technology emerged.[5]

The South Fails to Join the Manufacturing Belt

In contrast to the Midwest, which joined with the eastern anchor to form the manufacturing belt, the South failed to do so. Causes of this failure are complex, but a key factor must have been the slave plantation institution, with its capital invested in slaves rather than in land, infrastructure, or machinery; this institutional structure generated little demand for manufactures. Plantations internalized some craft manufactures and supplied some of their own food, and the inability of slaves to express their demands on open markets meant that rural densities of demand for manufactures in areas with high proportions of slaves were too low for many workshops and factories to meet market thresholds. This small commercial demand and investment of much capital in slaves produced few cities and towns across slave plantation areas of the South; therefore, urban infrastructure stimulated few manufactures. Outside of large processing factories, most firms remained tiny craft shops serving small market areas and a few substantial firms in regional market industrial sectors located in the small number of cities in the South as of 1860—Richmond, Virginia; Knoxville, Chattanooga, and Nashville in Tennessee; Atlanta and Augusta in Georgia; and New Orleans, Louisiana. By that year productivity of southern manufacturing firms compared quite favorably to midwestern firms, signifying that limited market demand retarded southern manufacturing, not the absence of entrepreneurial capacity or inadequate workers' skills.

Emancipation terminated the slave plantation institution in 1865, but its impacts on local and regional markets could not have been consequential until the 1870s, and even then southern industrial markets remained small. By that time numerous midwestern and eastern firms in many regional market industrial sectors possessed substantial market power, whereas the South had few such firms. National market manufactures, except textile firms utilizing large supplies of low-wage labor, rarely relocated or started outside regional industrial systems of the manufacturing belt, which housed networks of experienced industrial entrepreneurs, suppliers of capital attuned to industrial investment, inventors, and semiskilled and skilled manufacturing workers.[6]

Substantial evidence confirms the South's meager industrialization in 1860 and its failure to make significant gains by 1900. Its share of national value added in manufacturing remained about 8 percent between 1860 and 1900 (see table 9.1). By 1860 the South had significant shares of the nation's processing—tobacco, lumber, and chemicals—which continued to be important in 1900, and it housed small national shares of local and regional market manufactures from 1860 to 1900. Its share of national market manufactures stayed tiny throughout that period, with the notable exception of textiles, whose share rose to 10 percent by 1900. The almost total failure of southern factories to capture significant shares of interstate trade in manufacturing between 1860 and 1900, except in processing and in textiles, testifies to the limited markets in the South; local and regional markets were so small that few firms in those industries reached market areas outside their state (see table 9.2). Thus, the South failed to join the manufacturing belt, whereas the Midwest became an integral component of it.

Notes

CHAPTER ONE: The Puzzle of the Antebellum East

1. See chapter epigraph by Franklin, "Interest of Great Britain Considered, with Regard to Her Colonies and the Acquisitions of Canada and Guadaloupe," 19; Hamilton, "Report on Manufactures," 43–69. Also see Cooke, "Tench Coxe, Alexander Hamilton, and the Encouragement of American Manufactures"; Nelson, "Alexander Hamilton and American Manufacturing."

2. Gallman, "Commodity Output, 1839–1899," 24, 43, tables 3, A-1. The 1860 manufacturing employment data come from Niemi, *State and Regional Patterns in American Manufacturing, 1860–1900,* 125–29, app. 6, 7. The East includes New England (Massachusetts, Connecticut, Rhode Island, Maine, New Hampshire, and Vermont) and the Middle Atlantic (New York, New Jersey, Pennsylvania, Delaware, and Maryland). The Niemi data are more suitable than other sources because he eliminated nonmanufactures such as construction crafts and blacksmithing which the U.S. census included as manufactures.

3. North, *Economic Growth of the United States, 1790–1860,* 103. Classic statements include Callender, "Early Transportation and Banking Enterprises of States in Relation to the Growth of Corporations"; Schmidt, "Internal Commerce and the Development of a National Economy before 1860."

4. Fishlow, "Antebellum Interregional Trade Reconsidered"; Herbst, "Interregional Commodity Trade from the North to the South and American Economic Development in the Antebellum Period"; Kravis, "Role of Exports in Nineteenth-Century United States Growth," 399, table 5; Lindstrom, *Economic Development in the Philadelphia Region, 1810–1850,* 5–8; Uselding, "Note on the Inter-Regional Trade in Manufactures in 1840."

5. Bidwell and Falconer, *History of Agriculture in the Northern United States, 1620–1860;* Clark, *Roots of Rural Capitalism;* Danhof, *Change in Agriculture;* Dublin, *Transforming Women's Work;* Dublin, *Women at Work;* Field, "On the Unimportance of Machinery"; "Sectoral Shift in Antebellum Massachusetts"; Gates, *Farmer's Age;* Sokoloff, "Investment in Fixed and Working Capital during Early Industrialization"; "Productivity Growth in Manufacturing During Early Industrialization"; Ware, *Early New England Cotton Manufacture.*

6. Lindstrom, *Economic Development in the Philadelphia Region, 1810–1850.*

7. Brown, *Baldwin Locomotive Works, 1831–1915;* Clark, *History of Manufactures in the United States,* vol. 1; Davis, Easterlin, and Parker, *American Economic Growth,* 418–37; Dawley, *Class and Community;* Eggert, *Harrisburg Industrializes;* Fishlow, *American Railroads and the Transformation of the Ante-Bellum Economy;* Goodrich, *Canals and*

American Economic Development; Hounshell, *From the American System to Mass Production, 1800–1932;* Licht, *Industrializing America;* McGaw, *Most Wonderful Machine;* Paskoff, *Industrial Evolution;* Ransom, "Interregional Canals and Economic Specialization in the Antebellum United States"; Taylor, *Transportation Revolution, 1815–1860.*

8. Durkheim, *Division of Labor in Society;* Marx, *Capital;* Rueschemeyer, *Power and the Division of Labor;* Smith, *Inquiry into the Nature and Causes of the Wealth of Nations;* Spencer, *Principles of Sociology;* Tonnies, *Community and Society;* Weber, *Economy and Society.* The concepts of local and nonlocal exchange are from Meyer, "Division of Labor and the Market Areas of Manufacturing Firms."

9. Clark, "Economics and Culture"; Clark, *Roots of Rural Capitalism;* Henretta, *Origins of American Capitalism;* Kulikoff, *Agrarian Origins of American Capitalism;* Kulikoff, "Households and Markets"; Merrill, "Putting 'Capitalism' in Its Place"; Meyer, "Dynamic Model of the Integration of Frontier Urban Places into the United States System of Cities"; Post, "Agrarian Origins of US Capitalism"; Rothenberg, *From Market-Places to a Market Economy;* Vance, *Merchant's World;* Weiman, "Families, Farms and Rural Society in Preindustrial America."

10. Burt, *Structural Holes,* 8–49; Coleman, *Foundations of Social Theory,* 91–116, 175–96, 241–321; Granovetter, "Economic Action and Social Structure"; "Nature of Economic Relationships"; "Strength of Weak Ties"; Meyer, "Division of Labor and the Market Areas of Manufacturing Firms"; Meyer, *Hong Kong as a Global Metropolis,* chap. 2; Rothenberg, *From Market-Places to a Market Economy,* 24–55; Swedberg, "Markets as Social Structures."

11. Appleby, *Capitalism and a New Social Order; Inheriting the Revolution;* Blumin, *Emergence of the Middle Class;* Butler, *Becoming America;* De Vries, "Industrial Revolution and the Industrious Revolution"; McCusker and Menard, *Economy of British America, 1607–1789;* Newell, *From Dependency to Independence;* Wood, "Enemy Is Us"; *Radicalism of the American Revolution.*

CHAPTER TWO: Prosperous Farmers Energize the Economy

1. Nettels, *Emergence of a National Economy, 1775–1815,* 45–64; Pitkin, *Statistical View of the Commerce of the United States of America,* 30.

2. Nettels, *Emergence of a National Economy, 1775–1815,* 61, 89–108; Pitkin, *Statistical View of the Commerce of the United States of America,* 30; U.S. Bureau of the Census, *Historical Statistics of the United States,* ser. E52–63, 90–96, 97–111.

3. Goldin and Lewis, "Role of Exports in American Economic Growth during the Napoleonic Wars, 1793 to 1807"; Johnson, Metre, Huebner, and Hanchett, *History of Domestic and Foreign Commerce of the United States,* 2:18–25, 29–37; North, *Economic Growth of the United States, 1790–1860,* 221, 228–29, app. 1, tables B-III, C-III, E-III, F-III ; U.S. Bureau of the Census, *Historical Statistics of the United States,* ser. A7, 69, 93, 100, 195.

4. U.S. Bureau of the Census, *Historical Statistics of the United States,* ser. A7, 195. See table 1.2 for the composition of the East.

5. Weiss, "Economic Growth before 1860," 19, table 1.4.

6. Adams, "Earnings and Savings in the Early 19th Century." The 1814–20 savings rate was computed from Adams, 120–21, table 1. Whenever possible, end points of real-wage growth rates are based on three-year averages.

7. Gallman, "American Economic Growth before the Civil War," 80–81.

8. Haig, "Toward an Understanding of the Metropolis"; Lampard, "History of Cities in the Economically Advanced Areas"; Timmer, "Agricultural Transformation"; Williamson, "Antebellum Urbanization in the American Northeast."

9. Bjork, "Foreign Trade," 56, table 11; Pred, *Urban Growth and the Circulation of Information,* 189–94; Taylor, "American Urban Growth Preceding the Railway Age," 311–15, table 1, and 321–26. Territorial groupings in table 2.6 accord reasonably well with literature on trade reported in the sources listed here. Boundary shifts have little impact because New York State accounted for much of the difference in population; Meyer, "Dynamic Model of the Integration of Frontier Urban Places into the United States System of Cities"; Pred, *Urban Growth and the Circulation of Information;* Vance, *Merchant's World.*

10. Kirkland, *Men, Cities, and Transportation,* 1:66–70; Kistler, "Rise of Railroads in the Connecticut River Valley," 14–16; Livingood, *Philadelphia-Baltimore Trade Rivalry, 1780–1860,* 36–38; Ringwalt, *Development of Transportation Systems in the United States,* 12–13; Taylor, *Transportation Revolution, 1815–1860,* 32.

11. Durrenberger, *Turnpikes,* 26–75, 96–100; Jordan, *National Road,* 83–89; Ringwalt, *Development of Transportation Systems in the United States,* 31; Wood, *Turnpikes of New England.*

12. Durrenberger, *Turnpikes,* 26–75.

13. Durrenberger, *Turnpikes,* 96–129; Gerhold, "Productivity Change in Road Transport before and after Turnpiking, 1690–1840"; Kistler, "Rise of Railroads in the Connecticut River Valley," 93–100; Klein, "Voluntary Provision of Public Goods?"; Klein and Majewski, "Economy, Community, and Law"; Wood, *Turnpikes of New England.*

14. Barker and Gerhold, *Rise and Rise of Road Transport, 1700–1990,* 21–23; Szostak, *Role of Transportation in the Industrial Revolution,* 51–52.

15. Berry, *Western Prices before 1861,* 72–77; MacGill, *History of Transportation in the United States before 1860,* 77–84, 223 n. 1 ; Ringwalt, *Development of Transportation Systems in the United States,* 27, 33–34; Rothenberg, *From Market-Places to a Market Economy,* 94, table 4 ; Taylor, *Transportation Revolution, 1815–1860,* 133–34.

16. Baker and Izard, "New England Farmers and the Marketplace, 1780–1865"; Barker and Gerhold, *Rise and Rise of Road Transport, 1700–1990,* 6, table 1 ; Rothenberg, *From Market-Places to a Market Economy,* 82–95, 94, table 4.

17. Population concentrations are identified in sources such as: *Atlas of Pennsylvania,* 86; Hilliard, *Atlas of Antebellum Southern Agriculture,* 25–26, maps 21–24; Whitford, *History of the Canal System of the State of New York,* 1:917–19, table 21A-B.

18. Mathews, *Expansion of New England,* 139–95.

19. Grantham, "Agricultural Supply during the Industrial Revolution"; Hall, *Von Thunen's Isolated State;* McClelland, *Sowing Modernity,* 9, 222–35.

20. Rothenberg, *From Market-Places to a Market Economy,* 214–33.

21. Rothenberg, *From Market-Places to a Market Economy,* 85, 88–89, 157, tables 1, 3, 14; 84, 90, 97, 111, figs. 2, 3, 5, 6.

22. Karr, "Transformation of Agriculture in Brookline, 1770–1885."

23. Gross, "Culture and Cultivation."

24. Baker and Izard, "New England Farmers and the Marketplace, 1780–1865"; Clark, *Roots of Rural Capitalism,* 21–191; Martin, "Merchants and Trade of the Connecticut River Valley, 1750–1820," 11, 16, 58, 61–65, 55, chart 1, 265–66, app. A; Pabst, "Agricultural Trends in the Connecticut Valley Region of Massachusetts, 1800–1900," 49, table 12.

25. Fuller, "An Introduction to the History of Connecticut as a Manufacturing State," 2.

26. Davis, Easterlin, and Parker, *American Economic Growth*, 123, table 5.1; Mathews, *Expansion of New England*, 139–83; Purcell, *Connecticut in Transition*, 99–100; State of Connecticut, *Register and Manual, 1972*, 570–75; U.S. Bureau of the Census, *Historical Statistics of the United States*, ser. A7, 195.

27. Bushman, *From Puritan to Yankee*, 107–34; Daniels, *Connecticut Town*, 46–51, 140–51; Martin, "Merchants and Trade of the Connecticut River Valley, 1750–1820," 19–25.

28. Martin, "Merchants and Trade of the Connecticut River Valley, 1750–1820," 54–73, 265–66, app. A; Purcell, *Connecticut in Transition: 1775–1818*, 74–77, 103–11.

29. Dwight, *Travels in New-England and New-York*, 4 vols.

30. Dwight, *Travels in New-England and New-York*, 1:270–71.

31. See chapter epigraph by Dwight, *Travels in New-England and New-York*, 3:126.

32. Dwight, *Travels in New-England and New-York*, 2:352, 355, 3:499.

33. Pease and Niles, *Gazetteer of the States of Connecticut and Rhode Island.*

34. Pease and Niles, *Gazetteer of the States of Connecticut and Rhode Island;* State of Connecticut, *Register and Manual, 1972*, 570–75; Taylor, "American Urban Growth Preceding the Railway Age," 311–15, table 1.

35. Ellis, *Landlords and Farmers in the Hudson-Mohawk Region, 1790–1850*, 16–158; Wermuth, "New York Farmers and the Market Revolution."

36. Brooks, *Frontier Settlement and Market Revolution;* Wyckoff, *Developer's Frontier.*

37. Doerflinger, "Farmers and Dry Goods in the Philadelphia Market Area, 1750–1800"; Lemon, *Best Poor Man's Country*, 123–30, 150–83.

38. Lemon, *Best Poor Man's Country*, 184–217; Mancall, *Valley of Opportunity*, 160–216.

39. Lindstrom, *Economic Development in the Philadelphia Region, 1810–1850*, 31–33, 93–151; Livingood, *Philadelphia-Baltimore Trade Rivalry, 1780–1860*, 26, app. 3.

40. Wermuth, "New York Farmers and the Market Revolution."

41. Baker and Izard, "New England Farmers and the Marketplace, 1780–1865," 35–41; Christaller, *Central Places in Southern Germany*, 16–21; Wood, *New England Village*, 88–113.

42. Clark, *Roots of Rural Capitalism*, 156–70; Martin, "Merchants and Trade of the Connecticut River Valley, 1750–1820," 16–17, 93–101.

43. Martin, "Merchants and Trade of the Connecticut River Valley, 1750–1820," 102–69; Pease and Niles, *Gazetteer of the States of Connecticut and Rhode Island;* State of Connecticut, *Register and Manual, 1972*, 570–75.

44. Pease and Niles, *Gazetteer of the States of Connecticut and Rhode Island.*

45. See chapter epigraph by Dwight, *Travels in New-England and New-York*, 4:119.

46. Schein, "Urban Origin and Form in Central New York."

47. Lemon, *Best Poor Man's Country*, 118–49.

CHAPTER THREE: Bursting through the Bounds of Local Markets

1. Meyer, "Dynamic Model of the Integration of Frontier Urban Places into the United States System of Cities"; "Emergence of the American Manufacturing Belt."

2. Cain, "From Mud to Metropolis"; Losch, *Economics of Location;* Meyer, "Rise of the Industrial Metropolis"; Pred, *Spatial Dynamics of U. S. Urban-Industrial Growth, 1800–1914,* 167–77; Weber, *Theory of the Location of Industries.*

3. Meyer, "Emergence of the American Manufacturing Belt," 152–55.

4. Meyer, "Emergence of the American Manufacturing Belt," 155–57.

5. Meyer, "Emergence of the American Manufacturing Belt," 157–59; Pred, *City-Systems in Advanced Economies.*

6. Bush, "Return of Manufactures near Wilmington, Delaware"; Clark, *Roots of Rural Capitalism,* 95–100; Pease and Niles, *Gazetteer of the States of Connecticut and Rhode Island,* 14–16 (see chap. epigraph).

7. Field, "On the Unimportance of Machinery"; Porter and Livesay, *Merchants and Manufacturers,* 69–72; Sokoloff, "Investment in Fixed and Working Capital during Early Industrialization."

8. Dalzell, *Enterprising Elite,* 26–44; Gallatin, "Manufactures," 130; Homer and Sylla, *History of Interest Rates,* 292, chart 32; Porter and Livesay, *Merchants and Manufacturers,* 13–22, 62–78.

9. Coleman, *Foundations of Social Theory,* 91–116, 175–96, 241–99; Hannay, "Chronicle of Industry on the Mill River," 25–40; Martin, "Merchants and Trade of the Connecticut River Valley, 1750–1820," 184–89; Rothenberg, *From Market-Places to a Market Economy,* 112–47.

10. Lamoreaux, *Insider Lending,* 1–30, 52–83; Wright, "Bank Ownership and Lending Patterns in New York and Pennsylvania, 1781–1831."

11. Hammond, *Banks and Politics in America,* 144–45; U.S. Bureau of the Census, *Historical Statistics of the United States,* ser. A195.

12. Discussion and computations based on data underlying fig. 3.1; see Fenstermaker, *Development of American Commercial Banking,* 13, 77, 80, tables 4, 12, 13.

13. The ten banks were: Hartford (Hartford Bank, 1792; Phoenix Bank, 1814); New Haven (New Haven Bank, 1795; Eagle Bank, 1811); Middletown (Middletown Bank, 1801); New London (Union Bank, 1792; New London Bank, 1807); Norwich (Norwich Bank, 1796); Bridgeport (Bridgeport Bank, 1806); and Derby (Derby Bank, 1809). The nine marine and fire insurance companies were: Hartford (Hartford Company, 1810; Aetna, 1819); New Haven (New Haven Marine Insurance Company, 1797; New Haven Company, 1813); Middletown (Middletown Marine Insurance Company, 1803; Middletown Company, 1813); New London (Union Insurance Company, 1805); and Norwich (Mutual Assurance Company of Norwich, 1795; Norwich Marine Insurance Company, 1803). Purcell, *Connecticut in Transition,* 65–74.

14. Martin, "Merchants and Trade of the Connecticut River Valley, 1750–1820," 170–222.

15. Feller, "Determinants of the Composition of Urban Inventions"; "Invention, Diffusion and Industrial Location"; "Urban Location of United States Invention, 1860–1910"; Higgs, "Urbanization and Inventiveness in the United States, 1870–1920"; Pred, *Spatial Dynamics of U.S. Urban-Industrial Growth, 1800–1914,* 24–41, 86–142, 106–8, tables 3.1–3.3; Rosenberg, "Factors Affecting the Diffusion of Technology"; *Technology and American Economic Growth;* Schmookler, *Invention and Economic Growth.*

16. Sokoloff, "Inventive Activity in Early Industrial America," 829, 848–49, tables 3, 7; 832–35, figs. 2–5. Definitions: *large cities*—county with a city of fifty thousand or

more people by 1840; *small cities*—county with a city of ten thousand at any time or adjacent to large-city county.

17. Sokoloff and Khan, "Democratization of Invention during Early Industrialization," 367, 369, tables 1–2.

18. Commons, "American Shoemakers, 1648–1895," 45–59; Hazard, *Organization of the Boot and Shoe Industry in Massachusetts before 1875*, 3–23; Tryon, *Household Manufactures in the United States, 1640–1860*, 197–202.

19. Faler, *Mechanics and Manufacturers in the Early Industrial Revolution*, 11–15; Gray, *History of Agriculture in the Southern United States to 1860*, 2:673–95, 752–54; Hilliard, *Atlas of Antebellum Southern Agriculture*, 27–30, 67, maps 25–30, 94; U.S. Bureau of the Census, *Historical Statistics of the United States*, ser. A200, K554.

20. Bishop, *History of American Manufactures from 1608 to 1860*, 2:147; Dawley, *Class and Community*, 16–20; Faler, *Mechanics and Manufacturers in the Early Industrial Revolution*, 17–19; Hazard, *Organization of the Boot and Shoe Industry in Massachusetts before 1875*, 24–64; Thomson, *Path to Mechanized Shoe Production in the United States*, 18, table 2.4.

21. Thomson, *Path to Mechanized Shoe Production in the United States*, 67, table 6.1.

22. Grant, *Yankee Dreamers and Doers;* Higgs, "Urbanization and Inventiveness in the United States, 1870–1920," 255, table 8.2; Hoke, *Ingenious Yankees;* Roe, *Connecticut Inventors*, 1–4; *English and American Tool Builders*, 109.

23. Cole, *Industrial and Commercial Correspondence of Alexander Hamilton Anticipating His Report on Manufactures*, 1–52; Pease and Niles, *Gazetteer of the States of Connecticut and Rhode Island;* Purcell, *Connecticut in Transition*, 80–81.

24. Brown, *Knowledge Is Power*, 65–109, app., 297–99; Daniels, *Connecticut Town*, 108–9; Dwight, *Travels in New-England and New-York*, 4:284–87 (quote, 287), 292–93; Frost, *Connecticut Education in the Revolutionary Era*, 17–18, 54; Pease and Niles, *Gazetteer of the States of Connecticut and Rhode Island;* Purcell, *Connecticut in Transition: 1775–1818*.

25. Brown, *Knowledge Is Power*, 132–59; Daniels, *Connecticut Town*, 150; Dwight, *Travels in New-England and New-York*, 2:347, 3:127; Pease and Niles, *Gazetteer of the States of Connecticut and Rhode Island.*

26. Martin, "Merchants and Trade of the Connecticut River Valley, 1750–1820," 18–73; Mathews, *Expansion of New England*, 108–95; Morison, *Maritime History of Massachusetts, 1783–1860*, 41–118, 134–55; Pease and Niles, *Gazetteer of the States of Connecticut and Rhode Island;* Pred, *Urban Growth and the Circulation of Information;* Price, "Economic Function and the Growth of American Port Towns in the Eighteenth Century"; Purcell, *Connecticut in Transition*, 74–77.

27. Bushman, *From Puritan to Yankee*, 113; Dwight, *Travels in New-England and New-York*, 2:43–45; Gilmore, "Peddlers and the Dissemination of Printed Material in Northern New England, 1780–1840"; Jaffee, "Peddlers of Progress and the Transformation of the Rural North, 1760–1860"; Keir, "Tin Peddler"; Kline, "New Light on the Yankee Peddler."

28. Bailey, *History of Danbury, Conn., 1684–1896*, 176, 199–203, 226; Cole, *Industrial and Commercial Correspondence of Alexander Hamilton Anticipating His Report on Manufactures*, 19–23, 30, 32–35, 40–41; Pease and Niles, *Gazetteer of the States of Connecticut and Rhode Island*, 176–78.

29. Bailey, *History of Danbury, Conn., 1684–1896*, 215–19; Pease and Niles, *Gazetteer of the States of Connecticut and Rhode Island.*

30. Clouette and Roth, *Bristol, Connecticut,* 51; Colt, "Peter Colt to John Chester, On Manufactures in Connecticut," 7; DeVoe, *Tinsmiths of Connecticut,* 3–7, 31, 35–36, 40–42; Gillespie, *Century of Meriden,* pt. 1: 346–61.

31. DeVoe, *Tinsmiths of Connecticut,* 21–59.

32. DeVoe, *Tinsmiths of Connecticut,* 5–6, 13–17, 51; Dwight, *Travels in New-England and New-York,* 2:43–45, vols. 3–4; Gillespie, *Century of Meriden,* pt. 1: 347; Gray, *History of Agriculture in the Southern United States to 1860,* 2:673–95, 752–54; Keir, "Tin Peddler"; Pease and Niles, *Gazetteer of the States of Connecticut and Rhode Island.*

33. Bishop, *History of American Manufactures from 1608 to 1860,* 2:218–19; Clouette and Roth, *Bristol, Connecticut,* 50; Gillespie, *Century of Meriden,* pt. 1: 346–47, 353–54; pt. 3: 37; Lathrop, *Brass Industry in the United States,* 35.

34. Anderson, *Town and City of Waterbury,* 2:233–318; Dwight, *Travels in New-England and New-York,* 2:45; Lathrop, *Brass Industry in the United States,* 39; Pease and Niles, *Gazetteer of the States of Connecticut and Rhode Island.*

35. Anderson, *Town and City of Waterbury,* 2:258–59; Hoke, *Ingenious Yankees,* 47–52; Jerome, *History of the American Clock Business for the Past Sixty Years,* 35–36; Murphy, "Entrepreneurship in the Establishment of the American Clock Industry."

36. Anderson, *Town and City of Waterbury,* 1:618–19, 2:259, 275, 546; Hoke, *Ingenious Yankees,* 53–59; Jerome, *History of the American Clock Business for the Past Sixty Years,* 17, 36; Murphy, "Entrepreneurship in the Establishment of the American Clock Industry," 173–74.

37. Clouette and Roth, *Bristol, Connecticut,* 51, 57; Jerome, *History of the American Clock Business for the Past Sixty Years,* 37; Murphy, "Entrepreneurship in the Establishment of the American Clock Industry," 174–75; Roberts, *Contributions of Joseph Ives to Connecticut Clock Technology, 1810–1862,* 17, 27, tables 7, 8.

38. Hoke, *Ingenious Yankees,* 53–59; Jerome, *History of the American Clock Business for the Past Sixty Years,* 39–44; Murphy, "Entrepreneurship in the Establishment of the American Clock Industry," 175–77; Pease and Niles, *Gazetteer of the States of Connecticut and Rhode Island.*

CHAPTER FOUR: The Foundation of the Eastern Textile Cores

1. Tryon, *Household Manufactures in the United States, 1640–1860,* 127–30, 202–18.

2. Gibb, *Saco-Lowell Shops,* 29–32; Jeremy, *Transatlantic Industrial Revolution,* 20–30; Ware, *Early New England Cotton Manufacture,* 24–25.

3. Bagnall, *Textile Industries of the United States,* 89–108, 178–82; Bishop, *History of American Manufactures from 1608 to 1860,* 2:31, 60; Cabot, "George Cabot to Hamilton, on Cotton Manufacture at Beverly, Mass."; Colt, "Elisha Colt to John Chester, on Woolen Manufacture in Hartford, Ct."; "Peter Colt to John Chester, on Manufactures in Connecticut"; Hamilton, "Report on Manufactures," 66 (chap. epigraph); Jeremy, *Transatlantic Industrial Revolution,* 8–49, 83, 86; Martin, "Merchants and Trade of the Connecticut River Valley, 1750–1820," 188; "Prospectus of the Society for Establishing Useful Manufactures"; and "Projected Agreement for Binding the Subscribers to the S.U.M. Pending Incorporation of the Society"; Shelton, *Mills of Manayunk,* 7–25; Ware, *Early New England Cotton Manufacture,* 20.

4. Hunter, *Steamboats on the Western Rivers,* 21–27; MacGill, *History of Transportation in the United States before 1860,* 3–160; Tryon, *Household Manufactures in the United States, 1640–1860,* 169–82, table 12; U.S. Bureau of the Census, *Historical Statis-*

tics of the United States, ser. A172, 178, 195, 202; Zevin, "Growth of Cotton Textile Production after 1815," 128. For this period the West comprises the North Central census region, plus Kentucky and Tennessee.

5. Jeremy, *Transatlantic Industrial Revolution,* 36–140; Zevin, "Growth of Cotton Textile Production after 1815," 145–46.

6. Zevin, "Growth of Cotton Textile Production after 1815," 136.

7. Bagnall, *Textile Industries of the United States,* 135–375; Coleman, *Transformation of Rhode Island, 1790–1860,* 86–87, tables 5–6; Coxe, "Digest of Manufactures, 1810," pt. 3: 190; Day, "Early Development of the American Cotton Manufacture," 464–65; Gallatin, "Manufactures," 125. The embargo actually started on December 22, 1807, and was repealed on March 1, 1809; see Nettels, *Emergence of a National Economy, 1775–1815,* 135.

8. Data on number of factories established were computed from McLane, *Documents Relative to the Manufactures in the United States;* procedures followed Day, "Early Development of the American Cotton Manufacture." Results in table 4.1 diverge somewhat from Day (452), due to differences in interpretation of some founding dates and distinctions between firms. Details in McLane on ownership changes are only given for some cotton mills; therefore, for consistency all mills listed were included, regardless of ownership changes or bankruptcies, as long as the mill operated in 1832. Thus, the data are crude surrogates for competitive mills.

9. Gallatin, "Manufactures," 125; Nettels, *Emergence of a National Economy, 1775–1815,* 327–28.

10. Table 4.1 and Tryon, *Household Manufactures in the United States, 1640–1860,* 166, table 11.

11. Day, "Early Development of the American Cotton Manufacture," 465–66; Lebergott, *Americans,* 126–29.

12. Bagnall, *Textile Industries of the United States,* 96, 144–60; Coleman, *Transformation of Rhode Island, 1790–1860,* 26–70; Hedges, *Browns of Providence Plantations: The Nineteenth Century;* Jeremy, *Transatlantic Industrial Revolution,* 82–84; Pitkin, *Statistical View of the Commerce of the United States of America,* chap. 3, table 1; Tucker, *Samuel Slater and the Origins of the American Textile Industry, 1790–1860,* 49–51, 55; U.S. Bureau of the Census, *Historical Statistics of the United States,* ser. A195; White, *Memoir of Samuel Slater,* 71–76.

13. Bagnall, *Textile Industries of the United States;* Hedges, *Browns of Providence Plantations: Colonial Years,* 187; Jeremy, *Transatlantic Industrial Revolution,* 86–88; Tucker, *Samuel Slater and the Origins of the American Textile Industry, 1790–1860,* 51–58, 71–86; Vance, *Merchant's World;* Ware, *Early New England Cotton Manufacture,* 32, 161–66; White, *Memoir of Samuel Slater,* 102–12.

14. Bagnall, *Textile Industries of the United States,* 213–17, 251–54; Tucker, *Samuel Slater and the Origins of the American Textile Industry, 1790–1860,* 61; Ware, *Early New England Cotton Manufacture,* 32–33, 166–68.

15. Bagnall, *Textile Industries of the United States;* Brown, *Innovation Diffusion,* 50–86; Burt, *Structural Holes,* 18–30; Erickson, "Culture, Class, and Connections"; Granovetter, "Strength of Weak Ties"; "Economic Action and Social Structure"; Lawler and Yoon, "Commitment in Exchange Relations"; Uzzi, "Sources and Consequences of Embeddedness for the Economic Performance of Organizations."

16. Bagnall, *Textile Industries of the United States,* 394–99, 404–12, 433–35, 440–43,

451–54, 524–30; Hedges, *Browns of Providence Plantations: The Nineteenth Century,* 183–84.

17. Bagnall, *Textile Industries of the United States,* 444–46, 541–49.

18. Bagnall, *Textile Industries of the United States,* 416–21, 536–39, 593–96; Pease and Niles, *Gazetteer of the States of Connecticut and Rhode Island,* 139–67, 202–28.

19. Bagnall, *Textile Industries of the United States,* 373–75; Mathews, *Expansion of New England,* 153; Ware, *Early New England Cotton Manufacture,* 32.

20. Bagnall, *Textile Industries of the United States,* 368–72, 460–61, 473–80; Gregory, *Nathan Appleton,* 5–18; Jeremy, *Transatlantic Industrial Revolution,* 89; Ware, *Early New England Cotton Manufacture,* 32.

21. Bagnall, *Textile Industries of the United States,* 276–77, 376–78, 390–92, 501–9; Hedges, *Browns of Providence Plantations: The Nineteenth Century,* 179.

22. McLane, *Documents Relative to the Manufactures of the United States,* docs. 3, 9.

23. McLane, *Documents Relative to the Manufactures of the United States,* docs. 4, 5.

24. The counties are Columbia, Rensselaer, Saratoga, Schenectady, and Washington. McLane, *Documents Relative to the Manufactures of the United States,* doc. 10.

25. The counties are Cayuga, Chenango, Herkimer, Jefferson, Oneida, Onondaga, and Otsego. McLane, *Documents Relative to the Manufactures of the United States,* doc. 10.

26. For Boston Manufacturing Company sales, see Ware, *Early New England Cotton Manufacture,* 70. The date that Almy and Brown took control and minimum estimates of spindles for each firm around 1819 are: Almy, Brown, and Slater in Pawtucket (1793; 1,150 spindles); Warwick Spinning Mill (1799; 1,500); Almy, Brown, and Slaters in Slatersville (1807; 5,170); and Warwick Manufacturing Company (1807; 2,700). See Bagnall, *Textile Industries of the United States,* 153–60, 213–19, 396–99, 433–35; Gallatin, "Manufactures," pt. C: 133.

27. Bishop, *History of American Manufactures from 1608 to 1860,* 2:111; Hedges, *Browns of Providence Plantations: The Nineteenth Century,* 175–78; Ware, *Early New England Cotton Manufacture,* 167–69, 309–10, app. F-G.

28. Bagnall, *Textile Industries of the United States;* Conrad, "Drive That Branch"; Jeremy, *Transatlantic Industrial Revolution,* 93–103, 204–6; Lozier, *Taunton and Mason,* 66–68, 71, 90; Pease and Niles, *Gazetteer of the States of Connecticut and Rhode Island,* 336–38; U.S. Bureau of the Census, *Historical Statistics of the United States,* ser. E126; Ware, *Early New England Cotton Manufacture,* 309–10, app. F-G; Zevin, "Growth of Cotton Textile Production after 1815," 140.

29. Gibb, *Saco-Lowell Shops,* 32–35; Jeremy, *Transatlantic Industrial Revolution,* 96–98, 181–82; Temin, "Product Quality and Vertical Integration in the Early Cotton Textile Industry"; Zevin, "Growth of Cotton Textile Production after 1815," 141.

30. Shelton, *Mills of Manayunk,* 33–37; Ware, *Early New England Cotton Manufacture,* 309, app. F.

31. Bagnall, *Textile Industries of the United States,* 581–84; Scranton, *Proprietary Capitalism,* 81–82; Shelton, *Mills of Manayunk,* 37–46.

32. Jeremy, *Transatlantic Industrial Revolution,* 164–68, 174, 166, table 9.4; Scranton, *Proprietary Capitalism,* 95–104; Shelton, *Mills of Manayunk,* 47–51.

33. The appellation *new industrial form* comes from Ware, *Early New England Cotton Manufacture,* 60–78.

34. Albion, *Rise of New York Port, 1815–1860;* Browne, *Baltimore in the Nation, 1789–*

1861; Hedges, *Browns of Providence Plantations: The Nineteenth Century;* Lindstrom, *Economic Development in the Philadelphia Region, 1810–1850;* Porter and Livesay, *Merchants and Manufacturers;* Pred, *Urban Growth and the Circulation of Information.*

35. Dalzell, *Enterprising Elite,* 27; Ware, *Early New England Cotton Manufacture,* 60–118; Zevin, "Growth of Cotton Textile Production after 1815," 144, table 5. The term *Boston Associates* was coined in Shlakman, "Economic History of a Factory Town," 31.

36. Bagnall, *Textile Industries of the United States,* 320–25, 368–72, 396–99, 473–78, 524–30; Conrad, "Drive That Branch"; Gibb, *Saco-Lowell Shops,* 4–8, 23; Gregory, *Nathan Appleton,* 1–71, 114–46, 164; Hedges, *Browns of Providence Plantations: The Nineteenth Century,* 104, 183–84; McLane, *Documents Relative to the Manufactures in the United States,* doc. 3; Morison, *Maritime History of Massachusetts, 1783–1860,* 52–78; Porter, *Jacksons and the Lees,* 1:3–98; Pred, *Urban Growth and the Circulation of Information;* Ware, *Early New England Cotton Manufacture,* 32, 166, 309, app. F; Weil, "Capitalism and Industrialization in New England, 1815–1845."

37. Bagnall, *Textile Industries of the United States,* 153–60, 213–17, 396–99, 433–35; Gibb, *Saco-Lowell Shops,* 58–62; Gregory, *Nathan Appleton,* 146–55.

38. Bagnall, *Textile Industries of the United States,* 524–30; Gibb, *Saco-Lowell Shops,* 25–26, 50–51; Gregory, *Nathan Appleton,* 137–51, 150, table 3; Hedges, *Browns of Providence Plantations: The Nineteenth Century,* 183–84; Jackson, "Letter from Patrick Tracy Jackson"; Morison, *Maritime History of Massachusetts, 1783–1860,* 70–71; Ware, *Early New England Cotton Manufacture,* 138–40.

39. Bagnall, *Textile Industries of the United States,* 146, 152, 320–25, 404–9; Bathe and Bathe, *Jacob Perkins,* 1–55, 200, app. 1; Bishop, *History of American Manufactures from 1608 to 1860,* 2:227; Dalzell, *Enterprising Elite,* 32–36; Gibb, *Saco-Lowell Shops,* 10–50; Gregory, *Nathan Appleton,* 150, table 3, 151–53, 160–62; Hayes, *American Textile Machinery,* 27–28; Jeremy, *Transatlantic Industrial Revolution,* 98–99; Lozier, *Taunton and Mason,* 66; Mathews, *Expansion of New England,* 139–95; Ware, *Early New England Cotton Manufacture,* 200–201, 227–28; Zevin, "Growth of Cotton Textile Production after 1815," 140–41.

40. Bishop, *History of American Manufactures from 1608 to 1860,* 2:111, 226–27; Gibb, *Saco-Lowell Shops,* 61; Gregory, *Nathan Appleton,* 107, 164–67; Hedges, *Browns of Providence Plantations: The Nineteenth Century,* 175–78; Temin, "Product Quality and Vertical Integration in the Early Cotton Textile Industry," 897; Ware, *Early New England Cotton Manufacture,* 65–66, 70–72, 167–69, 178–79.

41. Bagnall, *Textile Industries of the United States,* 153–60, 213–17, 396–99, 433–35, 529–30; Dalzell, *Enterprising Elite,* 3–44; Gibb, *Saco-Lowell Shops,* 4–14, 23–39, 62–63, 738 n. 11; Gregory, *Nathan Appleton,* 141–72; Jeremy, *Transatlantic Industrial Revolution,* 194, table 10.2; Ware, *Early New England Cotton Manufacture,* 60–78; Zevin, "Growth of Cotton Textile Production after 1815."

42. Bagnall, *Textile Industries of the United States,* 122–27, 146, 178–94; Coxe, "Digest of Manufactures, 1810," pt. 4: 276, 284; Gallatin, "Manufactures," pt. B: 132.

43. Bishop, *History of American Manufactures from 1608 to 1860,* 2:206–8; Clark, *History of Manufactures in the United States,* 1:538; Jeremy, *Transatlantic Industrial Revolution,* 101–2; Lozier, *Taunton and Mason,* 32; McLane Report, *Documents Relative to the Manufactures of the United States,* doc. 10, nos. 2, 20; doc. 11, nos. 11, 13, 21–23.

CHAPTER FIVE: Tightening Ties That Bound the East

1. Harley, "Antebellum Tariff: Food Exports and Manufacturing"; Kravis, "Role of Exports in Nineteenth-Century United States Growth," 393, 397, 399, tables 2, 3, 5, and 404–5; U.S. Bureau of the Census, *Historical Statistics of the United States,* ser. A7, C89; Wahl, "New Results on the Decline in Household Fertility in the United States from 1750 to 1900."

2. Wage data for the East include the New England states, New York, New Jersey, and Pennsylvania. Margo, *Wages and Labor Markets in the United States, 1820–1860,* 71–73, tables 3A.9–3A.11; Weiss, "Economic Growth before 1860," 19, table 1.4.

3. Gallman, "Commodity Output, 1839–1899," 43, table A-1; "Gross National Product in the United States, 1834–1909," 26–27, tables A-1, A-2; Smith and Cole, *Fluctuations in American Business, 1790–1860,* 53, 60, 122–23, charts 12, 17, 41–43; 120–21, 185, tables 32, 72; and 53–54; Temin, *Jacksonian Economy,* 71, 159, tables 3.3, 5.2, and 59–171.

4. Bodenhorn, *History of Banking in Antebellum America;* Bodenhorn and Rockoff, "Regional Interest Rates in Antebellum America"; Davis and Gallman, "Capital Formation in the United States during the Nineteenth Century," 2, table 1, and 10–13; Gallman, "American Economic Growth before the Civil War," 94, table 2.8, and 95.

5. Atack, "Returns to Scale in Antebellum United States Manufacturing"; Field, "On the Unimportance of Machinery"; Sokoloff, "Investment in Fixed and Working Capital during Early Industrialization"; "Was the Transition from the Artisanal Shop to the Nonmechanized Factory Associated with Gains in Efficiency?"; "Productivity Growth in Manufacturing during Early Industrialization"; "Invention, Innovation, and Manufacturing Productivity Growth in the Antebellum Northeast."

6. Margo, *Wages and Labor Markets in the United States, 1820–1860,* 76–94; Sokoloff, "Invention, Innovation, and Manufacturing Productivity Growth in the Antebellum Northeast"; Sokoloff, "Productivity Growth in Manufacturing during Early Industrialization," 724; Sokoloff and Villaflor, "Market for Manufacturing Workers during Early Industrialization."

7. Williamson, "Antebellum Urbanization in the American Northeast," 600, table 1.

8. Meyer, "Dynamic Model of the Integration of Frontier Urban Places into the United States System of Cities"; Pred, *Urban Growth and the Circulation of Information;* Vance, *Merchant's World.* The Midwest comprises the North Central census region and Kentucky. U.S. Bureau of the Census, *Historical Statistics of the United States,* ser. A172, 195.

9. Rubin, "Canal or Railroad?"; Whitford, *History of the Canal System of the State of New York,* 1:48–130.

10. Railroads in figure 5.3: New York Central, New York and Erie, Pennsylvania, and Baltimore and Ohio.

11. Albion, *Rise of New York Port, 1815–1860,* 12–13; *Square-Riggers on Schedule,* 20–76, 274, app. 1; Pitkin, *Statistical View of the Commerce of the United States of America,* 53–54, 57–83, tables 2, 4, and 5; Pred, *Urban Growth and the Circulation of Information,* 189–94; Seybert, *Statistical Annals,* 425–37.

12. Pred, *Urban Growth and the Circulation of Information,* 32–35, 43–48, 202–27, 115–16, 124–25, tables 4.3–4.4, 4.9–4.10.

13. Hutchinson and Williamson, "Self-Sufficiency of the Antebellum South"; Kohl-

meier, *Old Northwest as the Keystone of the Arch of American Federal Union,* 7; Lindstrom, "Southern Dependence upon Interregional Grain Supplies"; Ransom, "Interregional Canals and Economic Specialization in the Antebellum United States," 15, table 2; Reiser, *Pittsburgh's Commercial Development, 1800–1850,* 76–78.

14. Fishlow, "Antebellum Interregional Trade Reconsidered," 363; Ransom, "Interregional Canals and Economic Specialization in the Antebellum United States."

15. Cranmer, "Improvements without Public Funds," 157–59; Fishlow, *American Railroads and the Transformation of the Ante-Bellum Economy,* 23–32; Fogel, *Railroads and American Economic Growth,* 19–22; Goodrich, *Government Promotion of American Canals and Railroads, 1800–1890,* 3–16.

16. Albion, *Rise of New York Port, 1815–1860,* 143–64; Kirkland, *Men, Cities, and Transportation,* 1:66–75; Morrison, *History of American Steam Navigation,* 539–44; Livingood, *Philadelphia-Baltimore Trade Rivalry, 1780–1860,* 36–38; Ringwalt, *Development of Transportation Systems in the United States,* 12–13.

17. Cranmer, "Canal Investment, 1815–1860," 553, 564, app. C; Goodrich, *Government Promotion of American Canals and Railroads, 1800–1890,* 56–59; Harlow, *Old Towpaths,* 295–307; Miller, *Enterprise of a Free People,* 107–8; Roberts, *Middlesex Canal, 1793–1860,* 65–100; Segal, "Cycles of Canal Construction," 189–205, 242, table 12; Taylor, *Transportation Revolution, 1815–1860,* 54; U.S. Bureau of the Census, *Tenth Census of the United States, 1880,* vol. 4: *Transportation,* 725–64, tables 1–2; Whitford, *History of the Canal System of the State of New York,* vol. 1.

18. Roberts, *Middlesex Canal, 1793–1860,* 28–45, 65, 114, 124–35, 176–87, 200, 227, app. C, L.

19. Roberts, *Middlesex Canal, 1793–1860,* 166, 209, app. H.

20. Cranmer, "Canal Investment, 1815–1860," 564, app. C; Hedges, *Browns of Providence Plantations: The Nineteenth Century,* 209–16; Kirkland, *Men, Cities, and Transportation,* 1:71–83; Kistler, "Rise of Railroads in the Connecticut River Valley," 18–28.

21. Aitken, *Welland Canal Company,* 18–21; Brooks, *Frontier Settlement and Market Revolution;* Cranmer, "Improvements without Public Funds," 157–58; Creighton, *Commercial Empire of the St. Lawrence, 1760–1850;* Harlow, *Old Towpaths,* 48; MacGill, *History of Transportation in the United States before 1860,* 170–79; McIlwraith, "Freight Capacity and Utilization of the Erie and Great Lakes Canals before 1850"; McNall, *Agricultural History of the Genesee Valley, 1790–1860,* 99–101; Miller, *Enterprise of a Free People,* 40, 61–73; Rubin, "Innovating Public Improvement," 23–55; Segal, "Canals and Economic Development," 216–48; U.S. Bureau of the Census, *Tenth Census of the United States, 1880,* vol. 4: *Transportation,* 733; Wyckoff, *Developer's Frontier.*

22. Bagnall, *Textile Industries of the United States,* 501–9; Filante, "Note on the Economic Viability of the Erie Canal, 1825–1860," 96, table 1; Hedges, *Browns of Providence Plantations: The Nineteenth Century,* 210; Kirkland, *Men, Cities, and Transportation,* 1:71, 76; Miller, *Enterprise of a Free People,* 115–18; Miller, *City and Hinterland,* 21–36; Whitford, *History of the Canal System of the State of New York,* 1:84–123, 917–19, table 21-A, 947–51, tables 41–42.

23. Bishop, "State Works of Pennsylvania," 157–200, 247, 278–79; Cranmer, "Canal Investment, 1815–1860," 564, app. C; Hartz, *Economic Policy and Democratic Thought,* 42–51, 131–42; Rubin, "Canal or Railroad?" 15–62.

24. Bishop, "State Works of Pennsylvania," 213–59, 277–86, app. 6; Ransom, "Inter-

regional Canals and Economic Specialization in the Antebellum United States," 19–21, tables 4, 6; Rubin, "Imitative Public Improvement," 107–8.

25. Browne, *Baltimore in the Nation, 1789–1861,* 82–86; Cranmer, "Canal Investment, 1815–1860," 564, app. C; Durrenberger, *Turnpikes,* 65–70; Livingood, *Philadelphia-Baltimore Trade Rivalry, 1780–1860,* 54–80; Sanderlin, *Great National Project,* 22–44, 61–160, 305–8, app. tables 3–4; Ward, *Early Development of the Chesapeake and Ohio Canal Project,* 9–83. Also see chapter epigraph.

26. Jones, *Economic History of the Anthracite-Tidewater Canals,* 36; Reizenstein, *Economic History of the Baltimore and Ohio Railroad, 1827–1853,* 85, app.; Sanderlin, *Great National Project,* 305–8, 312–14, app. tables 3, 4, 7. Conversion rate for wheat to flour is 4.5 bushels of wheat equivalent to 1 barrel of flour; see Berry, *Western Prices before 1861,* 146; Bidwell and Falconer, *History of Agriculture in the Northern United States, 1620–1860,* 493, table 66. Counties bordering the canal between Georgetown and Hancock included Frederick, Montgomery, and Washington in Maryland; Loudoun in Virginia; and Berkeley and Jefferson in West Virginia. They housed 126,196 people and produced 2,923,867 bushels of wheat; consumption, directly as flour or indirectly as feed, was estimated at 4.95 bushels of wheat per capita based on total production of wheat in the United States in 1840 and population size that year. U.S. Bureau of the Census, *Compendium of the Sixth Census, 1840; Historical Statistics of the United States,* ser. A7.

27. Cranmer, "Canal Investment, 1815–1860," 564, app. C. At a total cost of $3,260,000, a 6 percent interest charge required $195,000 in revenue just to cover that. During 1840–60 toll revenue totaled $2,839,132, whereas estimated operating expenses reached $803,691 and interest costs, assuming no debt amortization, would have hit $4,107,600; accumulated deficit, including construction costs, was $5,332,159. Livingood, *Philadelphia-Baltimore Trade Rivalry, 1780–1860,* 54–78, 79, app. 6, 80, app. 9. Operating expenditures were extrapolated from an annual average for the Eastern Division of the Pennsylvania Mainline, a forty-three-mile canal; Bishop, "State Works of Pennsylvania," 278, app. 6. MacGill, *History of Transportation in the United States before 1860,* 226, table 34.

28. Cranmer, "Canal Investment, 1815–1860," 564, app. C; "Improvements without Public Funds," 147–56; Gray, *National Waterway,* 1–137; Livingood, *Philadelphia-Baltimore Trade Rivalry, 1780–1860,* 81–91, 98–99, app. 10–11; MacGill, *History of Transportation in the United States before 1860,* 227–28, 233–34, tables 36–37.

29. Cranmer, "Canal Investment, 1815–1860," 564, app. C; Jones, *Economic History of the Anthracite-Tidewater Canals.*

30. Larson, *Internal Improvement;* Ransom, "Canals and Development."

31. Fishlow, *American Railroads and the Transformation of the Ante-Bellum Economy,* 18–95.

32. Coleman, *Transformation of Rhode Island, 1790–1860,* 71–160; Fishlow, *American Railroads and the Transformation of the Ante-Bellum Economy,* 99–160, 257–59, 252, table 32; MacGill, *History of Transportation in the United States before 1860,* 322–30; Shlakman, "Economic History of a Factory Town," 39–42, table 1; Withington, *First Twenty Years of Railroads in Connecticut.*

33. Bogen, *Anthracite Railroads,* 14–15; Cranmer, "Canal Investment, 1815–1860," 564, app. C; Dalzell, *Enterprising Elite,* 85–92; Johnson and Supple, *Boston Capitalists*

and Western Railroads, 33–59; Livingood, *Philadelphia-Baltimore Trade Rivalry, 1780–1860,* 146–47; Pierce, *Railroads of New York,* 9–11, 176, chart 2, 193, table 2; Reizenstein, *Economic History of the Baltimore and Ohio Railroad, 1827–1853,* 15; Salsbury, *State, the Investor, and the Railroad,* 93–111, 155, table 1; Stevens, *Beginnings of the New York Central Railroad,* 24–25; Taylor, *Transportation Revolution, 1815–1860,* 92–94.

34. Homer and Sylla, *History of Interest Rates,* 293–317; Kistler, "Rise of Railroads in the Connecticut River Valley," 80–126; Livingood, *Philadelphia-Baltimore Trade Rivalry, 1780–1860,* 116–60; Majewski, *House Dividing,* 72–84; Pierce, *Railroads of New York,* 3–25, 41–59; Reizenstein, *Economic History of the Baltimore and Ohio Railroad, 1827–1853,* 18–19, 46; Wilson, *History of the Pennsylvania Railroad Company,* 1:1–169.

35. Baer, *Canals and Railroads of the Mid-Atlantic States, 1800–1860;* Kirkland, *Men, Cities, and Transportation,* 1:158–222.

36. Baer, *Canals and Railroads of the Mid-Atlantic States, 1800–1860;* Burgess and Kennedy, *Centennial History of the Pennsylvania Railroad Company,* 35–100; Dilts, *Great Road,* 314–36; Stevens, *Beginnings of the New York Central Railroad;* Ward, *J. Edgar Thomson,* 24–90.

37. Those business elite were familiar with Philadelphia and Reading Railroad during the 1840s, when it attracted soaring volumes of low-value coal in competition with Schuylkill Navigation Canal, yet the railroad almost went bankrupt. See Bogen, *Anthracite Railroads,* 19–40.

CHAPTER SIX: Agriculture Augments Regional Industrial Systems

1. Bronson, "Report of Alvin Bronson to the Secretary of the Treasury, on the Manufactures of New York."

2. Steckel, "Economic Foundations of East-West Migration during the 19th Century," 15–17, table 1.

3. McClelland, *Sowing Modernity.*

4. Baker and Izard, "New England Farmers and the Marketplace, 1780–1865"; Clark, *Roots of Rural Capitalism,* 59–155, 195–227; Gross, "Culture and Cultivation"; Karr, "Transformation of Agriculture in Brookline, 1770–1885"; Rothenberg, *From Market-Places to a Market Economy,* 167–74.

5. Karr, "Transformation of Agriculture in Brookline, 1770–1885," 38–39; U.S. Bureau of the Census, *Twelfth Census, 1900,* 430–33, table 6; *Historical Statistics of the United States,* ser. A202.

6. Baker and Izard, "New England Farmers and the Marketplace, 1780–1865"; Gross, "Culture and Cultivation"; Smith and Bridges, "Brighton Market"; U.S. Bureau of the Census, *Compendium of the Sixth Census, 1840.*

7. Clark, *Roots of Rural Capitalism,* 59–155, 331, app.; Smith and Bridges, "Brighton Market"; U.S. Bureau of the Census, *Twelfth Census, 1900,* 430–33, table 6.

8. U.S. Bureau of the Census, *Compendium of the Sixth Census, 1840; Historical Statistics of the United States,* ser. A202; *Twelfth Census, 1900,* 430–33, table 6.

9. Baker and Izard, "New England Farmers and the Marketplace, 1780–1865"; Gross, "Culture and Cultivation"; Karr, "Transformation of Agriculture in Brookline, 1770–1885."

10. Clark, *Roots of Rural Capitalism*, 273–313; Ramsey, "History of Tobacco Production in the Connecticut Valley," 125–46.

11. Linder and Zacharias, *Of Cabbages and Kings County*, 24–44; U.S. Bureau of the Census, *Compendium of the Sixth Census, 1840*.

12. Troy is in Rensselaer County, and Saratoga County is north of the Mohawk and west of the Hudson.

13. McMurry, *Transforming Rural Life*, 6–61; U.S. Bureau of the Census, *Compendium of the Sixth Census, 1840*.

14. McNall, *Agricultural History of the Genesee Valley, 1790–1860*, 96–131, 156–57; U.S. Bureau of the Census, *Compendium of the Sixth Census, 1840*.

15. Ellis, *Landlords and Farmers in the Hudson-Mohawk Region, 1790–1850*, 190–200; U.S. Bureau of the Census, *Compendium of the Sixth Census, 1840*. The nine barley counties from east to west were: Albany, Schenectady, Montgomery, Schoharie, Otsego, Herkimer, Oneida, Madison, and Onondaga.

16. Ellis, *Landlords and Farmers in the Hudson-Mohawk Region, 1790–1850*, 193–208.

17. McMurry, *Transforming Rural Life*, 6–71; McNall, *Agricultural History of the Genesee Valley, 1790–1860*, 147–254; Miller, *City and Hinterland*, 80–95; Parkerson, *Agricultural Transition in New York State*, 55–102.

18. Easterlin, "Interregional Differences in Per Capita Income, Population, and Total Income, 1840–1950," 97–98, app. table A-1; Lindstrom, *Economic Development in the Philadelphia Region, 1810–1850*, 140–45, 104, table 4.3, 194–98, app. table B.2; U.S. Bureau of the Census, *Compendium of the Sixth Census, 1840; Twelfth Census, 1900*, 430–33, table 6.

19. Location quotients (in parentheses) based on per farmer production are as follows. Baltimore County: vegetables (17.7), nursery and florists (8.6), and dairy (1.8); Anne Arundel County: vegetables (6.1); Frederick County: dairy (2.2) and fruit (2.7). U.S. Bureau of the Census, *Compendium of the Sixth Census, 1840; Twelfth Census, 1900*, 430–33, table 6. Estimates of wheat (flour) sent on the Baltimore and Ohio Railroad come from Reizenstein, *Economic History of the Baltimore and Ohio Railroad, 1827–1853*, 88, app. Foreign export computations are based on production and consumption figures cited previously; and Rutter, *South American Trade of Baltimore*, 18.

20. Atack and Bateman, *To Their Own Soil*, 201–46; Lebergott, "Demand for Land," 189–92.

21. In 1840 per-worker agricultural income and nonagricultural income (current dollars), respectively, were $182 and $465 in New England, and $225 and $462 in the Middle Atlantic. Easterlin, "Interregional Differences in Per Capita Income, Population, and Total Income, 1840–1950," 97–98, app. table A-1; Atack and Bateman, *To Their Own Soil*, 243, table 13.8.

22. Clark, *Roots of Rural Capitalism*, 169–84, 229–32, 243–51, 261–63; Hannay, "Chronicle of Industry on the Mill River," 70–89; Schein, "Urban Origin and Form in Central New York."

23. Porter and Livesay, *Merchants and Manufacturers*, 65–77; Williamson, "Money and Commercial Banking, 1789–1865," 233, fig. 1. Bank money equals circulation plus deposits minus notes of other banks, and bank credit equals loans and discounts plus bills of exchange. Bodenhorn, *History of Banking in Antebellum America*, 63–64, tables 2.1–2.2.

24. Bodenhorn and Rockoff, "Regional Interest Rates in Antebellum America"; Fenstermaker, *Development of American Commercial Banking,* 111–84, app. A.

25. Sokoloff, "Inventive Activity in Early Industrial America," 829, 831, 848–49, tables 3, 4, 7; 832–35, figs. 2–5; "Invention, Innovation, and Manufacturing Productivity Growth in the Antebellum Northeast," 368 n. 17.

26. Sokoloff and Khan, "Democratization of Invention during Early Industrialization," 367, 369, tables 1–2.

27. Khan and Sokoloff, "Schemes of Practical Utility."

CHAPTER SEVEN: Metropolises Lead the Regional Industrial Expansion

1. Atack and Bateman, *To Their Own Soil,* 150–51, table 9.1, fig. 9.1; Ellis, *Landlords and Farmers in the Hudson-Mohawk Region, 1790–1850,* 189, 198; McNall, *Agricultural History of the Genesee Valley, 1790–1860,* 140–41, 153–54; Tryon, *Household Manufactures in the United States, 1640–1860,* 308–9, table 17.

2. U.S. Bureau of the Census, *Compendium of the Sixth Census, 1840.* Agricultural employment also includes mining, but that constituted only 0.4 percent of the total for "agriculture." Employment figures are conservative estimates of manufacturing because data were limited to workers employed in thirty manufactures for which the census listed employment data. The total of 387,303 contrasts with the category of manufactures and trades used in the census, which totaled 791,739.

3. Boston city had major manufacturing firms by 1832. See chapter epigraph by Tyler, "Report of John S. Tyler to the Secretary of the Treasury, on the Manufactures of the County of Suffolk, Massachusetts."

4. Thiel, *Economics and Information Theory,* 316–18. The Herfindahl index of diversification is computed as sum of squares of each ratio of employment in industry i to total industrial employment in the areal unit. Squared ratios sum to 1.0 if all employment in the areal unit concentrates in one industry, and they sum to $1/n$ ($1/30 = 0.03$) if all industries have equal employment in the areal unit. Maximum diversification occurs at an index value of 0.03, and maximum concentration (all employment in one industry) at an index of 1.0.

5. The location quotient is computed as the ratio of the share of each manufacture in total manufacturing employment in an areal unit to the share of that manufacture in total national manufacturing employment. A location quotient of 1.0 indicates the areal unit has the same share of manufacturing employment in an industry as in the nation; values above 1.0 indicate greater specialization than the nation, and values below 1.0 indicate less specialization.

6. Zakim, "Ready-Made Business."

7. U.S. Bureau of the Census, *Twelfth Census, 1900,* 430–33, table 6.

8. Bishop, *History of American Manufactures from 1608 to 1860,* 3:178–81, 188–90.

9. Hoffecker, *Wilmington, Delaware,* 15–35; Lindstrom, *Economic Development in the Philadelphia Region, 1810–1850,* 142–43, table 5.8; U.S. Bureau of the Census, *Twelfth Census, 1900,* 430–33, table 6.

10. U.S. Bureau of the Census, *Twelfth Census, 1900,* 430–33, table 6.

11. Deyrup, "Arms Makers of the Connecticut Valley"; Hannay, "Chronicle of Industry on the Mill River," 61–64; Prude, *Coming of Industrial Order,* 208, table 7.4; Shlak-

man, "Economic History of a Factory Town," 39–42, table 1, 64; Taber, "History of the Cutlery Industry in the Connecticut Valley."

12. McGaw, *Most Wonderful Machine,* 15–186; U.S. Bureau of the Census, *Compendium of the Sixth Census, 1840.*

13. Bishop, *History of American Manufactures From 1608 to 1860,* 3:206–23; Stevens, *Beginnings of the New York Central Railroad,* 115–46, 350–87; U.S. Bureau of the Census, *Compendium of the Sixth Census, 1840;* Walkowitz, *Worker City, Company Town,* 48–51.

14. Ardrey, *American Agricultural Implements,* 224–27; Bishop, *History of American Manufactures from 1608 to 1860,* 3:229–31; Hatch, *Remington Arms,* 28–61.

15. McKelvey, *Rochester,* 205–41.

16. Bishop, *History of American Manufactures from 1608 to 1860,* 3:232–39.

17. Hilliard, *Atlas of Antebellum Southern Agriculture,* 32, maps 33–34; U.S. Bureau of the Census, *Twelfth Census, 1900,* 430–33, table 6. Employment in milling: Lancaster (323), York (260), and Frederick (139). U.S. Bureau of the Census, *Compendium of the Sixth Census, 1840.*

CHAPTER EIGHT: Building Competitive National Market Industries

1. Ellsworth, "Report of H. L. Ellsworth to the Secretary of the Treasury, on the Manufactories in Connecticut"; Foster, "Explanations, Remarks, and Additional Facts"; Mudge, "Explanations, Remarks, and Additional Facts."

2. Meyer, "Emergence of the American Manufacturing Belt"; "Midwestern Industrialization and the American Manufacturing Belt in the Nineteenth Century"; Muller, "Selective Urban Growth in the Middle Ohio Valley, 1800–1860"; Thomson, *Path to Mechanized Shoe Production in the United States,* 19.

3. Field, "Sectoral Shift in Antebellum Massachusetts"; Goldin and Sokoloff, "Women, Children, and Industrialization in the Early Republic"; "Relative Productivity Hypothesis of Industrialization"; U.S. Bureau of the Census, *Historical Statistics of the United States,* ser. A203, 210–63.

4. Commons, "American Shoemakers, 1648–1895"; Dawley, *Class and Community,* 25–32; Faler, *Mechanics and Manufacturers in the Early Industrial Revolution,* 12–22, 154, table 7; Hazard, *Organization of the Boot and Shoe Industry in Massachusetts before 1875,* 58–63; Pred, *Urban Growth and City-Systems in the United States, 1840–1860,* 65–70, 181–82, tables A.7, A.8; Thomson, *Path to Mechanized Shoe Production in the United States,* 18, table 2.4; U.S. Bureau of the Census, *Manufactures of the United States in 1860, Eighth Census,* lxxii.

5. Dawley, *Class and Community,* 29; Faler, *Mechanics and Manufacturers in the Early Industrial Revolution,* 61; Hazard, *Organization of the Boot and Shoe Industry in Massachusetts before 1875,* 42–96; Christiansen and Philips, "Transition from Outwork to Factory Production in the Boot and Shoe Industry, 1830–1880," 24–29; Lamoreaux, *Insider Lending,* 14, table 1.1; Thomson, *Path to Mechanized Shoe Production in the United States,* 22–33.

6. Cooper, *Shaping Invention,* 171–74; Hazard, *Organization of the Boot and Shoe Industry in Massachusetts before 1875,* 65–96; Thomson, *Path to Mechanized Shoe Production in the United States,* 34–37, 49–58; U.S. Bureau of the Census, *Manufactures of the United States in 1860, Eighth Census,* lxx–lxxi.

7. Sokoloff, "Productivity Growth in Manufacturing during Early Industrialization," 698, 719, tables 13.6, 13.11.

8. Faler, *Mechanics and Manufacturers in the Early Industrial Revolution,* 58–76.

9. Jeremy, *Transatlantic Industrial Revolution,* 276, 279, app. D, tables D.1, D.5.

10. Harley, "Antebellum Tariff: Different Products or Competing Sources?"; "International Competitiveness of the Antebellum American Cotton Textile Industry"; Irwin and Temin, "Antebellum Tariff on Cotton Textiles Revisited"; Temin, "Product Quality and Vertical Integration in the Early Cotton Textile Industry"; Zevin, "Growth of Cotton Textile Production after 1815," 129–33.

11. McGouldrick, *New England Textiles in the Nineteenth Century,* 18–20; Sokoloff, "Productivity Growth in Manufacturing during Early Industrialization," 698, table 13.6; Zevin, "Growth of Cotton Textile Production after 1815," 137–46.

12. Jeremy, *Transatlantic Industrial Revolution,* 276, app. D, table D.1; U.S. Bureau of the Census, *Report on the Manufactures of the United States at the Tenth Census, 1880.*

13. Several data values in table 8.6 differ slightly from comparable numbers in tables 8.3 and 8.4 because the latter two tables used the 1880 U.S. census values for 1840, whereas table 8.6 is based directly on the 1840 census. Some values in the text were computed from census data underlying the tables.

14. Bodenhorn and Rockoff, "Regional Interest Rates in Antebellum America," 167–69, table 5.2.

15. Dalzell, *Enterprising Elite,* 45–73; Gregory, *Nathan Appleton;* Meyer, *Hong Kong as a Global Metropolis,* 5–27; Shlakman, "Economic History of a Factory Town"; Ware, *Early New England Cotton Manufacture.*

16. Bagnall, *Textile Industries of the United States,* 473–83; Gibb, *Saco-Lowell Shops,* 23–112; Jeremy, *Transatlantic Industrial Revolution,* 20–26; Zevin, "Growth of Cotton Textile Production after 1815," 143–45.

17. McGouldrick, *New England Textiles in the Nineteenth Century,* 18–20.

18. Chandler, *Visible Hand;* Gibb, *Saco-Lowell Shops,* 58–62.

19. Davis, "Stock Ownership in the Early New England Textile Industry," 221, table 6; Gregory, *Nathan Appleton,* 214–51; McGouldrick, *New England Textiles in the Nineteenth Century,* 42–43; Porter and Livesay, *Merchants and Manufacturers,* 22–27; Shlakman, "Economic History of a Factory Town," 38, 42; Ware, *Early New England Cotton Manufacture,* 161–97.

20. Davis, "Stock Ownership in the Early New England Textile Industry," 220–22, tables 4, 6, 8; Gregory, *Nathan Appleton,* 1–71, 114–37; Porter, *Jacksons and the Lees,* 1:3–98; Ware, *Early New England Cotton Manufacture,* 145–51. The count of seventy-seven Boston Associates for 1813–65 comes from Dalzell, *Enterprising Elite,* 233–38, app.

21. Anonymous, *Manchester,* 269–302; Bishop, *History of American Manufactures from 1608 to 1860,* 2:283; Browne, *Amoskeag Manufacturing Co. of Manchester, New Hampshire;* Clark, *History of Manufactures in the United States,* 1:546–51; Cudd, *Chicopee Manufacturing Company, 1823–1915,* 16–30, 229–30, app. 7; Davis and Stettler, "New England Textile Industry, 1825–60," 234–37, table A-1; Gibb, *Saco-Lowell Shops,* 55–73; Gregory, *Nathan Appleton,* 197–200; McGouldrick, *New England Textiles in the Nineteenth Century,* 248–50, app. D, table 48.

22. Bagnall, *Textile Industries of the United States,* 399, 529–30; McGouldrick, *New England Textiles in the Nineteenth Century,* 248–50, app. D, table 48; White, *Memoir of Samuel Slater,* 256.

23. Bagnall, *Textile Industries of the United States;* Coleman, *Transformation of Rhode Island, 1790–1860,* 92; McLane, *Documents Relative to the Manufactures in the United States,* doc. 8, no. 41; Meyer, "Dynamic Model of the Integration of Frontier Urban Places into the United States System of Cities"; Ware, *Early New England Cotton Manufacture,* 127–38.

24. Bagnall, *Textile Industries of the United States,* 414–15, 448–49, 532.

25. Bagnall, *Textile Industries of the United States,* 396–403, 444–57, 524–32, 541–51, 559–77.

26. Bagnall, *Textile Industries of the United States,* 376–83, 394–96, 404–23, 433–39, 451–57, 524–32; Coleman, *Transformation of Rhode Island, 1790–1860,* 91; Gibb, *Saco-Lowell Shops,* 32–35; Jeremy, *Transatlantic Industrial Revolution,* 96–98, 181–82; Tucker, *Samuel Slater and the Origins of the American Textile Industry, 1790–1860,* 89–124, 189–206.

27. Bagnall, *Textile Industries of the United States;* Dublin, *Women at Work;* McGouldrick, *New England Textiles in the Nineteenth Century,* 34–42; Hedges, *Browns of Providence Plantations: The Nineteenth Century,* 252; Prude, *Coming of Industrial Order,* 34–157; Tucker, *Samuel Slater and the Origins of the American Textile Industry, 1790–1860,* 139–62, 207–13; Ware, *Early New England Cotton Manufacture,* 176–78, 198–235.

28. Bagnall, *Textile Industries of the United States,* 559–77; Jeremy, *Transatlantic Industrial Revolution,* 216.

29. Bagnall, *Textile Industries of the United States,* 501–23.

30. Scranton, *Proprietary Capitalism,* 80–93, 135–76. Philadelphia also contained woolen, mixed goods, hosiery, and carpet sectors, whose characteristics resembled the cotton sector, the largest textile group.

31. McGouldrick, *New England Textiles in the Nineteenth Century,* 248–50, app. D, table 48; Scranton, *Proprietary Capitalism,* 133–34; Shelton, *Mills of Manayunk,* 54–75; U.S. Bureau of the Census, *Compendium of the Sixth Census, 1840.*

32. Shelton, *Mills of Manayunk;* Wallace, *Rockdale.*

33. Sokoloff, "Inventive Activity in Early Industrial America," 832, fig. 2.

34. See chapter epigraph by Ellsworth, "Report of H. L. Ellsworth to the Secretary of the Treasury, on the Manufactories in Connecticut."

35. Anderson, *Town and City of Waterbury,* 2:265, 275–76; Bishop, *History of American Manufactures from 1608 to 1860,* 2:228, 242; Lathrop, *Brass Industry in the United States,* 73.

36. Anderson, *Town and City of Waterbury,* 2:275–331; Lathrop, *Brass Industry in the United States,* 11, 52–53; McLane, *Documents Relative to the Manufactures in the United States,* doc. 9, nos. 1, 32; Purcell, *Connecticut in Transition: 1775–1818,* 65–74; Woodbury, *Studies in the History of Machine Tools.*

37. Anderson, *Town and City of Waterbury,* 2:298; Bishop, *History of American Manufactures from 1608 to 1860,* 2:395, 415; Cleland, *History of Phelps Dodge, 1834–1950,* 3–35; Jerome, *History of the American Clock Business for the Past Sixty Years,* 51–96; Lathrop, *Brass Industry in the United States,* 58–63; Lubar, "Culture and Technological Design in the 19th-Century Pin Industry."

38. Anderson, *Town and City of Waterbury,* 2:275–487; U.S. Bureau of the Census, *Report of the Manufactures of the United States at the Tenth Census, 1880,* 9, 96, tables 2, 4.

39. Anderson, *Town and City of Waterbury,* 2:275–406.

40. Anderson, *Town and City of Waterbury,* 2:275–406.

CHAPTER NINE: The East Anchors the Manufacturing Belt

1. Value added in manufacturing is the gross value of output minus the cost of materials.

2. Meyer, "Emergence of the American Manufacturing Belt"; "Midwestern Industrialization and the American Manufacturing Belt in the Nineteenth Century."

3. Meyer, "Midwestern Industrialization and the American Manufacturing Belt in the Nineteenth Century."

4. Meyer, "Emergence of the American Manufacturing Belt"; "Midwestern Industrialization and the American Manufacturing Belt in the Nineteenth Century," 924, 931, tables 1, 4; "National Integration of Regional Economies, 1860–1920"; "Rise of the Industrial Metropolis"; U.S. Bureau of the Census, *Historical Statistics of the United States,* ser. A57, 178, 202. Also see chapter epigraph by De Geer, "American Manufacturing Belt," 247.

5. Meyer, "Emergence of the American Manufacturing Belt"; Winder, "Before the Corporation and Mass Production"; "North American Manufacturing Belt in 1880."

6. Meyer, "Industrial Retardation of Southern Cities, 1860–1880"; Sokoloff and Tchakerian, "Manufacturing Where Agriculture Predominates"; Tchakerian, "Productivity, Extent of Markets, and Manufacturing in the Late Antebellum South and Midwest."

Bibliography

Adams, Donald R., Jr. "Earnings and Savings in the Early 19th Century." *Explorations in Economic History* 17 (April 1980): 118–34.

———. "Prices and Wages in Maryland, 1750–1850." *Journal of Economic History* 46 (September 1986): 625–45.

———. "The Standard of Living during American Industrialization: Evidence from the Brandywine Region, 1800–1860." *Journal of Economic History* 42 (December 1982): 903–17.

———. "Wage Rates in the Early National Period: Philadelphia, 1785–1830." *Journal of Economic History* 28 (September 1968): 404–26.

Aitken, Hugh G. J. *The Welland Canal Company: A Study in Canadian Enterprise.* Cambridge, Mass.: Harvard University Press, 1954.

Albion, Robert G. *The Rise of New York Port, 1815–1860.* New York: Charles Scribner's Sons, 1939.

———. *Square-Riggers on Schedule: The New York Sailing Packets to England, France, and the Cotton Ports.* Princeton, N.J.: Princeton University Press, 1938.

Anderson, Joseph, ed. *Town and City of Waterbury.* 3 vols. New Haven, Conn.: Price and Lee Co., 1896.

Anonymous. *Manchester: A Brief Record of Its Past and a Picture of Its Present.* Manchester, N.H.: John B. Clarke, 1875.

Appleby, Joyce. *Capitalism and a New Social Order: The Republican Vision of the 1790s.* New York: New York University Press, 1984.

———. *Inheriting the Revolution: The First Generation of Americans.* Cambridge, Mass.: Belknap Press of Harvard University Press, 2000.

Ardrey, Robert L. *American Agricultural Implements.* Chicago, 1894.

Atack, Jeremy. "Returns to Scale in Antebellum United States Manufacturing." *Explorations in Economic History* 14 (October 1977): 337–59.

Atack, Jeremy, and Fred Bateman. *To Their Own Soil: Agriculture in the Antebellum North.* Ames: Iowa State University Press, 1987.

The Atlas of Pennsylvania. Philadelphia: Temple University Press, 1989.

Baer, Christopher T. *Canals and Railroads of the Mid-Atlantic States, 1800–1860.* Wilmington, Del.: Regional Economic History Research Center, Eleutherian Mills–Hagley Foundation, 1981.

Bagnall, William R. *The Textile Industries of the United States.* Cambridge, Mass.: Riverside Press, 1893.

Bailey, James M. *History of Danbury, Conn., 1684–1896,* ed. Susan B. Hill. New York: Burr Printing House, 1896.

Baker, Andrew H., and Holly V. Izard. "New England Farmers and the Marketplace, 1780–1865: A Case Study." *Agricultural History* 65 (summer 1991): 29–52.

Barker, Theo, and Dorian Gerhold. *The Rise and Rise of Road Transport, 1700–1990.* Cambridge, Eng.: Cambridge University Press, 1995.

Bathe, Greville, and Dorothy Bathe. *Jacob Perkins: His Inventions, His Times, and His Contemporaries.* Philadelphia: Historical Society of Pennsylvania, 1943.

Berry, Thomas S. *Western Prices before 1861: A Study of the Cincinnati Market.* Cambridge, Mass.: Harvard University Press, 1943.

Bidwell, Percy W., and John I. Falconer. *History of Agriculture in the Northern United States, 1620–1860.* Washington, D.C.: Carnegie Institution of Washington, 1925.

Bishop, Avard L. "The State Works of Pennsylvania." *Transactions of the Connecticut Academy of Arts and Sciences* 13 (1907–8).

Bishop, J. Leander. *A History of American Manufactures from 1608 to 1860.* 3 vols. Philadelphia: Edward Young and Co., 1866.

Bjork, Gordon C. "Foreign Trade." In *The Growth of the Seaport Cities, 1790–1825,* ed. David T. Gilchrist, 54–61. Charlottesville: University Press of Virginia, 1967.

Blumin, Stuart M. *The Emergence of the Middle Class: Social Experience in the American City, 1760–1900.* Cambridge, Eng.: Cambridge University Press, 1989.

Bodenhorn, Howard. *A History of Banking in Antebellum America: Financial Markets and Economic Development in an Era of Nation-Building.* Cambridge, Eng.: Cambridge University Press, 2000.

Bodenhorn, Howard, and Hugh Rockoff. "Regional Interest Rates in Antebellum America." In *Strategic Factors in Nineteenth Century American Economic History,* ed. Claudia Goldin and Hugh Rockoff, 159–87. Chicago: University of Chicago Press, 1992.

Bogen, Jules I. *The Anthracite Railroads: A Study in American Railroad Enterprise.* New York: Ronald Press, 1927.

Bronson, Alvin. "Report of Alvin Bronson to the Secretary of the Treasury, on the Manufactures of New York." Albany, New York, April 17, 1832. In Louis McLane, *Documents Relative to the Manufactures in the United States Collected and Transmitted to the House of Representatives, 1832, by the Secretary of the Treasury.* House Document no. 308, 1st sess., 22d Congress, doc. 10, no. 1. Washington, D.C.: Duff Green, 1833.

Brooks, Charles E. *Frontier Settlement and Market Revolution: The Holland Land Purchase.* Ithaca, N.Y.: Cornell University Press, 1996.

Brown, John K. *The Baldwin Locomotive Works, 1831–1915.* Baltimore, Md.: Johns Hopkins University Press, 1995.

Brown, Lawrence A. *Innovation Diffusion: A New Perspective.* London: Methuen, 1981.

Brown, Moses. "Moses Brown to John Dexter, On Manufactures in Rhode Island." May 1791. In *Industrial and Commercial Correspondence of Alexander Hamilton Anticipating His Report on Manufactures,* ed. Arthur H. Cole, 71–80. Chicago: A. W. Shaw Co., 1928.

Brown, Richard D. *Knowledge Is Power: The Diffusion of Information in Early America, 1700–1865.* New York: Oxford University Press, 1989.

Browne, Gary L. *Baltimore in the Nation, 1789–1861.* Chapel Hill: University of North Carolina Press, 1980.

Browne, George W. *The Amoskeag Manufacturing Co. of Manchester, New Hampshire.* Manchester, N.H.: Amoskeag Manufacturing Co., 1915.

Burgess, George H., and Miles C. Kennedy. *Centennial History of the Pennsylvania Railroad Company.* Philadelphia: Pennsylvania Railroad Co., 1949.

Burt, Ronald S. *Structural Holes: The Social Structure of Competition.* Cambridge, Mass.: Harvard University Press, 1992.

Bush, George. "Return of Manufactures near Wilmington, Delaware." November 28, 1791. In *Industrial and Commercial Correspondence of Alexander Hamilton Anticipating His Report on Manufactures,* ed. Arthur H. Cole, 53. Chicago: A. W. Shaw Co., 1928.

Bushman, Richard L. *From Puritan to Yankee: Character and the Social Order in Connecticut, 1690–1765.* Cambridge, Mass.: Harvard University Press, 1967.

Butler, Jon. *Becoming America: The Revolution before 1776.* Cambridge, Mass.: Harvard University Press, 2000.

Cabot, George. "George Cabot to Hamilton, on Cotton Manufacture at Beverly, Mass." September 6, 1791. In *Industrial and Commercial Correspondence of Alexander Hamilton Anticipating His Report on Manufactures,* ed. Arthur H. Cole, 61–65. Chicago: A. W. Shaw Co., 1928.

Cain, Louis P. "From Mud to Metropolis: Chicago before the Fire." *Research in Economic History* 10 (1986): 93–129.

Callender, Guy S. "The Early Transportation and Banking Enterprises of States in Relation to the Growth of Corporations." *Quarterly Journal of Economics* 17 (November 1902): 111–62.

Chandler, Alfred D., Jr. *The Visible Hand: The Managerial Revolution in American Business.* Cambridge, Mass.: Belknap Press of Harvard University Press, 1977.

Christaller, Walter. *Central Places in Southern Germany,* trans. Carlisle W. Baskin. Englewood Cliffs, N.J.: Prentice-Hall, 1966.

Christiansen, Jens, and Peter Philips. "The Transition from Outwork to Factory Production in the Boot and Shoe Industry, 1830–1880." In *Masters to Managers: Historical and Comparative Perspectives on American Employers,* ed. Sanford M. Jacoby, 21–42. New York: Columbia University Press, 1991.

Clark, Christopher. "Economics and Culture: Opening Up the Rural History of the Early American Northeast." *American Quarterly* 43 (June 1991): 279–301.

———. *The Roots of Rural Capitalism: Western Massachusetts, 1780–1860.* Ithaca, N.Y.: Cornell University Press, 1990.

Clark, Victor S. *History of Manufactures in the United States.* 3 vols. New York: McGraw-Hill Book Co., 1929.

Cleland, Robert G. *A History of Phelps Dodge, 1834–1950.* New York: Alfred A. Knopf, 1952.

Clouette, Bruce, and Matthew Roth. *Bristol, Connecticut: A Bicentennial History, 1785–1985.* Canaan, N.H.: Phoenix Publishing, 1984.

Cole, Arthur H., ed. *Industrial and Commercial Correspondence of Alexander Hamilton Anticipating His Report on Manufactures.* Chicago: A. W. Shaw Co., 1928.

Coleman, James S. *Foundations of Social Theory.* Cambridge, Mass.: Belknap Press of Harvard University Press, 1990.

Coleman, Peter J. *The Transformation of Rhode Island, 1790–1860.* Providence, R.I.: Brown University Press, 1963.

Colt, Elisha. "Elisha Colt to John Chester, on Woolen Manufacture in Hartford, Ct." August 20, 1791. In *Industrial and Commercial Correspondence of Alexander Hamil-*

ton Anticipating His Report on Manufactures, ed. Arthur H. Cole, 7–11. Chicago: A. W. Shaw Co., 1928.

Colt, Peter. "Peter Colt to John Chester, On Manufactures in Connecticut." July 21, 1791. In *Industrial and Commercial Correspondence of Alexander Hamilton Anticipating His Report on Manufactures,* ed. Arthur H. Cole, 2–7. Chicago: A. W. Shaw Co., 1928.

Commons, John R. "American Shoemakers, 1648–1895: A Sketch of Industrial Evolution." *Quarterly Journal of Economics* 24 (November 1909): 39–84.

Conrad, James L., Jr. "'Drive That Branch': Samuel Slater, the Power Loom, and the Writing of America's Textile History." *Technology and Culture* 36 (January 1995): 1–28.

Cooke, Jacob E. "Tench Coxe, Alexander Hamilton, and the Encouragement of American Manufactures." *William and Mary Quarterly,* 3d ser., 32 (July 1975): 369–92.

Cooper, Carolyn C. *Shaping Invention: Thomas Blanchard's Machinery and Patent Management in Nineteenth-Century America.* New York: Columbia University Press, 1991.

Coxe, Tench. "Digest of Manufactures, 1810." Communicated to the Senate, January 5, 1814, 13th Cong., 2d sess., *New American State Papers, Manufactures,* 1:160–410. Wilmington, Del.: Scholarly Resources, 1972.

Cranmer, H. Jerome. "Canal Investment, 1815–1860." *Trends in the American Economy in the Nineteenth Century.* Studies in Income and Wealth, 24:547–70. Princeton, N.J.: Princeton University Press, 1960.

———. "Improvements without Public Funds: The New Jersey Canals." In *Canals and American Economic Development,* ed. Carter Goodrich, 115–66. New York: Columbia University Press, 1961.

Creighton, Donald G. *The Commercial Empire of the St. Lawrence, 1760–1850.* Toronto: Ryerson Press, 1937.

Cudd, John M. *The Chicopee Manufacturing Co., 1823–1915.* Wilmington, Del.: Scholarly Resources, 1974.

Dalzell, Robert F., Jr. *Enterprising Elite: The Boston Associates and the World They Made.* Cambridge, Mass.: Harvard University Press, 1987.

Danhof, Clarence H. *Change in Agriculture: The Northern United States, 1820–1870.* Cambridge, Mass.: Harvard University Press, 1969.

Daniels, Bruce C. *The Connecticut Town: Growth and Development, 1635–1790.* Middletown, Conn.: Wesleyan University Press, 1979).

David, Paul A., and Peter Solar. "A Bicentenary Contribution to the History of the Cost of Living in America." *Research in Economic History* 2 (1977): 1–80.

Davis, Lance E. "Stock Ownership in the Early New England Textile Industry." *Business History Review* 32 (summer 1958): 204–22.

Davis, Lance E., Richard A. Easterlin, and William N. Parker. *American Economic Growth.* New York: Harper and Row, 1972.

Davis, Lance E., and Robert E. Gallman. "Capital Formation in the United States during the Nineteenth Century." In *The Cambridge Economic History of Europe,* ed. Peter Mathias and M. M. Postan, vol. 7, pt. 2: 1–69. Cambridge, Eng.: Cambridge University Press, 1978.

Davis, Lance E., and H. Louis Stettler III. "The New England Textile Industry, 1825–60: Trends and Fluctuations." *Output, Employment, and Productivity in the United*

States after 1800. Studies in Income and Wealth, 30:213–42. New York: National Bureau of Economic Research, 1966.

Dawley, Alan. *Class and Community: The Industrial Revolution in Lynn*. Cambridge, Mass.: Harvard University Press, 1976.

Day, Clive. "The Early Development of the American Cotton Manufacture." *Quarterly Journal of Economics* 39 (May 1925): 450–68.

De Vries, Jan. "The Industrial Revolution and the Industrious Revolution." *Journal of Economic History* 54 (June 1994): 249–70.

De Geer, Sten. "The American Manufacturing Belt." *Geografiska Annaler* 9 (1927): 233–359.

DeVoe, Shirley S. *The Tinsmiths of Connecticut*. Middletown, Conn.: Wesleyan University Press, 1968.

Deyrup, Felicia J. "Arms Makers of the Connecticut Valley: A Regional Study of the Economic Development of the Small Arms Industry, 1798–1870." *Smith College Studies in History* 33 (1948).

Dilts, James D. *The Great Road: The Building of the Baltimore and Ohio, the Nation's First Railroad, 1828–1853*. Stanford, Calif.: Stanford University Press, 1993.

Doerflinger, Thomas M. "Farmers and Dry Goods in the Philadelphia Market Area, 1750–1800." In *The Economy of Early America: The Revolutionary Period, 1763–1790*, ed. Ronald Hoffman, John J. McCusker, Russell R. Menard, and Peter J. Albert, 166–95. Charlottesville: University Press of Virginia, 1988.

Dublin, Thomas. *Transforming Women's Work: New England Lives in the Industrial Revolution*. Ithaca, N.Y.: Cornell University Press, 1994.

———. *Women at Work: The Transformation of Work and Community in Lowell, Massachusetts, 1826–1860*. New York: Columbia University Press, 1979.

Durkheim, Emile. *The Division of Labor in Society*, trans. George Simpson. New York: Free Press, 1964.

Durrenberger, Joseph A. *Turnpikes: A Study of the Toll Road Movement in the Middle Atlantic States and Maryland*. Cos Cob, Conn.: John E. Edwards, 1968.

Dwight, Timothy. *Travels in New-England and New-York*. 4 vols. London: Charles Wood, 1823.

Easterlin, Richard A. "Interregional Differences in Per Capita Income, Population, and Total Income, 1840–1950." *Trends in the American Economy in the Nineteenth Century*. Studies in Income and Wealth, 24:73–140. Princeton, N.J.: Princeton University Press, 1960.

Eggert, Gerald G. *Harrisburg Industrializes: The Coming of Factories to an American Community*. University Park: Pennsylvania State University Press, 1993.

Ellis, David M. *Landlords and Farmers in the Hudson-Mohawk Region, 1790–1850*. Ithaca, N.Y.: Cornell University Press, 1946.

Ellsworth, H. L. "Report of H. L. Ellsworth to the Secretary of the Treasury, on the Manufactories in Connecticut." Hartford, Connecticut, April 16, 1832. In Louis McLane, *Documents Relative to the Manufactures in the United States Collected and Transmitted to the House of Representatives, 1832, by the Secretary of the Treasury*. House Document no. 308, 1st sess., 22d Cong., doc. 9, no. 1. Washington, D.C.: Duff Green, 1833.

Erickson, Bonnie H. "Culture, Class, and Connections." *American Journal of Sociology* 102 (July 1996): 217–51.

Faler, Paul G. *Mechanics and Manufacturers in the Early Industrial Revolution: Lynn, Massachusetts, 1780–1860.* Albany: State University of New York Press, 1981.

Feller, Irwin. "Determinants of the Composition of Urban Inventions." *Economic Geography* 49 (January 1973): 47–58.

———. "Invention, Diffusion and Industrial Location." In *Locational Dynamics of Manufacturing Activity,* ed. Lyndhurst Collins and David F. Walker, 83–107. London: John Wiley, 1975.

———. "The Urban Location of United States Invention, 1860–1910." *Explorations in Economic History* 8 (spring 1971): 285–303.

Fenstermaker, J. Van. *The Development of American Commercial Banking: 1782–1837,* Bureau of Economic and Business Research Monographs, no. 5. Kent, Ohio: Kent State University, 1965.

Field, Alexander J. "On the Unimportance of Machinery." *Explorations in Economic History* 22 (October 1985): 378–401.

———. "Sectoral Shift in Antebellum Massachusetts: A Reconsideration." *Explorations in Economic History* 15 (April 1978): 146–71.

Filante, Ronald W. "A Note on the Economic Viability of the Erie Canal, 1825–1860." *Business History Review* 48 (spring 1974): 95–102.

Fishlow, Albert. *American Railroads and the Transformation of the Ante-Bellum Economy.* Cambridge, Mass.: Harvard University Press, 1965.

———. "Antebellum Interregional Trade Reconsidered." *American Economic Review* 54 (May 1964): 352–64.

Fogel, Robert W. *Railroads and American Economic Growth: Essays in Econometric History.* Baltimore, Md.: Johns Hopkins Press, 1964.

Foster, William H. "Explanations, Remarks, and Additional Facts." Wales, Massachusetts, in Louis McLane, *Documents Relative to the Manufactures in the United States Collected and Transmitted to the House of Representatives, 1832, by the Secretary of the Treasury.* House Document no. 308, 1st sess., 22d Cong., doc. 3, no. 107. Washington, D.C.: Duff Green, 1833.

Franklin, Benjamin. "The Interest of Great Britain Considered, with Regard to Her Colonies and the Acquisitions of Canada and Guadaloupe." 1760. In *The Works of Benjamin Franklin,* ed. Jared Sparks, 4:1–53. Boston: Tappan and Whittemore, 1837.

Frost, J. William. *Connecticut Education in the Revolutionary Era: "For God and Country."* Connecticut Bicentennial Series, no. 7. Chester, Conn.: Pequot Press, 1974.

Fuller, Grace P. "An Introduction to the History of Connecticut as a Manufacturing State." *Smith College Studies in History* 1, no. 1 (October 1915).

Gallatin, Albert. "Manufactures." Communicated to the House of Representatives, April 19, 1810, 11th Cong., 2d sess. *New American State Papers, Manufactures,* 1:124–42. Wilmington, Del.: Scholarly Resources, 1972.

Gallman, Robert E. "American Economic Growth before the Civil War: The Testimony of the Capital Stock Estimates." In *American Economic Growth and Standards of Living before the Civil War,* ed. Robert E. Gallman and John J. Wallis, 79–120. Chicago: University of Chicago Press, 1992.

———. "Commodity Output, 1839–1899." *Trends in the American Economy in the Nineteenth Century,* Studies in Income and Wealth, 24:13–71. Princeton, N.J.: Princeton University Press, 1960.

———. "Gross National Product in the United States, 1834–1909." *Output, Employment, and Productivity in the United States after 1800.* Studies in Income and Wealth, 30:3–90. New York: National Bureau of Economic Research, 1966.

Gates, Paul W. *The Farmer's Age: Agriculture, 1815–1860.* New York: Holt, Rinehart and Winston, 1960.

Geib-Gundersen, Lisa, and Elizabeth Zahrt. "A New Look at U.S. Agricultural Productivity Growth, 1800–1910." *Journal of Economic History* 56 (September 1996): 679–86.

Gerhold, Dorian. "Productivity Change in Road Transport before and after Turnpiking, 1690–1840." *Economic History Review* 49 (August 1996): 491–515.

Gibb, George S. *The Saco-Lowell Shops: Textile Machinery Building in New England, 1813–1949.* Cambridge, Mass.: Harvard University Press, 1950.

Gillespie, C. Bancroft. *A Century of Meriden: "The Silver City."* 3 pts. Meriden, Conn.: Journal Publishing Co., 1906.

Gilmore, William J. "Peddlers and the Dissemination of Printed Material in Northern New England, 1780–1840." In *Itinerancy in New England and New York,* ed. Peter Benes and Jane M. Benes, 76–89. Boston: Boston University, 1986.

Goldin, Claudia D., and Frank D. Lewis. "The Role of Exports in American Economic Growth during the Napoleonic Wars, 1793 to 1807." *Explorations in Economic History* 17 (January 1980): 6–25.

Goldin, Claudia D., and Kenneth L. Sokoloff. "The Relative Productivity Hypothesis of Industrialization: The American Case, 1820 to 1850." *Quarterly Journal of Economics* 99 (August 1984): 461–87.

———. "Women, Children, and Industrialization in the Early Republic: Evidence from the Manufacturing Censuses." *Journal of Economic History* 42 (December 1982): 741–74.

Goodrich, Carter. *Government Promotion of American Canals and Railroads, 1800–1890.* New York: Columbia University Press, 1960.

———, ed., *Canals and American Economic Development.* New York: Columbia University Press, 1961.

Granovetter, Mark. "Economic Action and Social Structure: The Problem of Embeddedness." *American Journal of Sociology* 91 (November 1985): 481–510.

———. "The Nature of Economic Relationships." In *Explorations in Economic Sociology,* ed. Richard Swedberg, 3–41. New York: Russell Sage Foundation, 1993.

———. "The Strength of Weak Ties." *American Journal of Sociology* 78 (May 1973): 1360–80.

Grant, Ellsworth S. *Yankee Dreamers and Doers.* Chester, Conn.: Pequot Press, 1975.

Grantham, George. "Agricultural Supply during the Industrial Revolution: French Evidence and European Implications." *Journal of Economic History* 49 (March 1989): 43–72.

Gray, Lewis C. *History of Agriculture in the Southern United States to 1860.* 2 vols. Washington, D.C.: Carnegie Institution of Washington, 1933.

Gray, Ralph D. *The National Waterway: A History of the Chesapeake and Delaware Canal, 1769–1965.* Urbana: University of Illinois Press, 1967.

Gregory, Frances W. *Nathan Appleton: Merchant and Entrepreneur, 1779–1861.* Charlottesville: University Press of Virginia, 1975.

Gross, Robert A. "Culture and Cultivation: Agriculture and Society in Thoreau's Concord." *Journal of American History* 69 (June 1982): 42–61.

Haig, Robert M. "Toward an Understanding of the Metropolis: I. Some Speculations Regarding the Economic Basis of Urban Concentration." *Quarterly Journal of Economics* 40 (February 1926): 179–208.

Haites, Erik F., James Mak, and Gary M. Walton. *Western River Transportation: The Era of Early Internal Development, 1810–1860.* Baltimore, Md.: Johns Hopkins University Press, 1975.

Hall, Peter, ed. *Von Thunen's Isolated State.* Oxford: Pergamon Press, 1966.

Hamilton, Alexander. "Report on Manufactures." Communicated to the House of Representatives, December 5, 1791, 2d Cong., 1st sess. *New American State Papers, Manufactures,* 1:43–69. Wilmington, Del.: Scholarly Resources, 1972.

Hammond, Bray. *Banks and Politics in America: From the Revolution to the Civil War.* Princeton, N.J.: Princeton University Press, 1957.

Hannay, Agnes. "A Chronicle of Industry on the Mill River." *Smith College Studies in History* 21, nos. 1–4 (October 1935–July 1936).

Harley, C. Knick. "The Antebellum Tariff: Different Products or Competing Sources? A Comment on Irwin and Temin." *Journal of Economic History* 61 (September 2001): 799–805.

———. "The Antebellum Tariff: Food Exports and Manufacturing." *Explorations in Economic History* 29 (October 1992): 375–400.

———. "International Competitiveness of the Antebellum American Cotton Textile Industry." *Journal of Economic History* 52 (September 1992): 559–84.

Harlow, Alvin F. *Old Towpaths: The Story of the American Canal Era.* New York: D. Appleton and Co., 1926.

Hartz, Louis. *Economic Policy and Democratic Thought: Pennsylvania, 1776–1860.* Cambridge, Mass.: Harvard University Press, 1948.

Hatch, Alden. *Remington Arms: An American History.* New York: Rinehart and Co., 1956.

Hayes, John L. *American Textile Machinery.* Cambridge, Mass.: University Press, John Wilson and Son, 1879.

Hazard, Blanche E. *The Organization of the Boot and Shoe Industry in Massachusetts before 1875.* Cambridge, Mass.: Harvard University Press, 1921.

Hedges, James B. *The Browns of Providence Plantations: Colonial Years.* Cambridge, Mass.: Harvard University Press, 1952.

———. *The Browns of Providence Plantations: The Nineteenth Century.* Providence, R.I.: Brown University Press, 1968.

Henretta, James A. *The Origins of American Capitalism.* Boston: Northeastern University Press, 1991.

Herbst, Lawrence A. "Interregional Commodity Trade from the North to the South and American Economic Development in the Antebellum Period." *Journal of Economic History* 35 (March 1975): 264–70.

Higgs, Robert. "Urbanization and Inventiveness in the United States, 1870–1920." In *The New Urban History,* ed. Leo F. Schnore, 247–59. Princeton, N.J.: Princeton University Press, 1975.

Hilliard, Sam B. *Atlas of Antebellum Southern Agriculture.* Baton Rouge: Louisiana State University Press, 1984.

Hoffecker, Carol E. *Wilmington, Delaware: Portrait of an Industrial City, 1830–1910.* Charlottesville: University Press of Virginia, 1974.

Hoke, Donald R. *Ingenious Yankees: The Rise of the American System of Manufactures in the Private Sector.* New York: Columbia University Press, 1990.

Homer, Sidney, and Richard Sylla. *A History of Interest Rates.* 3d rev. ed. New Brunswick, N.J.: Rutgers University Press, 1996.

Hounshell, David A. *From the American System to Mass Production, 1800–1932: The Development of Manufacturing Technology in the United States.* Baltimore, Md.: Johns Hopkins University Press, 1984.

Hunter, Louis C. *Steamboats on the Western Rivers.* Cambridge, Mass.: Harvard University Press, 1949.

Hutchinson, William K., and Samuel H. Williamson. "The Self-Sufficiency of the Antebellum South: Estimates of the Food Supply." *Journal of Economic History* 31 (September 1971): 591–612.

Irwin, Douglas, and Peter Temin. "The Antebellum Tariff on Cotton Textiles Revisited." *Journal of Economic History* 61 (September 2001): 777–98.

Jackson, Patrick T. "Letter from Patrick Tracy Jackson, Boston, April 30, 1810, to Edward A. Newton, Madras, with Instructions for a Voyage from India." Jackson-Lee Papers, P. T. Jackson—Letter Book A, 56–59. Reprinted in Kenneth W. Porter, *The Jacksons and the Lees: Two Generations of Massachusetts Merchants, 1765–1844,* 1:646–50. 2 vols. Cambridge, Mass.: Harvard University Press, 1937.

Jaffee, David. "Peddlers of Progress and the Transformation of the Rural North, 1760–1860." *Journal of American History* 78 (September 1991): 511–35.

Jeremy, David J. *Transatlantic Industrial Revolution: The Diffusion of Textile Technologies between Britain and America, 1790–1830s.* Cambridge, Mass.: MIT Press, 1981.

Jerome, Chauncey. *History of the American Clock Business for the Past Sixty Years, and Life of Chauncey Jerome.* New Haven, Conn.: F. C. Dayton Jr., 1860.

Johnson, Arthur M., and Barry E. Supple. *Boston Capitalists and Western Railroads: A Study in the Nineteenth-Century Railroad Investment Process.* Cambridge, Mass.: Harvard University Press, 1967.

Johnson, Emory R., T. W. Van Metre, G. G. Huebner, and D. S. Hanchett. *History of Domestic and Foreign Commerce of the United States.* 2 vols. Washington, D.C.: Carnegie Institution of Washington, 1915.

Jones, Chester L. *The Economic History of the Anthracite-Tidewater Canals.* University of Pennsylvania Series on Political Economy and Public Law, no. 22. Philadelphia: John C. Winston, 1908.

Jordan, Philip D. *The National Road.* Indianapolis, Ind.: Bobbs-Merrill, 1948.

Karr, Ronald D. "The Transformation of Agriculture in Brookline, 1770–1885." *Historical Journal of Massachusetts* 15 (January 1987): 33–49.

Keir, R. Malcolm. "The Tin Peddler." *Journal of Political Economy* 21 (March 1913): 255–58.

Khan, B. Zorina, and Kenneth L. Sokoloff. "'Schemes of Practical Utility': Entrepreneurship and Innovation among 'Great Inventors' in the United States, 1790–1865." *Journal of Economic History* 53 (June 1993): 289–307.

Kirkland, Edward C. *Men, Cities, and Transportation: A Study in New England History, 1820–1900.* 2 vols. Cambridge, Mass.: Harvard University Press, 1948.

Kistler, Thelma M. "The Rise of Railroads in the Connecticut River Valley." *Smith College Studies in History* 23 (October 1937–July 1938).

Klein, Daniel B. "The Voluntary Provision of Public Goods? The Turnpike Companies of Early America." *Economic Inquiry* 28 (October 1990): 788–812.
Klein, Daniel B., and John Majewski. "Economy, Community, and Law: The Turnpike Movement in New York, 1797–1845." *Law and Society Review* 26, no. 3 (1992): 469–512.
Kline, Priscilla C. "New Light on the Yankee Peddler." *New England Quarterly* 12 (March 1939): 80–98.
Kohlmeier, A. L. *The Old Northwest as the Keystone of the Arch of American Federal Union.* Bloomington, Ind.: Principia Press, 1938.
Kravis, Irving B. "The Role of Exports in Nineteenth-Century United States Growth." *Economic Development and Cultural Change* 20 (April 1972): 387–405.
Kulikoff, Allan. *The Agrarian Origins of American Capitalism.* Charlottesville: University Press of Virginia, 1992.
———. "Households and Markets: Toward a New Synthesis of American Agrarian History." *William and Mary Quarterly,* 3d ser., 50 (April 1993): 342–55.
Lamoreaux, Naomi R. *Insider Lending: Banks, Personal Connections, and Economic Development in Industrial New England.* Cambridge, Eng.: Cambridge University Press, 1994.
Lampard, Eric E. "The History of Cities in the Economically Advanced Areas." *Economic Development and Cultural Change* 3 (January 1955): 81–136.
Larson, John L. *Internal Improvement: National Public Works and the Promise of Popular Government in the Early United States.* Chapel Hill: University of North Carolina Press, 2001.
Lathrop, William G. *The Brass Industry in the United States,* rev. ed. Mount Carmel, Conn.: privately published, 1926.
Lawler, Edward J., and Jeongkoo Yoon. "Commitment in Exchange Relations: Test of a Theory of Relational Cohesion." *American Sociological Review* 61 (February 1996): 89–108.
Lebergott, Stanley. *The Americans: An Economic Record.* New York: W. W. Norton, 1984.
———. "The Demand for Land: The United States, 1820–1860." *Journal of Economic History* 45 (June 1985): 181–212.
———. "Labor Force and Employment, 1800–1960." *Output, Employment, and Productivity in the United States after 1800.* Studies in Income and Wealth, 30:117–210. New York: National Bureau of Economic Research, 1966.
Lemon, James T. *The Best Poor Man's Country: A Geographical Study of Early Southeastern Pennsylvania.* Baltimore, Md.: Johns Hopkins University Press, 1972.
Licht, Walter. *Industrializing America: The Nineteenth Century.* Baltimore, Md.: Johns Hopkins University Press, 1995.
Linder, Marc, and Lawrence S. Zacharias. *Of Cabbages and Kings County: Agriculture and the Formation of Modern Brooklyn.* Iowa City: University of Iowa Press, 1999.
Lindstrom, Diane. *Economic Development in the Philadelphia Region, 1810–1850.* New York: Columbia University Press, 1978.
———. "Southern Dependence upon Interregional Grain Supplies: A Review of the Trade Flows, 1840–1860." *Agricultural History* 44 (January 1970): 101–13.
Livingood, James W. *The Philadelphia-Baltimore Trade Rivalry, 1780–1860.* Harrisburg: Pennsylvania Historical and Museum Commission, 1947.

Losch, August. *The Economics of Location.* New Haven, Conn.: Yale University Press, 1954.
Lozier, John W. *Taunton and Mason: Cotton Machinery and Locomotive Manufacture in Taunton, Massachusetts, 1811–1861.* New York: Garland Publishing, 1986.
Lubar, Steven. "Culture and Technological Design in the 19th-Century Pin Industry: John Howe and the Howe Manufacturing Co." *Technology and Culture* 28 (April 1987): 253–82.
McClelland, Peter D. *Sowing Modernity: America's First Agricultural Revolution.* Ithaca, N.Y.: Cornell University Press, 1997.
McCusker, John J., and Russell R. Menard. *The Economy of British America, 1607–1789.* Chapel Hill: University of North Carolina Press, 1985.
McGaw, Judith A. *Most Wonderful Machine: Mechanization and Social Change in Berkshire Paper Making, 1801–1885.* Princeton, N.J.: Princeton University Press, 1987.
MacGill, Caroline E. *History of Transportation in the United States before 1860.* Washington, D.C.: Carnegie Institution of Washington, 1917.
McGouldrick, Paul F. *New England Textiles in the Nineteenth Century: Profits and Investment.* Cambridge, Mass.: Harvard University Press, 1968.
McIlwraith, Thomas F. "Freight Capacity and Utilization of the Erie and Great Lakes Canals before 1850." *Journal of Economic History* 36 (December 1976): 852–77.
McKelvey, Blake. *Rochester: The Water-Power City, 1812–1854.* Cambridge, Mass.: Harvard University Press, 1945.
McLane, Louis. *Documents Relative to the Manufactures in the United States Collected and Transmitted to the House of Representatives, 1832, by the Secretary of the Treasury.* House Document no. 308, 1st sess., 22d Cong. Washington, D.C.: Duff Green, 1833.
McMurry, Sally. *Transforming Rural Life: Dairying Families and Agricultural Change, 1820–1885.* Baltimore, Md.: Johns Hopkins University Press, 1995.
McNall, Neil A. *An Agricultural History of the Genesee Valley, 1790–1860.* Philadelphia: University of Pennsylvania Press, 1952.
Majewski, John. *A House Dividing: Economic Development in Pennsylvania and Virginia before the Civil War.* Cambridge, Eng.: Cambridge University Press, 2000.
Mancall, Peter C. *Valley of Opportunity: Economic Culture along the Upper Susquehanna, 1700–1800.* Ithaca, N.Y.: Cornell University Press, 1991.
Margo, Robert A. *Wages and Labor Markets in the United States, 1820–1860.* Chicago: University of Chicago Press, 2000.
Martin, Margaret E. "Merchants and Trade of the Connecticut River Valley, 1750–1820." *Smith College Studies in History* 24, nos. 1–4 (October 1938–July 1939).
Marx, Karl. *Capital.* 3 vols. New York: International Publishers, 1967.
Mathews, Lois K. *The Expansion of New England.* Boston, Mass.: Houghton Mifflin Co., 1909.
Merrill, Michael. "Putting 'Capitalism' in Its Place: A Review of Recent Literature." *William and Mary Quarterly,* 3d ser., 52 (April 1995): 315–26.
Meyer, David R. "The Division of Labor and the Market Areas of Manufacturing Firms." *Sociological Forum* 3 (summer 1988): 433–53.
———. "A Dynamic Model of the Integration of Frontier Urban Places into the United States System of Cities." *Economic Geography* 56 (April 1980): 120–40.
———. "Emergence of the American Manufacturing Belt: An Interpretation." *Journal of Historical Geography* 9 (April 1983): 145–74.

——. *Hong Kong as a Global Metropolis.* Cambridge, Eng.: Cambridge University Press, 2000.

——. "The Industrial Retardation of Southern Cities, 1860–1880." *Explorations in Economic History* 25 (October 1988): 366–86.

——. "Midwestern Industrialization and the American Manufacturing Belt in the Nineteenth Century." *Journal of Economic History* 49 (December 1989): 921–37.

——. "The National Integration of Regional Economies, 1860–1920." In *North America: The Historical Geography of a Changing Continent,* ed. Thomas F. McIlwraith and Edward K. Muller, 307–31. 2d ed. Lanham, Md.: Rowman and Littlefield, 2001.

——. "The Rise of the Industrial Metropolis: The Myth and the Reality." *Social Forces* 68 (March 1990): 731–52.

Miller, Nathan. *The Enterprise of a Free People: Aspects of Economic Development in New York State during the Canal Period, 1792–1838.* Ithaca, N.Y.: Cornell University Press, 1962.

Miller, Roberta B. *City and Hinterland: A Case Study of Urban Growth and Regional Development.* Westport, Conn.: Greenwood Press, 1979.

Morison, Samuel E. *The Maritime History of Massachusetts, 1783–1860.* Boston: Houghton Mifflin Co., 1921.

Morrison, John H. *History of American Steam Navigation.* New York: W. F. Sametz and Co., 1903.

Mudge, Benjamin. "Explanations, Remarks, and Additional Facts." Marblehead, Massachusetts. In Louis McLane, *Documents Relative to the Manufactures in the United States Collected and Transmitted to the House of Representatives, 1832, by the Secretary of the Treasury.* House Document no. 308, 1st sess., 22d Cong., doc. 3, no. 84. Washington, D.C.: Duff Green, 1833.

Muller, Edward K. "Selective Urban Growth in the Middle Ohio Valley, 1800–1860." *Geographical Review* 66 (April 1976): 178–99.

Murphy, John J. "Entrepreneurship in the Establishment of the American Clock Industry." *Journal of Economic History* 26 (June 1966): 169–86.

Nelson, John R., Jr. "Alexander Hamilton and American Manufacturing: A Reexamination." *Journal of American History* 65 (March 1979): 971–95.

Nettels, Curtis P. *The Emergence of a National Economy, 1775–1815.* New York: Holt, Rinehart and Winston, 1962.

Newell, Margaret E. *From Dependency to Independence: Economic Revolution in Colonial New England.* Ithaca, N.Y.: Cornell University Press, 1998.

Niemi, Albert W., Jr. *State and Regional Patterns in American Manufacturing, 1860–1900.* Westport, Conn.: Greenwood Press, 1974.

North, Douglass C. *The Economic Growth of the United States, 1790–1860.* Englewood Cliffs, N.J.: Prentice-Hall, 1961.

Pabst, Margaret R. "Agricultural Trends in the Connecticut Valley Region of Massachusetts, 1800–1900." *Smith College Studies in History* 26, nos. 1–4 (October 1940–July 1941).

Parkerson, Donald H. *The Agricultural Transition in New York State: Markets and Migration in Mid-Nineteenth-Century America.* Ames: Iowa State University Press, 1995.

Paskoff, Paul F. *Industrial Evolution: Organization, Structure, and Growth of the Penn-*

sylvania Iron Industry, 1750–1860. Baltimore, Md.: Johns Hopkins University Press, 1983.

Pease, John C., and John M. Niles. *The Gazeteer of the States of Connecticut and Rhode Island.* Hartford, Conn.: privately published, 1819.

Pierce, Harry H. *Railroads of New York: A Study of Government Aid, 1826–1875.* Cambridge, Mass.: Harvard University Press, 1953.

Pitkin, Timothy. *A Statistical View of the Commerce of the United States of America.* New Haven, Conn.: Durrie and Peck, 1835.

Porter, Glenn, and Harold C. Livesay. *Merchants and Manufacturers: Studies in the Changing Structure of Nineteenth-Century Marketing.* Baltimore, Md.: Johns Hopkins Press, 1971.

Porter, Kenneth W. *The Jacksons and the Lees: Two Generations of Massachusetts Merchants, 1765–1844.* 2 vols. Cambridge, Mass.: Harvard University Press, 1937.

Post, Charles. "The Agrarian Origins of U.S. Capitalism: The Transformation of the Northern Countryside before the Civil War." *Journal of Peasant Studies* 22 (April 1995): 389–445.

Pred, Allan R. *City-Systems in Advanced Economies.* London: Hutchinson and Co., 1977.

———. *The Spatial Dynamics of U.S. Urban-Industrial Growth, 1800–1914.* Cambridge, Mass.: MIT Press, 1966.

———. *Urban Growth and the Circulation of Information: The United States System of Cities, 1790–1840.* Cambridge, Mass.: Harvard University Press, 1973.

———. *Urban Growth and City-Systems in the United States, 1840–1860.* Cambridge, Mass.: Harvard University Press, 1980.

Price, Jacob M. "Economic Function and the Growth of American Port Towns in the Eighteenth Century." *Perspectives in American History* 8 (1974): 123–86.

Proceedings of the Chesapeake and Ohio Canal Convention, Washington, 1823, 4. Reprinted in George W. Ward. *The Early Development of the Chesapeake and Ohio Canal Project.* Johns Hopkins University Studies in Historical and Political Science, ser. 17, nos. 9–11. Baltimore, Md.: Johns Hopkins Press, 1899, 50.

"Prospectus of the Society for Establishing Useful Manufactures" and "Projected Agreement for Binding the Subscribers to the S.U.M. Pending Incorporation of the Society." In *Industrial and Commercial Correspondence of Alexander Hamilton Anticipating His Report on Manufactures,* ed. Arthur H. Cole, 191–200. Chicago: A. W. Shaw Co., 1928.

Prude, Jonathan. *The Coming of Industrial Order: Town and Factory Life in Rural Massachusetts, 1810–1860.* Cambridge, Eng.: Cambridge University Press, 1983.

Purcell, Richard J. *Connecticut in Transition: 1775–1818.* Washington, D.C.: American Historical Association, 1918.

Ramsey, Elizabeth. "The History of Tobacco Production in the Connecticut Valley." *Smith College Studies in History* 15, nos. 3–4 (April–July 1930).

Ransom, Roger L. "Canals and Development: A Discussion of the Issues." *American Economic Review* 54 (May 1964): 365–76.

———. "Interregional Canals and Economic Specialization in the Antebellum United States." *Explorations in Entrepreneurial History* 5, no. 1 (1967–68): 12–35.

Reiser, Catherine E. *Pittsburgh's Commercial Development, 1800–1850.* Harrisburg: Pennsylvania Historical and Museum Commission, 1951.

Reizenstein, Milton. *The Economic History of the Baltimore and Ohio Railroad, 1827–1853.* Johns Hopkins University Studies in Historical and Political Science, ser. 15, nos. 7–8. Baltimore, Md.: Johns Hopkins Press, 1897.

Ringwalt, J. L. *Development of Transportation Systems in the United States.* Philadelphia: Railway World Office, 1888.

Roberts, Christopher. *The Middlesex Canal, 1793–1860.* Cambridge, Mass.: Harvard University Press, 1938.

Roberts, Kenneth D. *The Contributions of Joseph Ives to Connecticut Clock Technology, 1810–1862.* Bristol, Conn.: American Clock and Watch Museum, 1970.

Roe, Joseph W. *Connecticut Inventors,* Tercentary Commission of the State of Connecticut, Committee on Historical Publications, no. 33. New Haven, Conn.: Yale University Press, 1934.

———. *English and American Tool Builders.* New York: McGraw-Hill, 1916.

Rosenberg, Nathan. "Factors Affecting the Diffusion of Technology." *Explorations in Economic History* 10 (fall 1972): 3–33.

———. *Technology and American Economic Growth.* New York: Harper and Row, 1972.

Rothenberg, Winifred B. *From Market-Places to a Market Economy: The Transformation of Rural Massachusetts, 1750–1850.* Chicago: University of Chicago Press, 1992.

Rubin, Julius. "Canal or Railroad? Imitation and Innovation in the Response to the Erie Canal in Philadelphia, Baltimore, and Boston." *Transactions of the American Philosophical Society,* n.s., 51, pt. 7 (1961): 1–106.

———. "An Imitative Public Improvement: The Pennsylvania Mainline." In *Canals and American Economic Development,* ed. Carter Goodrich, 67–114. New York: Columbia University Press, 1961.

———. "An Innovating Public Improvement: The Erie Canal." In *Canals and American Economic Development,* ed. Carter Goodrich, 15–66. New York: Columbia University Press, 1961.

Rueschemeyer, Dietrich. *Power and the Division of Labor.* Stanford, Calif.: Stanford University Press, 1986.

Rutter, Frank R. *South American Trade of Baltimore.* Johns Hopkins University Studies in Historical and Political Science, ser. 15, no. 9. Baltimore, Md.: Johns Hopkins Press, 1897.

Salsbury, Stephen. *The State, the Investor, and the Railroad: The Boston & Albany, 1825–1867.* Cambridge, Mass.: Harvard University Press, 1967.

Sanderlin, Walter S. *The Great National Project: A History of the Chesapeake and Ohio Canal.* Johns Hopkins University Studies in Historical and Political Science, ser. 64, no. 1. Baltimore, Md.: Johns Hopkins Press, 1946.

Schein, Richard H. "Urban Origin and Form in Central New York." *Geographical Review* 81 (January 1991): 52–69.

Schmidt, Louis B. "Internal Commerce and the Development of a National Economy before 1860." *Journal of Political Economy* 47 (December 1939): 798–822.

Schmookler, Jacob. *Invention and Economic Growth.* Cambridge, Mass.: Harvard University Press, 1966.

Scranton, Philip. *Proprietary Capitalism: The Textile Manufacture at Philadelphia, 1800–1885.* Cambridge, Eng.: Cambridge University Press, 1983.

Secretary of the Interior. *Abstract of the Statistics of Manufactures, Seventh Census, 1850.* Washington, D.C.: Government Printing Office, 1859.

Segal, Harvey H. "Canals and Economic Development." In *Canals and American Economic Development,* ed. Carter Goodrich, 216–48. New York: Columbia University Press, 1961.
———. "Cycles of Canal Construction." In *Canals and American Economic Development,* ed. Carter Goodrich, 169–215. New York: Columbia University Press, 1961.
Seybert, Adam. *Statistical Annals: Embracing Views of the Population, Commerce, Navigation, . . . of the United States of America.* Philadelphia: Thomas Dobson and Son, 1818.
Shelton, Cynthia J. *The Mills of Manayunk: Industrialization and Social Conflict in the Philadelphia Region, 1787–1837.* Baltimore, Md.: Johns Hopkins University Press, 1986.
Shlakman, Vera. "Economic History of a Factory Town: A Study of Chicopee, Massachusetts." *Smith College Studies in History* 20, nos. 1–4 (October 1934–July 1935).
Smith, Adam. *An Inquiry into the Nature and Causes of the Wealth of Nations.* 1776. Reprint. Chicago: Encyclopedia Britannica, 1952.
Smith, David C., and Anne E. Bridges. "The Brighton Market: Feeding Nineteenth-Century Boston." *Agricultural History* 56 (January 1982): 3–21.
Smith, Walter B., and Arthur H. Cole. *Fluctuations in American Business, 1790–1860.* Cambridge, Mass.: Harvard University Press, 1935.
Sokoloff, Kenneth L. "Invention, Innovation, and Manufacturing Productivity Growth in the Antebellum Northeast." In *American Economic Growth and Standards of Living before the Civil War,* ed. Robert E. Gallman and John J. Wallis, 345–78. Chicago: University of Chicago Press, 1992.
———. "Inventive Activity in Early Industrial America: Evidence from Patent Records, 1790–1846." *Journal of Economic History* 48 (December 1988): 813–50.
———. "Investment in Fixed and Working Capital during Early Industrialization: Evidence from U.S. Manufacturing Firms." *Journal of Economic History* 44 (June 1984): 545–56.
———. "Productivity Growth in Manufacturing during Early Industrialization: Evidence from the American Northeast, 1820–1860." In *Long-Term Factors in American Economic Growth,* ed. Stanley L. Engerman and Robert E. Gallman. Studies in Income and Wealth, 51:679–729. Chicago: University of Chicago Press, 1986.
———. "Was the Transition from the Artisanal Shop to the Nonmechanized Factory Associated with Gains in Efficiency? Evidence from the U.S. Manufacturing Censuses of 1820 and 1850." *Explorations in Economic History* 21 (October 1984): 351–82.
Sokoloff, Kenneth L., and B. Zorina Khan. "The Democratization of Invention during Early Industrialization: Evidence from the United States, 1790–1846." *Journal of Economic History* 50 (June 1990): 363–78.
Sokoloff, Kenneth L., and Viken Tchakerian. "Manufacturing Where Agriculture Predominates: Evidence from the South and Midwest in 1860." *Explorations in Economic History* 34 (July 1997): 243–64.
Sokoloff, Kenneth L., and Georgia C. Villaflor. "The Market for Manufacturing Workers during Early Industrialization: The American Northeast, 1820 to 1860." In *Strategic Factors in Nineteenth Century American Economic History,* ed. Claudia Goldin and Hugh Rockoff, 29–65. Chicago: University of Chicago Press, 1992.
Spencer, Herbert. *The Principles of Sociology.* 3 vols. New York: Williams and Norgate, 1897–1906.

State of Connecticut. *Register and Manual: 1972*. Hartford: State of Connecticut, 1972.
Steckel, Richard H. "The Economic Foundations of East-West Migration during the 19th Century." *Explorations in Economic History* 20 (January 1983): 14–36.
Stevens, Frank W. *The Beginnings of the New York Central Railroad: A History.* New York: G. P. Putnam's Sons, 1926.
Swedberg, Richard. "Markets as Social Structures." In *The Handbook of Economic Sociology,* ed. Neil J. Smelser and Richard Swedberg, 255–82. Princeton, N.J.: Princeton University Press, 1994.
Szostak, Rick. *The Role of Transportation in the Industrial Revolution: A Comparison of England and France.* Montreal: McGill-Queen's University Press, 1991.
Taber, Martha. "A History of the Cutlery Industry in the Connecticut Valley." *Smith College Studies in History* 41 (1955).
Taylor, George R. "American Urban Growth Preceding the Railway Age." *Journal of Economic History* 27 (September 1967): 309–39.
———. "Comment." *Trends in the American Economy in the Nineteenth Century.* Studies in Income and Wealth, 24:524–44. Princeton, N.J.: Princeton University Press, 1960.
———. *The Transportation Revolution, 1815–1860.* New York: Holt, Rinehart and Winston, 1951.
Tchakerian, Viken. "Productivity, Extent of Markets, and Manufacturing in the Late Antebellum South and Midwest." *Journal of Economic History* 54 (September 1994): 497–525.
Temin, Peter. *The Jacksonian Economy.* New York: W. W. Norton, 1969.
———. "Product Quality and Vertical Integration in the Early Cotton Textile Industry." *Journal of Economic History* 48 (December 1988): 891–907.
Thiel, Henri. *Economics and Information Theory.* Amsterdam: North Holland, 1967.
Thomson, Ross. *The Path to Mechanized Shoe Production in the United States.* Chapel Hill: University of North Carolina Press, 1989.
Timmer, C. Peter. "The Agricultural Transformation." In *Handbook of Development Economics,* ed. Hollis Chenery and T. N. Srinivasan, 1:276–331. Amsterdam: North Holland, 1988.
Tonnies, Ferdinand. *Community and Society (Gemeinschaft and Gesellschaft),* ed. and trans. Charles P. Loomis. East Lansing: Michigan State University Press, 1957.
Tryon, Rolla M. *Household Manufactures in the United States, 1640–1860.* Chicago: University of Chicago Press, 1917.
Tucker, Barbara M. *Samuel Slater and the Origins of the American Textile Industry, 1790–1860.* Ithaca, N.Y.: Cornell University Press, 1984.
Tyler, Daniel P. *Statistics of the Condition and Products of Certain Branches of Industry in Connecticut, for the Year Ending October 1, 1845.* Hartford, Conn.: John L. Boswell, 1846.
Tyler, John S. "Report of John S. Tyler to the Secretary of the Treasury, on the Manufactures of the County of Suffolk, Massachusetts." Boston, Massachusetts, May 19, 1832. In Louis McLane, *Documents Relative to the Manufactures in the United States Collected and Transmitted to the House of Representatives, 1832, by the Secretary of the Treasury.* House Document no. 308, 1st sess., 22d Cong., doc. 3, no. 218. Washington, D.C.: Duff Green, 1833.
Uselding, Paul J. "A Note on the Inter-Regional Trade in Manufactures in 1840." *Journal of Economic History* 36 (June 1976): 428–35.

U.S. Bureau of the Census. *Compendium of the Sixth Census, 1840.* Washington, D.C.: Blair and Rives, 1841.

———. *Manufactures of the United States in 1860, Eighth Census.* Washington, D.C.: Government Printing Office, 1865.

———. *Report on the Manufactures of the United States at the Tenth Census, 1880.* Washington, D.C.: Government Printing Office, 1883.

———. *Tenth Census of the United States, 1880,* vol. 4: *Transportation.* Washington, D.C.: Government Printing Office, 1883.

———. *Twelfth Census, 1900,* vol. 1, pt. 1: *Population.* Washington, D.C.: Government Printing Office, 1901.

———. *Thirteenth Census, 1910,* vol. 1: *Population.* Washington, D.C.: Government Printing Office, 1913.

———. *Seventeenth Census, 1950,* vol. 1: *Number of Inhabitants.* Washington, D.C.: Government Printing Office, 1952.

———. *Historical Statistics of the United States, Colonial Times to 1970, Bicentennial Edition.* 2 pts. Washington, D.C.: Government Printing Office, 1975.

Uzzi, Brian. "The Sources and Consequences of Embeddedness for the Economic Performance of Organizations: The Network Effect." *American Sociological Review* 61 (August 1996): 674–98.

Vance, James E., Jr. *The Merchant's World: The Geography of Wholesaling.* Englewood Cliffs, N.J.: Prentice-Hall, 1970.

Wahl, Jenny B. "New Results on the Decline in Household Fertility in the United States from 1750 to 1900." In *Long-Term Factors in American Economic Growth,* ed. Stanley L. Engerman and Robert E. Gallman. Studies in Income and Wealth, 51:391–437. Chicago: University of Chicago Press, 1986.

Walkowitz, Daniel J. *Worker City, Company Town: Iron and Cotton-Worker Protest in Troy and Cohoes, New York, 1855–84.* Urbana: University of Illinois Press, 1978.

Wallace, Anthony F. C. *Rockdale: The Growth of an American Village in the Early Industrial Revolution.* New York: Alfred A. Knopf, 1978.

Ward, George W. *The Early Development of the Chesapeake and Ohio Canal Project.* Johns Hopkins University Studies in Historical and Political Science, ser. 17, nos. 9–11. Baltimore, Md.: Johns Hopkins Press, 1899.

Ward, James A. *J. Edgar Thomson: Master of the Pennsylvania.* Westport, Conn.: Greenwood Press, 1980.

Ware, Caroline F. *The Early New England Cotton Manufacture: A Study in Industrial Beginnings.* Boston: Houghton Mifflin, 1931.

Weber, Alfred. *Theory of the Location of Industries,* trans. by Carl J. Friedrich. Chicago: University of Chicago Press, 1929.

Weber, Max. *Economy and Society,* ed. Guenther Roth and Claus Wittich. 2 vols. Berkeley: University of California Press, 1978.

Weil, Francois. "Capitalism and Industrialization in New England, 1815–1845." *Journal of American History* 84 (March 1998): 1334–54.

Weiman, David F. "Families, Farms and Rural Society in Preindustrial America." *Research in Economic History,* supp. 5 (1989): 255–77.

Weiss, Thomas. "Economic Growth before 1860: Revised Conjectures." In *American Economic Development in Historical Perspective,* ed. Thomas Weiss and Donald Schaefer, 11–27. Stanford, Calif.: Stanford University Press, 1994.

———. "Long-Term Changes in U.S. Agricultural Output per Worker, 1800–1900." *Economic History Review* 46 (May 1993): 324–41.

———. "U.S. Labor Force Estimates and Economic Growth, 1800–1860." In *American Economic Growth and Standards of Living before the Civil War,* ed. Robert E. Gallman and John J. Wallis, 19–75. Chicago: University of Chicago Press, 1992.

Wermuth, Thomas S. "New York Farmers and the Market Revolution: Economic Behavior in the Mid–Hudson Valley, 1780–1830." *Journal of Social History* 32 (fall 1998): 179–96.

White, George S. *Memoir of Samuel Slater.* 2d ed. Philadelphia: privately printed, 1836.

Whitford, Noble E. *History of the Canal System of the State of New York.* 2 vols. Albany, N.Y.: Brandow Printing Co., 1906.

Williamson, Harold F. "Money and Commercial Banking, 1789–1865." In *The Growth of the American Economy,* ed. Harold F. Williamson, 227–55. 2d ed. New York: Prentice-Hall, 1951.

Williamson, Jeffrey G. "Antebellum Urbanization in the American Northeast." *Journal of Economic History* 25 (December 1965): 592–608.

Wilson, William B. *History of the Pennsylvania Railroad Co.* 2 vols. Philadelphia: Henry T. Coates and Co., 1895.

Winder, Gordon M. "Before the Corporation and Mass Production: The Licensing Regime in the Manufacture of North American Harvesting Machinery, 1830–1910." *Annals of the Association of American Geographers* 85 (September 1995): 521–52.

———. "The North American Manufacturing Belt in 1880: A Cluster of Regional Industrial Systems or One Large Industrial District." *Economic Geography* 75 (January 1999): 71–92.

Withington, Sidney. *The First Twenty Years of Railroads in Connecticut.* Tercentenary Commission of the State of Connecticut, Committee on Historical Publications, vol. 45. New Haven, Conn.: Yale University Press, 1935.

Wood, Frederic J. *The Turnpikes of New England.* Boston: Marshall Jones, 1919.

Wood, Gordon S. "The Enemy Is Us: Democratic Capitalism in the Early Republic." In *Wages of Independence: Capitalism in the Early Republic,* ed. Paul A. Gilje, 137–53. Madison, Wis.: Madison House, 1997.

———. *The Radicalism of the American Revolution.* New York: Alfred A. Knopf, 1992.

Wood, Joseph S. *The New England Village.* Baltimore, Md.: Johns Hopkins University Press, 1997.

Woodbury, Robert S. *Studies in the History of Machine Tools.* Cambridge, Mass.: MIT Press, 1972.

Wright, Robert E. "Bank Ownership and Lending Patterns in New York and Pennsylvania, 1781–1831." *Business History Review* 73 (spring 1999): 40–60.

Wyckoff, William. *The Developer's Frontier: The Making of the Western New York Landscape.* New Haven, Conn.: Yale University Press, 1988.

Zakim, Michael. "A Ready-Made Business: The Birth of the Clothing Industry in America." *Business History Review* 73 (spring 1999): 61–90.

Zevin, Robert B. "The Growth of Cotton Textile Production after 1815." In *The Reinterpretation of American Economic History,* ed. Robert W. Fogel and Stanley L. Engerman, 122–47. New York: Harper and Row, 1971.

Index

About the Author

David R. Meyer completed an M.S. degree in geography at Southern Illinois University and a doctorate in geography at the University of Chicago. From 1970 to 1981 he was an assistant and then associate professor of geography at the University of Massachusetts, Amherst, and in 1981 he joined the faculty at Brown University, where he is professor of sociology and urban studies. Meyer's research has focused on the urban and industrial growth of the nineteenth-century United States and on global cities and comparative economic development, especially in Asia. Meyer also served as coeditor of the *Journal of Historical Geography* from 1995 to 1999. His scholarly articles have appeared in the *Annals of the Association of American Geographers, Cities, Economic Geography, Explorations in Economic History, Journal of Economic History, Journal of Historical Geography, Social Forces, Sociological Forum, Studies in Comparative International Development, Urban Affairs Quarterly, Urban Geography,* and *World Development,* among other journals, and he has written numerous chapters for scholarly books and encyclopedias. Meyer is the author of *From Farm to Factory to Urban Pastoralism: Urban Change in Central Connecticut* (1976) and *Hong Kong as a Global Metropolis* (2000).

Related Books in the Series

Across This Land: A Regional Geography of the United States and Canada
John C. Hudson
Along the Ohio
Andrew Borowiec
Apostle of Taste: Andrew Jackson Downing, 1815–1852
David Schuyler
Boston's "Changeful Times": Origins of Preservation and Planning in America
Michael Holleran
Cities and Buildings: Skyscrapers, Skid Rows, and Suburbs
Larry R. Ford
The City Beautiful Movement
William H. Wilson
The Cotton Plantation South since 1865
Charles S. Aiken
Delta Sugar: Louisiana's Vanishing Plantations
John B. Rehder
Historic American Towns along the Atlantic Coast
Warren Boeschenstein
Invisible New York: The Hidden Infrastructure of the City
Stanley Greenberg, with an introductory essay by Thomas H. Garner
The Last Great Necessity: Cemeteries in American History
David Charles Sloane
Local Attachments: The Making of an American Urban Neighborhood, 1850 to 1920
Alexander von Hoffman
Manufacturing Montreal: The Making of an Industrial Landscape, 1850 to 1930
Robert Lewis
The New England Village
Joseph S. Wood
The North American Railroad: Its Origin, Evolution, and Geography
James E. Vance Jr.
The Pennsylvania Barn: Its Origin, Evolution, and Distribution in North America
Robert F. Ensminger
Petrolia: The Landscape of Pennsylvania's Oil Boom
Brian C. Black
Redevelopment and Race: Planning a Finer City in Postwar Detroit
June Manning Thomas

The Rough Road to Renaissance: Urban Revitalization in America, 1940–1985
Jon C. Teaford

The Sanitary City: Urban Infrastructure in America from Colonial Times to the Present
Martin V. Melosi

Suburban Landscapes: Culture and Politics in a New York Metropolitan Community
Paul H. Mattingly

Unplanned Suburbs: Toronto's American Tragedy, 1900 to 1950
Richard Harris